AF662056

Jharkhand

Land and People

Jharkhand

Land and People

V.N.P. Sinha
L.K.P. Singh • Manoj Kumar

RAJESH PUBLICATIONS
New Delhi - 110 002

Jharkhand
Land and People

Published by :
Sanjay Gupta
Rajesh Publications
1, Ansari Road, Darya Ganj, New Delhi - 110002
Website : www.rajeshpublications.com
Email : rajeshpublications@gmail.com,
rajeshpubdelhi@gmail.com

© Writers

First Edition : 2003
Second Edition : 2015
3rd Revised & Enlarged Edition : 2018
Reprint : 2024

ISBN : 978-81-85891-44-6 (PB)
ISBN : 978-93-83684-64-9 (HB)

Printed at :
R.K. Offset Process
Delhi - 110095

All rights reserved with the Publisher including to translate or to reproduce this book or parts of it except for brief quotations in critical articles or reviews.

Preface

The present book is an expression of keen interest, long association, love and affection to the people of Jharkhand. This region has been in focus of the authors throughout their academic life through research papers and publications. The formation of the state once again inspired us to write in detail about specialities of the land influencing human habitation, economic development and social and cultural traits.

This book, though, useful for people of the state as a whole, it has special importance for the students and teachers of geography and other disciplines. The book contains three broad aspects, the land, the economy and the people. In case of land the emphasis has been laid on geological formations, terrain characteristics, forest resources and climatic conditions. The economy of the region deals with land utilization, cropping pattern, transport net-work, mining and industrial development. Human resource has been analysed in much detail.

Most of the text is authors original writing. At a few places, however, there are some abstracts from their earlier publications for which they are indebted to original publishers. As far as possible attempts have been made to express their ideas in simple language. At places local terms have also been used.

We are obliged to the authors and scholars whose ideas have been incorporated in the book. We are highly obliged to our respected teacher Dr. P. Dayal, Ex-Vice-Chancellor, M.U. and former Head of Deptt. of Geography, P.U. for his inspirations. We are thankful to the publisher for publishing the book in time. Authors will highly appreciate the comments and suggestions for the improvement of the book in future.

Vishwanath Prasad Sinha
Lal Keshwar Prasad Singh
Manoj Kumar

Contents

List of Maps

Introduction

The landlocked State of Jharkhand endowed with heterogeneous landscape, huge natural resources, dominance of aboriginal's habitat and their culture is an unique geographical unit. Heterogeneity is observed in geological formations, physical appearance and the pattern of development. This part of the country is an integral portion of the Peninsular highland, part of ancient Gondwanaland, portrays areas formed of rock formations ranging from the earliest Archean Era to the latest Post-tertiary period. These rock formations have resulted into varied physical appearance on account of erosional processes. The area is also marked with uneven economic development. A few industrial centres have developed to such an extent that they are enlisted in the major industries of the world such as Bokaro Steel Plant, Heavy Engineering Corporation at Ranchi, Jamshedpur Industrial Complex, etc.

This area is store house of minerals and forest resources. It is the richest state in terms of mineral resources and plays leading role in production. It leads other states in the output of coal, mica, iron ore, copper, china clay, fire clay and kyanite, while it is one of the important producers of bauxite, asbestos, felspar, quartz, etc. In terms of forest products this state does not lack behind any state. It leads other states in the production of lac contributing about 40% of the total output of the country. This state also contributes in the production of sal, timber, bamboos and Sabai grass. With this limited area of the country there is good concentration of tribal people. There are 30 different tribes inhabiting the region and portray varied cultural traits which form anthropologist paradise.

Certainly in the process of economic development and imparting modern education the natural beauty of the region has been adversely affected and original cultural traits have been eroded. Virtually people coming from different parts of the country having varied cultural background have brought changes in the native culture of the region. However the rolling terrain, enveloped with green vegetal cover, rivers forming waterfalls and cascades, laterite soil, etc. describe the picturesque story of the land. Tribals living in extreme poverty even in inaccessible areas hardly look more discontented with pretty wealth and comfort. They still thrill with songs, dances and traditional musical instruments.

Shape and Extent :

The state of Jharkhand exhibits an amorphous shape. It has however, broad northern portion and narrow base in the south. A small projection lies in the north-eastern portion. It extends from 21°58′ N. Lat. to 25° 18′ N. Lat. and 83° 22′ E. Long. to 87° 57′ E. Long. Its maximum east-west length is about 457 kms. and north-south extent is about 310 kms. It comprises an area of 79716.00 Sq. kms. which is 2.34% of the area of the country and contains 3.67% population of the country. The state is bounded by Bihar in the north, Odisha in the south, West Bengal in the east and Chattisgarh and Uttar Pradesh in the west. This state comprises 5 divisions, 24 districts and 259 development blocks.

Historical Background : The historical accounts of the state of Jharkhand are interesting and full of dynamism. The inhabitants of this area, though, simple and innocent, have not remained aloof from the main stream of the country. The area provides evidences of very early habitation and cultural history. Vedic literature contains description of Asurs specializing in iron and copper smelting and trading of iron and copper articles with Aryans. These products used to be exported by the early civilized people to Athens and Rome.

The availability of copper and iron made goods and ornaments like knife, spear, axe, etc. in the coal field of Jharia, areas around Chaibasa and Chakradharpur attest the reality of the fact that there used to live some civilized people aware with the art of smelting iron and copper and leading civilized life. There are also evidences of early trade relations with outside countries because the coins of different countries have been also found in the process of archeological excavations. During early years this region was known as Pundru. It used to be called Ark-khand or Kark-khand. It might have got this nomenclature because of the tropic of cancer passing through the region. Even Pauranic Literature portrays description of Karusha and Pundru administration. They were known as Kharwars. The northern portion of this region provides ample evidences of its linkage with Magadh Empire and other areas of northern plains. The arrival of people having faith in Jainism is documented by the Nirman Site of 23rd Jain Tirthankar at Parasnath hilly area. It is also assumed that the impact of Magadh Empire must have exerted considerable impact on this region because of the proximity to this portion of the state with the administrative headquartors at Rajgirha (Rajgir) and Patallputra (Patna). Several old sites of habitation in the North Koel basin and Hazaribag plateau have preserved the statue of Lord Budha.

Table - 1
Administrative Divisions

S. No.	Division/ District	Area in Sq. Kms.	Population 2011	Population Growth 2001	Head-quarter
A.	**North Chotanagpur**				
1.	Hazaribagh	3,555.00	17,34,495	25.79	Hazaribagh
2.	Ramgarh	1,341.00	9,49,493	13.10	Ramgarh
3.	Chatra	3,718.00	10,42,886	29.05	Chatra
4.	Koderma	2,540.00	7,16,259	32.42	Koderma
5.	Giridih	4,962.00	24,45,474	21.78	Giridih
6.	Bokaro	2,833.00	20,62,330	16.01	Bokaro
7.	Dhanbad	2,040.00	26,84,487	11.99	Dhanbad
B.	**South Chotanagpur**				
8.	Ranchi	5,097.00	29,14,253	24.00	Ranchi
9.	Khunti	2,535.00	5,31,885	22.32	Khunti
10.	Simdega	3,774.00	5,99,578	16.58	Simdega
11.	Gumla	5,360.00	10,25,213	23.16	Gumla
12.	Lohardaga	1,502.00	4,61,790	26.68	Lohardaga
C.	**Palamu**				
13.	Palamu	4,393.00	19,39,869	26.17	Daltonganj
14.	Garhwa	4,093.00	13,22,784	27.75	Garhwa
15.	Latehar	4,291.00	7,26,978	29.61	Latehar
D.	**Santhal Pargana**				
16.	Sahebganj	2,063.00	11,50,567	35.74	Sahebganj
17.	Pakur	1,811.00	9,00,422	28.33	Pakur
18.	Jamtara	1,811.00	7,91,042	25.46	Jamtara
19.	Deoghar	2,477.00	14,92,073	28.03	Deoghar
20.	Dumka	3,761.00	13,21,442	19.42	Dumka

21.	Godda	2,266.00	13,13,551	25.35	Godda
E.	**Kolhan**				
22.	E. Singhbhum	3,562.00	22,93,919	15.68	Jamshedpur
23.	W. Singhbhum	7,224.00	15,02,338	21.75	Chaibasa
24.	Saraikela-Kharsawan	2,657.00	10,65,056	25.47	Saraikela
	Jharkhand	**7,97,16.00**	**3,29,88,134**	**22.42**	**Ranchi**

Source : Census of India 2001 & 2011.

Old legends of a few aboriginal tribes and old religious books and literature have enabled the historians and other social scientists to reveal the fact that in the beginning of the present Christian era there were habitations of Munda, Oraon and Nag tribes. Virtually there was rule of Nagbanshis and the centre of administration was Chutia, 3 kms. south of Ranchi. It is opined that this administrative headquarter of Nagbanshis was formed in 278 A.D. and this village was named after Chutia Munda. These scholars also hold the view that Chutia and his brother Nangu gave the development of nomenclature Chutia Nagpur. This name is also recorded in Rennel's Map of India of the year 1792. This name was further developed in the name of Chotanagpur by the British Parliament in 1872.

Table - 2
Jharkhand at a Glance (2011)

Items	Jharkhand State
Total Area	79716.00 Sq. Km.
Total Population	3,29,88,134 Persons
No. of Divisions	05
No. of Districts	24
No. of Sub-Division	38
No. of C.D. Block	260
No. of Panchayats	4423
No. of Villages	32,620
No. of Towns	228
Total Literacy Rate	66.41%
Male Literacy Rate	76.84%
Female Literacy Rate	55.42%
Population Density	414 Person / Sq. Km.
Sex Ratio	948 Females / 1000 Males

Urbanisation Level	24.05%
ST Population	26.21%
SC Population	12.08%
Decadal Population Growth (2001-2011)	22.42%

After the downfall of Buddhism, the Magadh Empire along with this region witnessed turmoils and the area lost peace and stable administration. During 6th century Sasank rule prevailed over the region and his administrative area was spread over Bihar, Bengal, Odisha and parts of Madhya Pradesh and Uttar Pradesh. Huen Sang, a Chinese traveller has described about Chotanagpur in his travel account. During 8th century this region was ruled by Pal Dynasty and at that time also the area of administration was almost the same as stated earlier. The period after 8th century ruler did not provide stronghold of any ruler who could keep the entire area intact. Different areas were being ruled by different rulers. This portion, however, provides evidence of rule of Nagbanshis from 8th century to the emergence of Muslim rule. In the year 1574 muslim ruler invaded Palamu area under the leadership of Mansingh. They remained confined to the accessible areas. They did not enter in interiors because of preponderence of thick forest cover and termed this area "Jharkhand". In the year 1589 Mansingh crossed through this area and gave the nomenclature of Manbhum and Singhbhum. In 1616, Ibrahim Khan invaded Chotanagpur. During the reign of Aurangzeb, this region was attacked several times. On 12th August, 1765, the East India Company got rule over this region including Bihar, Bengal and Odisha. In the year 1780. Ramgarh hill tract district was formed and this district remained upto 1783 which consisted of Hazaribag, Dhanbad and Palamu districts. At Hazaribag, Ramgarh battalion was stationed. During this period the local rulers, tribal leaders and people aligned with freedom movement frequently raised voices against the British rule. In the year 1831-32 Kol's revolted against the British rule followed by Santhal's rebels, Mundas and Chero's revolts, etc. As a result of these tribal revolts and freedom movements the British Government under the Bengal regulation (13) 1833 formed a separate independent province consisting of Chotanagpur and Santhal parganas under west frontier agency. It was formally inaugurated on 15th January, 1834. Capt. T. Wilkinson was appointed as first agent. Later on East India Company divided India into five parts like Calcutta Presidency, Madras Presidency,

Bombay Presidency, North - Western Provinces and South-West Frontier Agency and appointed governors for each presidency and provinces. In this process this region came under the rule of Governor of Calcutta Presidency. In 1857 Military revolted against the British rule and the people of this region took active part in support of military revolt. In the year 1881 the nation witnessed a farmers movement known as Sardar movement which motivated tribal people and they stopped payment of taxes to their Zamindars. In 1891, the people of Jharkhand raised strong protest against the British administration under the patronage of great Birsa Munda, whose contributions may not be ignored in the interest of the state as well as the nation. The district of Palamu was created in 1892. In 1899 the district of Lohardaga was degraded to the status of a subdivision and Ranchi belonging to Lohargada district became district. In the year 1912 Bihar was separated from Bengal and in 1936 Odisha was separated from Bihar. At this time Chotanagpur and Santhal Parganas remained with Bihar and districts of Keonjhar, Mayurbhanj, Sundargarh and Sambalpur were kept with Odisha. In the year 1956 the state reorganisation took place and at that time districts of Purulia and major parts of Manbhum were

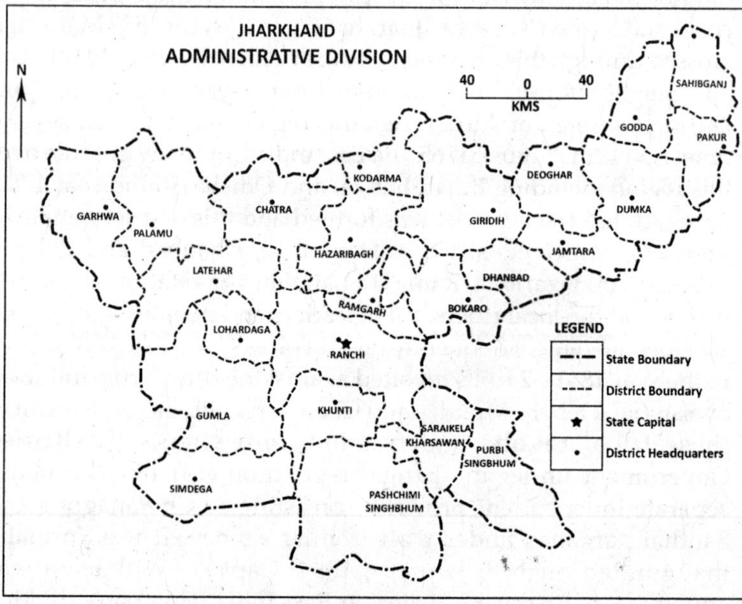

Fig. 1

transferred from Chotanagpur to Bengal. A small part of Manbhum became Dhanbad district of this state. At the time of state reorganisation also the demand for separate state for Jharkhand was raised but the same was not considered. In the year 1989 this area was facilitated by forming Jharkhand Kshetra Swashasi Parishad. But the struggle for separate state did not stop and this area got the statehood as Jharkhand on 15th November, 2000.

1

Geology & Structure

Geology

Jharkhand, well known as the Chotanagpur plateau is the north-eastern projection of the Peninsular table land which has remained part of the Gondwanaland. It, therefore, consists of geological and structural formations ranging from the earliest period to the recent. It, however, contains more than ninety percent of Archean formations dominated by granite and gneisses. No doubt, the area has been stable but minor structural changes have taken place in the form of faulting, warping, tilting, igneous intrusions, etc. Erosional agents have also made additions to geology in the area during recent geological periods. Marine encroachment has hardly influenced the formation of rocks except in the Subarnarekha Basin.

Geological formations are arranged from the youngest to the oldest (in order of age) in the table 1.1

Table - 1.1

	Eras	Periods and Systems	Formations
IV.	Cainozoic		
	Quarternary	Recent	Alluvium
		Pleistocene	Laterite
	Tertiary	Pliocene	High Level Laterite
		Miocene	Tertiary Grits and Gravels
		Oligocene	- do -
		Eocene	Deccan Lavas
III.	Mesozoic	Unconformity	
		Cretaceous	Upper Gondwanas, Rajmahal Series
		Jurassic	Unconformity- Rajmahal Lavas
		Triassic	Panchet beds of

			Raniganj, Karan- - pura, Bokaro, etc. Coal fields	
			Lower Gondwanas	
II.	Palaeozoic Stage	Permian	Damuda Series	Raniganj
		Carboni- ferous	Talchir Series Measures	Iron Stone Shales Barren
				Barakar Coal Measures
		Devonian Silurian Ordovician Cambrian		Unrepresented
I.	Pre-Cambrian or Archean	Vindhyan System		Kaimur Series Unconformity Kolhan Series
		Cuddapah System		Unrepresented Unconformity granites and Gneisses Dalma- Dhanjori Lavas
				Dhanjori Sand stone - Conglomerate
		Unconformity		
		Iron ore Series		Iron ore Stage (Series)
		Dharwarian System		Possible overlap Chaibasa Stage (Series)

The Archeans form the basement over which other geological formations have taken place. Virtually these rocks were formed in the process of solidification of the crust. It, therefore, constitutes the surface rocks over the greater part of the state. These rocks have largely been transformed on account of diastrophism and intrusion and it has become difficult to distinguish Archean formations from Dharwarian system.

The Archeans formations consist of : (i) sedimentary rocks and

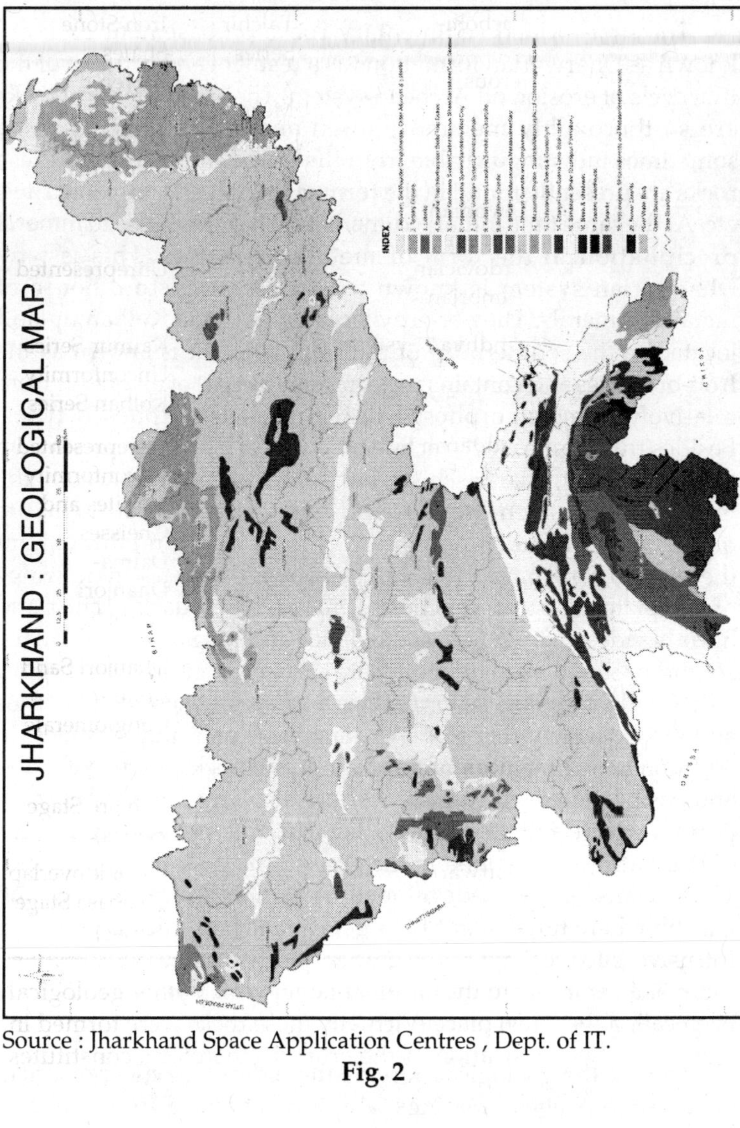

Source : Jharkhand Space Application Centres , Dept. of IT.

Fig. 2

(ii) wide spread basic intrusions and granite.

The sedimentary and basic intrusions are important in northern and southern parts of the Jharkhand state whereas the granite is dominant in central areas of the state. The oldest granite and gneiss rocks aging 1.5 Arab years are found in this area.

The first-formed sedimentary rocks on basement rock are known as Dharwarian formations as a result of end product of the first cycle of erosion on Archean system. These sedimentary rocks are so thoroughly metamorphosed and crystallined as to be sometimes indistinguishable from the basement Archeans. These rocks are presently found in the form of phyllites, slates, quarzites, etc. At the same time initial sedimentation has given rise to mineral precipitation in the form of metallic minerals. This is why Dharwarian system is known as stone house/store house of metallic minerals. They occupy large areas in the Kolhan upland located in the southern tip of the state where it is known as the Iron-ore series and contains rocks representative of all grades from relatively unmetamorphosed to highly metamorphosed types. Besides this region, such rocks also occur in Chaibasa area where the mica-schist, quartz-schists and horn-blende schist occupy the most part. The Iron-ore is found within the banded-hematite quartzite rocks. The second group of rocks are found throughout the plateau excepting a few patches.Towards the northern fringe of the plateau occur streches of mica-schist, quartz -schist and horn-blende-schist. A few isolated hills and ranges peep out in the southern Ganga plain of Bihar state such as Rajgir and Kharagpur hills. Of all the geological formations they are the most metalliferous and are extremely rich in iron, manganese ores and Kyanites. In addition to sedimentary and metamorphic rocks extrusive lavas and basic igneous intrusive rocks are also found in the southern portion which contain minerals like chromite and asbestos. Viens of Pegmatite are intruded in the granite and metamorphic rocks. These veins are well-developed in Kodarma, Hazaribag and Giridih where huge quantity of good quality of mica is obtained. Intrusive into the Singhbhum granite and also in the Iron ore series, there is a remarkable group of basic igneous dykes which have been called the 'Newer Dolerites'.

During the geological periods the oldest tectonic movement gave rise to Archean foldings which are alingned from north to south direction. As a result of this east-west extending hilly ranges have developed in the southern portion of Ranchi & Dhalbhum region of Singhbhum. At the same time Dalma and Dhanjori lava deposition took place which covered conglomerates of the region.

Archeans and Dharwar geological formations are followed by Cuddapah and Vindhyan formations. Cuddapah formations are not found in the state. Only part of this system is witnessed in Kolhan upland. Even Vindhyan formations are combined to a small area in the north-west particularly along the Son valley (Garhwa & Palamu districts) where crystalline limestone, dolomite and sandstone are found as outcrops over the granites at different places.

During Permo - Carboniferous period most parts of the earth surface experienced eiperiogenic (tectonic) movements resulting formation of a no. of mountains but this part of Gondwanaland remained aloof of mountain building. Certainly faults developed due to Hercynian movement which gave rise to Damodar Valley. Trough Basin became the area of sedimentation which took place from the last stage of Carboniferous period to first stage of Cretaceous period. This trough basin extends from Hutar Coal field in the west to the Jharia and Raniganj Coal field in the east. In this trough basin sedimentation process buried huge amount of flora and fauna which have turned into different type of coal seams. In this area there lies the best quality of coal in India. Most of the coal seams exist in between shale and sandstone. The second important mineral is fire-clay which is found beneath the coal seams and is used as refractory materials. It is used for firebricks, furnace lining, etc. The third important formation is Iron- stone in the form of limonite and siderite usually found in Dhanbad district especially in Raniganj Coal field. In addition to it there lies a very small area of about 10 sq. kms, south-west of Daltonganj having magnetite iron ore.The formation of sandstone is also found in Rajmahal Region especially in Lugo- Panchet and Dubraj. The sign of faulting is also witnessed in the Rajmahal highland. It also experienced igneous intrusions of volcanic eruptions known as Rajmahal lava eruption. Virtually the entire Rajmahal highland covers the Santhal Parganas having topography formed of lava eruptions except the south and south-western areas where Archeans rocks are found . This area contains varied types of basalts. In between layers of

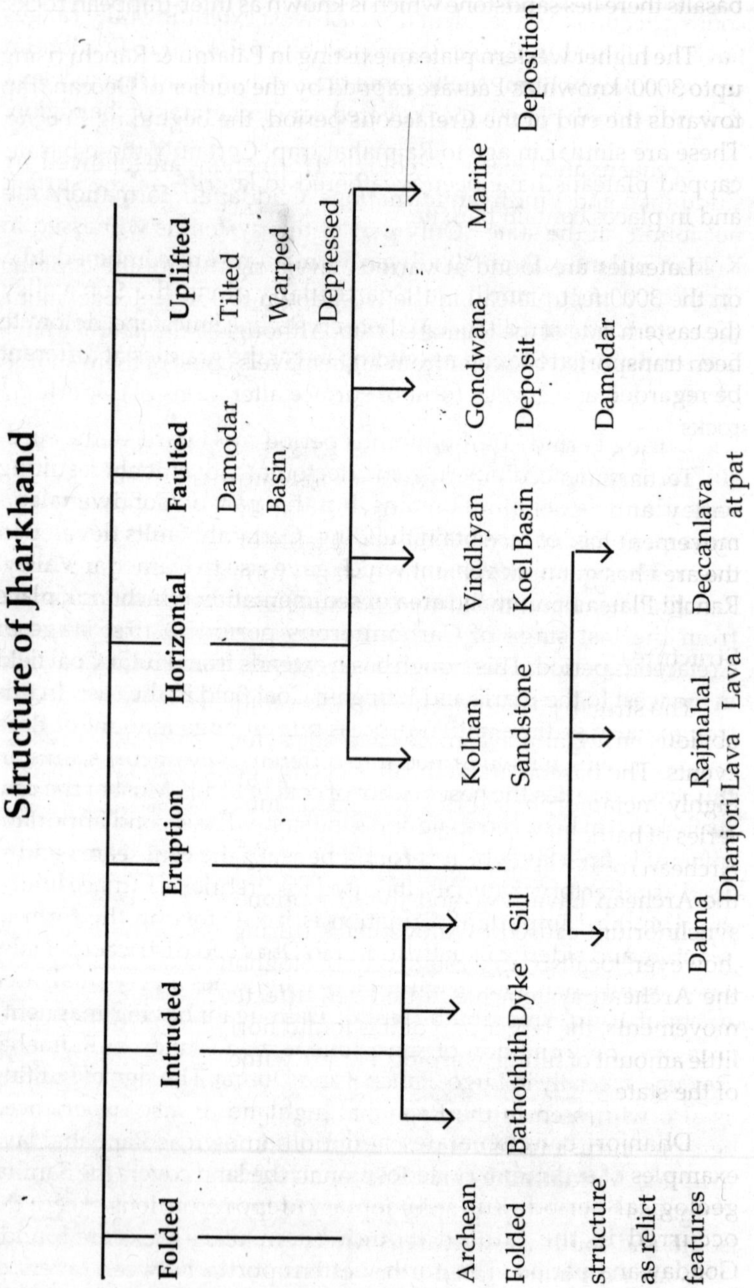
Structure of Jharkhand
Folded
Intruded
Eruption
Horizontal
Faulted
Uplifted
Tilted
Warped
Depressed
Damodar Basin
Archean Folded structure as relict features
Bathothith
Dyke
Sill
Dalma
Dhanjori Lava
Rajmahal Lava
Deccanlava at pat
Kolhan Sandstone
Vindhyan Koel Basin
Gondwana Deposit Damodar
Marine Deposition

basalts there lies sandstone which is known as Inter-trappean rocks.

The higher western plateau existing in Palamu & Ranchi rising upto 3000' known as Pats are capped by the outlier of Deccan trap towards the end of the Cretaceous period, the beginning Eocene. These are similar in age to Rajmahal trap. Certainly these basalts capped plateaus have been weathered to laterites at the surface and in places contain bauxite.

Laterites are found at various levels right from the capping on the 3000 feet plateau in the west down to 150 feet level along the eastern side of the Rajmahal trap. Although some laterites have been transported by water from higher levels, most of them should be regarded as a special form of surface alterations of underlying rocks.

Tertiary and post-tertiary depositions are only confined in river valley and other depressions. Certainly the Alpine Earth-movement has influenced the land table of Jharkhand state and the area has got uplifted in three stages forming High Pat region, Ranchi Plateau and the lower Plateau areas.

Structure

The structural complexity and the evolution of landscape are obvious on account of complex geological formations and tectonic events. The basement Archean rocks i.e. granite, gneiss, etc. are highly metamorphosed due to igneous intrusions in the form of series of batholiths and movements during Pre-Cambrian period. Archean rocks of Dalma and Dhanjori areas have been affected by the Archean Lava flow and the formation of anticlinorium and synclinorium caused by movements during Archean period are ,however, localised in western part of Singhbhum district. As such the Archean system has little been affected by major tectonic movements, the larger part of it is found horizontally arranged. A little amount of tilting, warping, etc. are witnessed in some pockets of the state.

Dhanjori conglomerates and Kolhan sandstone are good examples of sedimentary deposition in the southern part in early geological period. But sedimentary deposition in large scale occurred in the faulted trough basin of Damodar during Gondawana period. The north-western portion of this state also contains some amount of sedimentation over the eroded granites developed during Vindhyan period. The south-eastern portion i.e. Dhalbhum and north-eastern portion of Rajmahal highlands have also representation of sedimentation caused by marine transgression. These sedimentary rocks are highly metamorphosed and tilted.

During tertiary period this state has got records of successive uplifts, warps over its northern fringes and Subarnarekha Basin, tilts in the highlands of Rajmahal in relation to Apline folding.

2

Relief and Physiography

The evolution of landscape has been determined by prolonged erosion, deposition, intrusion, eruption, faulting, uplifting, tilting, etc. As a result the entire area is characterised by land of great inequalities, succession of plateaus, hills, valleys, scarps, ridges; etc. It is true that the earliest geological formations of the area had been reduced to peneplain but later structural disturbances have created unevenness in land surface.

The evolution of landscape concerns the basement Archean rocks which were formed due to solidification of molten masses, erosion of the primaveal sedimentary rocks deposition of earliest sediments containing metallic minerals, erosion of lava caps formed during Archean & Deccan eruptions, faulting, deposition of sediments and flora and fauna during Gondwana period, other sedimentary deposition due to marine transgression in limited areas during Vindhyan and Post-tertiary period. The stagewise upliftment of the region linked with Himalayan Orogen is also responsible for preserving the inequalities of land surface. As a result plateaus of different elevations , valley, scarps, ridges hillocks, etc. characterise the land surface. First upliftment is supposed to be of about 300m. This uplift permitted the development of a wide spread peneplain, 300m below the lavas in western portion and the remanants of the later were left as small plateau in the west. This peneplain was being formed throughout the greater part of Tertiary times. Next in later part of Tertiary period uplift was renewed to an extent of about 300m in western parts but in the form of block movement with sharp warping along a north-south line which is most represented by the eastern edge of the region. At the same time a trough developed along the valley of the Damodar due to faulting and cutting off the small Hazaribag plateau to the north. Along the northern side of the plateau the movement was more of the nature of a tilt, the surface sloping more gradually down towards Koderma. Since this late Tertiary differential block movement uplift of the whole western region has contributed to the extent of a further 300m elevation, the later

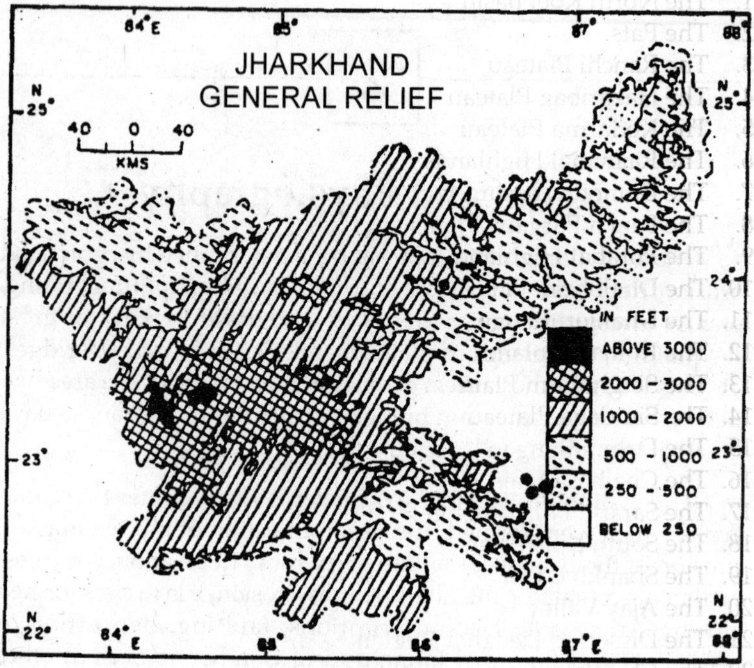

Fig. 3

movement has given rise to a tilting of the land surface.

Most of the area has an elevation exceeding 300m with a great part rising above about 650m above sea level. But while there are hills rising to about 1000m. There are also river valleys which are only 120m high. The heart of plateau consists of two large plateaus those of Hazaribag and Ranchi being separated by the Trough Basin of Damodar. On the western margin of the plateau, there is still higher plateau known as `Pats' having an elevation of about 1000m. On the eastern and south-eastern areas eroded plateaus having less then 300m elevation are found.

On the basis of contour forms, varied elevations and geomorphological features, the area can be divided into the following physiographic units:

This classification is almost similar to that of classification made by Dr. S.P. Chatterjee in National Atlas.

1. The North Koel Basin
2. The Pats
3. The Ranchi Plateau
4. The Hazaribag Plateau
5. The Kodarma Plateau
6. The Rajmahal Highlands
7. The Damodar Basin
8. The Panch Pargana Plains
9. The Porahat Highland
10. The Dhalbhum Plains
11. The Dhanjori Range
12. The Kolhan Upland
13. The Singhbhum Plain
14. The Simdega Plateau
15. The Dalma Range
16. The Chaibasa Plain
17. The Sarand Hilly Region
18. The South Koel Basin
19. The Shankh Basin
20. The Ajay Valley
21. The Dissected Deoghar Region
22. The Chatra Plateau
23. The Chattarpur Plain
24. The Northern Scarp

1. The North Koel Basin : This physiograpic unit, covering most areas of the districts of Garhwa and Palamu, has great physical inequalities from the steep scarps of the Pats in the south to the alluvial section of north consisting of Hussainabad, Bhawanathpur and Bishunpur Thanas. This unit has an average elevation of about 400 m. The general land slope is from south to north. The topography is highly dissected, partly due to the alternating arrangement of rocks of varied hardness. The rocks in the north - west are Vindhyans as against harder granites in the south. The main streams are the Son, the North Koel, the Amanat, the Gogra and the Kankar. All of them except the Son originate in the neighbouring highlands. The upper valleys of these streams are dissected and forested but the lower parts are more habitable. In upper areas of the basin the tribal people are in considerable percentage while the lower parts are dominated by non-tribals. Upper areas have considerable forest cover as agricultural activities are combined to limited areas but the lower portions exhibit wide range of cultivation.

2. **The Pats:** This region is the highest portion. It lies in the west of Ranchi Plateau and south of Garhwa and Palamu district. This region consists of several plateaus being separated by tributaries of the North Koel river, Sankh river, etc. but some how they have maintained elevation ranging between 1000m and 1200m. These smaller plateaus are often known as mesa & butte. These isolated pats are outliers of the Deccan lavas and have maintained flat tops.

After lava capping the region was elevated to about 300m in early tertiary period and another 300m in Middle tertiary period to attain the present elevation. Certainly the surface has been weathered into laterites. The western part of this plateau has steep escarpments of Bagru Pahar and Rudanipat Mesas. The highest waterfall is Burhaghagh on the river Burha in Chechari Basin has the fall of about 150m. The Ghoghri falls 6km north of Netarhat has a waterfall of 50m. Important higher Pats are Ganeshpur, Netarhat and Janira which have elevation of about 1250, 1260, and 1225m respectively. Important hills are Shanu and Saryu which are dissected by the rivers of Koel and Shankh.

The fundamental geomorphic features are the narrow erosional valleys, some of which are more than 150 m. deep, the deepest being that of Dhardhani Nadi. These areas are mostly inaccessible due to steep slope. The general slope of the Pat region is towards the north and the east. The highly inaccessible areas of the region are dotted with smaller nucleations of tribal people. Their economy is mostly based on forest products and isolated patches of agriculture.

3. **The Ranchi Plateau:** The Ranchi plateau is the largest plateau amongst the plateaus of Jharkhand state. It covers an area of about 4500 Sq. miles. It is bounded by Pats in the west , the Damodar Basin in the north, the Subarnarekha and the Kharkai rivers in the south-east. An everage elevation of this plateau is about 650m. It is gently undulating area intersected with numerous streams and rivers studded with low wooded rocky hills and isolated peaks which form a conspicuous feature of the landscape. These residual hills and ridges are made of resistant rock masses above the undulating gneiss surface. The rivers flowing over hard crystalline

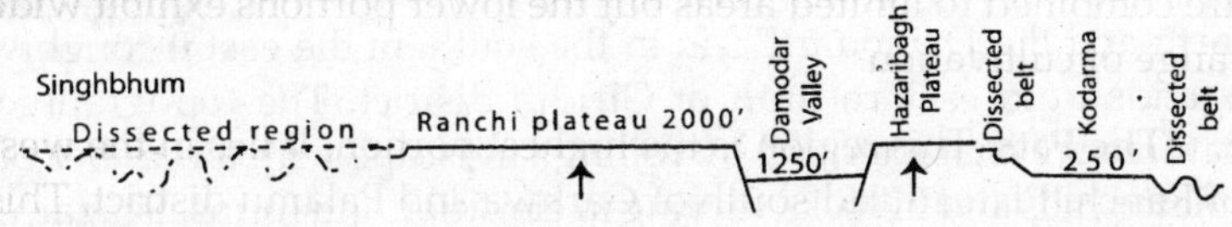

Cross Section : Showing Physiography of CNP (From North to South)

Fig. 4

*Arrows indicate direction of movement.

rocks and softer Dharwars with braided channels give picturesque view of peneplain. The monotony of undulating uplands and lowlands locally known as Tanrs and Dons is relieved only when a dome hummocky boulders of tors dot the landscape. Areas west of Ranchi reveal north to south warping whose extremities are characterized with radial drainage pattern.

The average slope of 1 to 3 degree is indicator of the flat character of the uplifted peneplain being evident from the uniformity of the plateau surface, low relief within the plateau surface, isolated residual low hills, high degree of truncation of structure, the rivers being indifferent to softer Dharwars and harder gneisses and granites. In the southern portion the relief reduces by 300m within the span of 8 km but in the eastern parts the relief decreases ranging between 150 and 300m within the span of 30 to 35 km. It indicates that the eastern scarp is relatively lesser sloping than the southern scarp. The southern scarp of the Ranchi plateau appears relatively higher than the Ranchi plateau as such it is expected that these scarps might be formed due to anticline and syncline which is parallel to border of Ranchi plateau.

Since this area consists of mostly homogeneous surface terraced agriculture has developed. Don areas are relatively more fertile than Tanr areas leading to growth of settlements, orchard and on Tanrs and cultivation in Don land. The entire plateau is dotted with compact settlements in smaller forms.

4. The Hazaribag Plateau: The Hazaribag plateau is relatively smaller than the Ranchi plateau being separated by the Trough Basin of Damodar but has almost the same altitude and morphological features. This physiographic unit is marked by the North-Koel Basin in the west and the south-west, mica field in the north and the Damodar Basin in the south, in the east it roughly extends upto eastern limit of Giridih district. The topographic features are characterized by rolling landscape with a number of isolated hills and hillocks varying in elevation and forms. Sitagarh, Bamhanbai, Silwar, Bhuswa, Banadag, Belian and Canary form a ring of seven hills around Hazaribag town. The western portion of the Hazaribag plateau constitutes a broad water-shed between the Damodar drainage to the south and the Lilajan and Mohane rivers in the north. The uplift of this plateau alongwith the Ranchi plateau marks clear steep scarp on the border and give rise to waterfalls in the rivers. The western part of the plateau has relatively coverage where sprinkled types of settlements are found while the eastern portion has relatively dispersed settlements because of better agricultural land. The tribal population is relatively more in the west and non-tribal people are in considerable percentage in the east. The region is rich in natural resources.

5. **The Kodarma Plateau:** This plateau extends from Barakar-

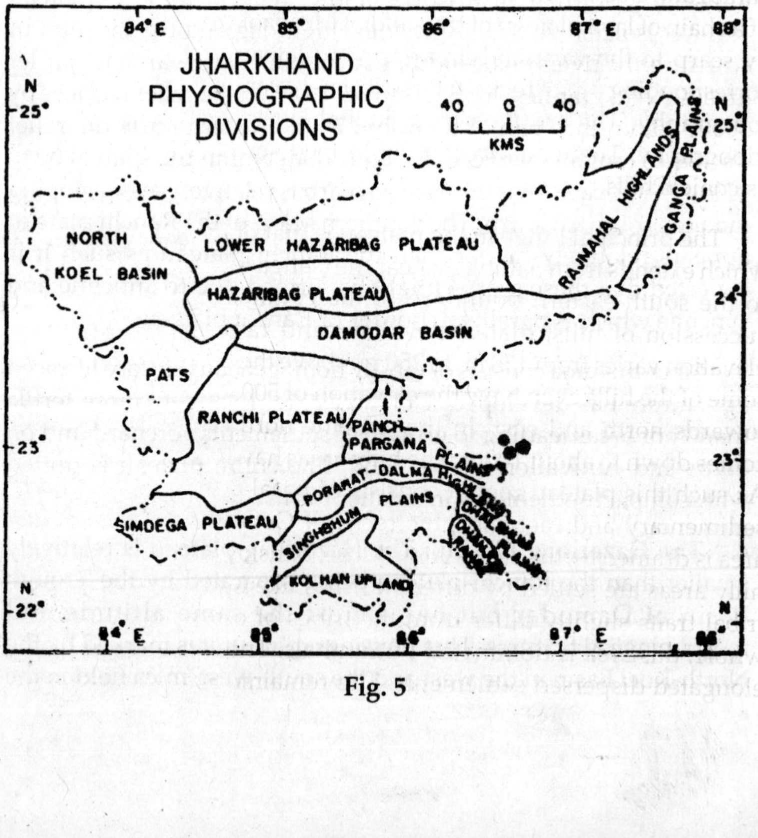

Fig. 5

Mohane divide in the west to the origin place of the rivers Kini and Ajay in the east. The northern and southern borders are demarcated by the Rajauli hills and Hazaribag plateau respectively. The average elevation is about 400m. Certainly the western portion is relatively undulating , northern portion has relatively more steep slope. This area has considerable rocks of Dharwar period like quartzite and schists where good quality mica is found into pegmatite veins. The hard quartzite rocks have been relatively less affected by erosional processes and form higher peaks. Areas having schist rocks have relatively lower relief because of more erosion. The area is inhabited by tribal and non-tribal people having agriculture, forest and mineral economy.

6. **The Rajmahal Highlands:** The Rajmahal Highlands and associated plains form the north-eastern part of the state. It is bounded by the alluvial tract of the Ganga plain in the north and east and the south and west are surrounded by Damodar valley and Kodarma plateau respectively. It includes the areas of Deoghar, Sahibganj, Godda, Dumka, Pakur districts. This highland consists of a chain of lava plateau of trap and of Mesozoic beds. It is marked by scarp to the west and lava capped plateau deeps to the east corresponding dip of rocks. It has extremely varied and picturesque topography. The eruption of Rajmahal lavas gave birth to such topography. There are several flat topped small plateaus as well as conical hills.

The principal range in the plateau is that of the Rajmahal hills which extends from Sahibganj near the Ganga to Nagalbanga close to the south-eastern boundary of the commissionary. There is succession of hills, plateaus, valleys and ravines. The general elevation varies from 150 m. to 250 m. above the sea level. Certainly some of the hills even have the elevation of 500 m. The land slopes towards north and east. In the extreme northern east the relief comes down to about 100m. The hilly areas have basaltic coverage. As such this plateau region consists of basal granite, gneiss and sedimentary and metamorphic rocks of Gondwana period. The area is drained by the rivers Ajay, More, Bansloy, Morang, etc. The hilly areas are settled by different tribes like Malpaharias whose tribal traits slightly differ from the tribes of other areas. On the whole, this area is dominated by Santhals. The hilly ranges have elongated dispersed settlements. The remaining hilly areas have relatively dispersed settlements. The adjoining plains have semi-compact and hamleted clusters. In southern portion coal is found.

7. **The Damodar Basin:** The structural basin of the Damodar formation lying along the zone of Gondwana formation occurs roughly in the middle and separates Ranchi and Hazaribag plateaus. In the west there lies North Koel valley while in the east it extends upto the boundary of this state. It covers most areas of Dhanbad, Bokaro, a small part of Hazaribag, Lohardaga, Ranchi, Palamu, etc. It covers an area of about 3000 Sq. km. Its upper parts lie in the state while the lower parts lie in the state of West-Bengal. The southern and northern boundaries are delimited by fault scarps. The valley has gentle slope. The width of valley varies from west to east. It contains thick deposits of sedimentary and metamorphic rocks of Gondwana period where huge reserve of good quality coal, china clay, fire clay, etc. are found. In western parts its relief is about 550m which comes down to 350m near Ramgarh and 150m near Dhanbad. The area is dotted with mining and industrial centres in the mineralised areas while forest and agricultural resource based activities have developed in other areas. The region has given rise to relatively dispersed settlement in the rural areas but the coal mining areas have given rise to bunches of urban agglomerations. This area has experienced considerable concentration of non-tribals; who have come from other areas of the country because of urban and industrial development.

8. **The Panch Pargana Plains:** This name has been used instead of lower Ranchi plateau by Prof. S.P. Chatterjee[1]. It is surrounded by Chotanagpur granites on three sides and terminates at the Dalma lavas in the south. As such it is bounded by Ranchi plateau in the west, Moryan in the north, Dalma in the south and state boundary in the east. A number of waterfalls including the Hundru mark the break in the Thalweg of the Subarnarekha and other streams causing these scarps. Its elevation varies between 150m and 300m. The western portion is relatively more uneven than the eastern portion. The land slopes from west to east. Compared with the plateau areas this eroded plateau has homogeneous land surface having more culturable land and habitation. The area is dotted with smaller nucleations having tribal and non-tribal population.

9. **The Porahat Highland:** This highland lies in the south of Ranchi plateau. This area begins from southern border of Ranchi

plateau and goes upto Sanjay-Karo water divide in the south. It comprises several hills, valleys and smaller plateaus. The entire region is formed of granite and gneiss. This region has also rock exposure of Dharwarian period like quartzites, schists and igneous intrusion. The landscape of the region is marked by dome-shaped hills and outliers of hard rock surfaces. The land has undulating surface. Forest and agricultural resources determine the economy of people. Relatively dispersed settlements dominate the area.

10. The Dhalbhum Plains: These plains represent the south-eastern portion of the state. It covers most areas of Purbi Singhbhum. It is formed by the river Subarnarekha and its tributaries and cover an area of 800 Sq.kms. It has gentle slope except in the south where the terrain is rugged and highlands are forested. The area provides a good size of cultivable land. Relatively homogeneous Plateau land has encouraged smaller nucleations of tribal and non-tribal settlements. The area has reserve of copper containing ores.

11. The Dhanjori Range: This range consists of picturesque scenery of the highland. It stands a great contrast to the mica-schist plain. The pre-cambrian fold mountains which have been denuded to a peneplain are composed of mainly mica-schist and phyllites of the iron ore series. Phyllites and schists bordering the range are its remains . It consists of group of continuous hills broken by a few streams. The south-eastern part of this range extends upto the Subarnarekha valley. The general elevation of this region is about 300m but there are peaks like Charai Pahar 700m, Mankaraora 775m which rise above the sky line. Some parts of this range are capped with laterite and bear evidence of tertiary deposition and subsequent uplift. The area is densely forested. The Gara river valley provides fertile land where agriculture is relatively better. The area is dotted with tribal settlements.

12. The Kolhan Upland: This upland lies in the southern part of the state. It is bounded by Saranda highland in the west, Chaibasa plain in the north and state boundary in the south and east. It has flat summit at 660m above sea level and it is supposed to be the southern extension of the Ranchi plateau.[1] This area is highly dissected and shows greater differentiation in settlement types and their distribution because of rough terrain. The residual hills are generally devoid of habitation due to dense forest cover but the surrounding plains have more settlements . This region has tribal population. This upland exhibits radial slope. This upland is composed of rocks of varying resistance which determine different form of landscapes.

13. The Singhbhum Plain: This plain lies in the south-east of the Chotanagpur plateau covering an area of about 2000 Sq. kms. It has an average elevation of 130m and is drained by the Sanjay, a tributary of Kharkai. The plain has formed indifferently over the softer mica-schist and shales as well as granites of Singhbhum and Chakradharpur.[2] It has less diverse slope and as a result the entire area is cultivated and the density of population is high in the Chakradharpur and Saraikela areas. There are smaller nucleations of settlements and isolated homestead. Both tribal and non-tribal population are present in this area.

14. TheSimdega Plateau: This plateau lies in the south-western parts of the state. The area is bounded by the state boundary in the west and south but in the north and east there lie Ranchi plateau, Porahat highland and Singhbhum plain. The area is highly rugged and hilly with turbulent streams, steep slopes, high cliffs and narrow valleys. It is drained by the South Koel and the Deo Nadi whose tributaries are spread over the entire area. Due to rugged terrain, it is still under forest cover but small fertile valleys occur at places where tribal population live in relatively dispersed settlements.

15. The Dalma Range: This range lies on the bordering areas of Ranchi and Purbi Singhbhum. This region is composed of older lavas and somehow remained less affected by erosional processes. Important ridges are Debar, Raisindri, Dalma and Chadari Pahar. This range has been cut near Chakulia by the river Subarnarekha. The western portion extending from North Karo river to Subarnarekha river has been in regular chain but the eastern portion is irregular. This range also forms water divide between Kharkai and North Karo and Sanjay rivers. This has developed into an uneven land surface due to prolonged erosion. Basalts and laterites dominate the land surface but other igneous intrusions have presented surface exposure of granites and gneisses. This area has got considerable forest cover and tribal habitations in the form of relatively dispersed settlements.

16. The Chaibasa Plain: This plain lies in the west of Subarnarekha valley and is probably an eroded remnant of the uplifted Ranchi plateau. Kharsawan and Saranda hilly regions form the northern and western boundary and Kolhan upland and Dhalbhum hilly area determine the southern and eastern boundary respectively. It has an area of about 3500 Sq. kms. The average height ranges between 150 and 300 m. Important rivers of the area are Sanjay, Kharkai and Rodi. Surrounding hilly areas are highly dissected. The landscape is formed of igneous intrusions and Dharwarian rocks. The land slopes from south-west to north-east direction. The land is fertile and has relatively dense population. Both tribal and non tribal population reside in this plain. Kyanite and China clay are found in this region.

17. The Sarand Hilly Region: This hilly region lies between Porahat hilly areas in the east and the south Koel valley in the west. The southern boundary is demarcated by the state boundary. This region is also formed of Dharwarian rocks like phyllite, lavas, quartzite, etc. The exposition of igneous rocks determine the landscape of the region. Certainly there are areas where igneous and Dharwarian rocks have been highly metamorphosed. The hills are of varied size. The area has considerable forest cover and patches of agricultural land on account of which tribal people reside and manage their living with the help of forest and agricultural products.

18. The South Koel Basin: This valley lies in the south of Ranchi Plateau. The river South Koel spans over large catchment area in the south-western parts of the state. This river flows through wide course and has made a large headward erosion in the Ranchi plateau. The landscape of the region has been determined by the erosional process of the tributaries of the river. The land slopes towards south. This region has both forest and agricultural economy. The people are tribal by and large and reside in relatively dispered settlements.

19. The Shankh Basin: In the south of Ranchi Plateau there lie drainage basins of two important rivers the Shankh and the South Koel. These two rivers form separate basin upto the boundary of the state but while leaving the state they merge together and form Brahamani river. The Sankh valley lies in the south of Raidy and the Shankh river originates in Barbe plain area. The river flows through narrow valley and steep slopes on both sides are present. This narrow valley is about 45 kms. in length and has straight course. This region is highly dissected. There are several peaks having the elevation of about 650m. These hills have flat tops and steep slopes. They look like remnants of the Ranchi Plateau. Along the state border there lies a narrow belt having Dharwarian schists which has experienced intense erosion. The average relief is about 300m. The area is forested but the homogeneous areas have agricultural land and habitaion of tribal people. Settlements are relatively dispersed.

20. The Ajay Valley: It starts from the Kodarma plateau and forms valley east of Barakar-Ajay water divide. The region has been formed by the main river Ajay and its tributaries. These rivers flow through gneissic area where excessire erosion has developed lower valley in comparison to the neighbouring valleys. This valley is characterized by isolated hills and ridges. The lower parts of the valley has got gentle slope and the relief is about 160m. The upper areas of the region have relatively more forest cover but the lower areas have more agricultural land. This valley has also tribal and non-tribal population living in smaller nucleations and semi-sprinkled settlements. This area has got coal resource.

21. The Dissected Deoghar Region: This dissected plateau area extends between Kodarma Plateau in the west and Rajmahal hills in the east. The average relief is much lower than the neighbouring highlands. This region has experienced relatively more erosion and some hills are, however, above 650 m. The landscape exhibits undulating terrain. The higher areas have vegetational cover in the form of bushes and lower areas have more agricultural uses. The area is dominated by non-tribal population. Settlements of the lower areas look like semi-compact but the higher areas have relatively dispered settlements.

22. The Chatra Plateau : This plateau region extends between Amanat river valley in the west and Mohane river valley in the east. This plateau region has relatively rough terrain characterized by inter-locking spurs, high peaks and river valleys. The Western portion is more dissected especially west of Simaria and the lower areas having sedimentary depositions have better growth of forest. It comprises of tribal and non-tribal population. The settlements are, by and large, dispersed.

23. The Chattarpur Plain : This area is the upper portion of the river Punpun. It is part of the northern flank of the Chotanagpur Plateau. The average elevation ranges between 150 and 300 metres. This plain has wide river valleys being separated by ridges at some places. These ridges look more dissected. Sandy areas also dominate the region. The land slopes from south to north. The upper areas have considerable forest cover but the lower areas have more culturable land. Both tribal and non-tribal population live in this area. The settlements are relatively dispersed.

24. The Northern Scarp : This scarp region exists in the northern flank of the Chotanagpur Plateau. It has an average elevation 150 to 300 m. and has landscapes different from the southern highlands and northern plains. It has several hilly ranges extending from south-west to north-east direction. It has been formed by the recession of scarps. The scarp near Mohane has witnessed more recession and has reached upto Hazaribag and Ranchi plateau west of Simaria. This recession has given rise to the formation of Chattarpur Plain and the North Koel Basin. This area also exhibits dissected land surface. Both tribal and non-tribal have settled in this area. Settlements are relatively dispered.

3

Drainage

The drainage refers to flow of water received from natural precipitation and other sources. Such water flows through either rivers or get accumulated in low lying areas in the form of ponds, lakes, etc. These flow of water are categorised as surface drainage. In addition to the surface flow or surface accumulation of water, considerable amount of water gets percolated beneath the surface and such drainage is known as underground drainage.

The drainage of this region is basically rain-fed and there is hardly any other sources of water giving rise to surface flow of water. At the same time this region does not get water through rivers from outside except little area of the Son valley in the north west. The flow of water is more influenced by the variation in amount of rainfall as well as the diversity of terrain. The diversified areas have drainage of different nature than that of the relatively homogeneous regions. Underground drainage mainly depends upon the composition of rocks and alignment of beds which determine the percolation as well as underground storage of water. It may be noted that this region comprises varied type of rocks whose percolating capacity differ. This region slopes in all directions except in the west because the land has been uplifted more in the west. It is, therefore, the flow of river is in all directions except in the west. Local slopes have also determined the pattern of drainage in some parts of the region. Since the region has considerable slope it leads to quick run off rainwater and hardly allow water stagnation in the form of water bodies. Certainly artificial reservoirs and water bodies have been made to store rain water or river water for different uses. This stagnation of water also helps in the formation of underground drainage. Areas having joints, faults, cracks in rocks, etc. in the region have generally allowed the percolation of water. The vegetal cover also influences the surface drainage of water.

In addition to the aforesaid facts the drainage of this region has also been considerably influenced by the scarps formed during different upliftments. As a result most rivers have given rise to a number of waterfalls. The varied hardness of rocks found in the river courses also determines the vertical and horizontal processes of erosion and flow of water.

The rivers of this region have some special characteristics. Except a few rivers most of them remain dry except during rainy season. Certainly these rivers become turbulent during rainy season particularly at the time of heavy rainfall. Areas having high slope provide quick run off water and for few hours the rivers become furious. The rivers have been flowing in wider courses and these rivers form drainage pattern of different types. Most of the rivers of this state feed the major rivers discharging into the Bay of Bengal. To the north the main drainage axis runs from west to east before turning to north-east in the Rajmahal highlands, forming a divide between north and south-east flowing streams[1]. The centre of the drainage axis lies on an elevated tracts of the plateau south of Piska

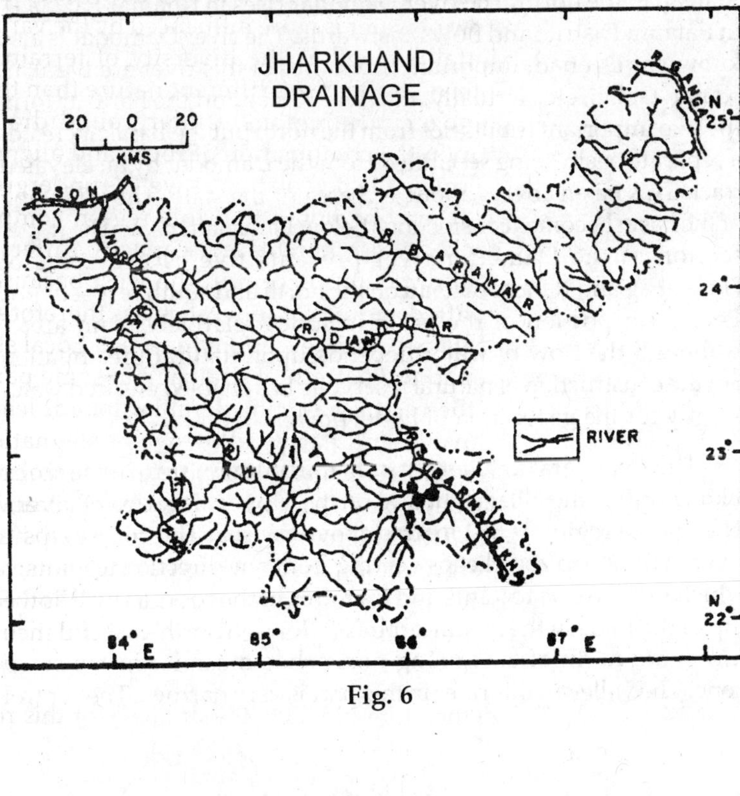

Fig. 6

near Ranchi. Three drainage axis radiate from this elevated tract of the state which divides the main river basins.

The eastward drainage axis forms watershed between the Damodar and the Subarnarekha Basins. The westward drainage axis forms a semi-circular water divide till it takes a south eastern loop to turn west and reaches eastern limit of the state. A narrow neck of highland links this drainage axis with the drainage axis following the northern flank of the highland[2].

Drainage Systems

This plateau region gives rise to several river basins which flow in different directions mainly in south, east and north. These river basins include Damodar, Barakar, Subarnarekha, North Koel, South Koel, etc.

1. The Damodar System : This system lies in the trough basin existing between the Ranchi Plateau in the south and the Hazaribag Plateau in the north. The river Damodar rises in Khamar Pat, 3504' in Palamu District and flows eastwards. The river Damodar is also known as Deonad. Important tributaries of this river are Barakar, Konar, Gobai, etc. Virtually, the river Konar and the Barakar form the two important tributaries from the north but the Barakar creates a separate basin being separated from the Damodar by an elevated tract. This basin covers about 1/3 area of the state. The widening of the basin becomes conspicuous below Parasnath peak. The upper section of the Damodar river presents five escarpments of Hazaribag and Ranchi plateau on the north and south respectively. The lower portion of the basin includes relatively plain areas. Although the flow of water depends upon natural precipitation but the construction of natural reservoirs has enabled ensured water supply for the generation of hydel power.

The river Barakar emerges from the southern scarp zone about 1 km. north to the village Bhuswa on the Upper Hazaribag Plateau. The source region is 582 metres above the sea. The course of the river in the scarp zone keeps veering from one direction to another which may be due to joints and fractures in the rocks as well as the dip of the folds in the metamorphic. It flows to north-east and then turns due north after forming horse-shoe meander in the scarp zone. The valley of this river in this part is very narrow. The control of structure on the flow of the river is apparent. The gorge, it has notched the northern escarpment of the plateau ,has reached a point only 9 kms. from the source near Padama. This river receives its leading tributary known as Kewta near Majhgawan. The length of this tributary is about 25 kms. and originates at an elevation of 590 metres. Its upper course is known as Asna river.

Another important tributary of the river Barakar is the Nadhadwa Nadi. It originates from the scarp of Mahabar Pahar. The next tributary is the Nonia Nadi which drains the area lying between the Nadhadwa and the Barakar river. It rises in the Bangh Pahar. Anjawadi Nadi is also an important tributary of Barakar. It originates on the outlying scarp ridge at an elevation of.640 m. Leriya Nadi is a slow moving sluggish river. It originates close to the western boundary on the Barakar Mohane divide. The Margoja Nadi is the only major tributary of the Barakar flowing from north to south across the Dharwar schist.

The first tributary from the northern side is the Garhi or Tandwa river which rises near Kariatu hill and drains the western portion of Barkagaon development block. Next comes the Haharo on the east of the Mahudi hill bearing the drainage of the eastern Karampura valley. The southern tributaries of Damodar have developed characteristic valleys incised the edges of the escarpment sending long arm into parallel ridges. Maikari Nadi has developed the largest number of such valleys. The tributaries of the Bokaro and Konar drain the south-western portion of the Hazaribag Plateau and join the river Damodar near Bermo. The river Sironi is the tributary of the Konar.

2. **The North-Eastern Drainage Basin** : There are several rivers such as the Ajay, the Mor, the Brahmani, the Gumani, the Bansloi, etc. which drain the eastern portion of the Rajmahal Highland and join the river Ganga. The most important river in the north of the Rajmahal highlands is the Gumani which rises in the Rajmahal hills in the extreme east of Godda district and makes its way north-east through the gorge which it has scoured out for itself. At Burhaigt it is joined by the Morel river coming down from the north and from this point the Gumani flows a short distance to the east and then turns sharply to the south. After a winding course of about 40 to 60 kms. it emerges from the hills and flowing eastwards

makes its way through the plain and joins the river Ganga.

The river Bansloi rises at a hill known as Bans Pahar in the Godda district and flows in easternly direction. It forms the northern boundary of the Dumka district separating it from the Godda and Pakur districts. This river emerges into the Dumka Damin through the Pachwara Pass and meanders along its northern boundary.

The river Brahmani rises in the west of Dudhua Hills in the north of the Dumka district and flowing through Parasemul and Sankora forms the Southern boundary of the Dumka Damin. Its main tributaries are the Gumro and Ero which drain the watershed between the Ramgarh and the Damin Hills.

The river Mero rises in the Tiur Hills at the extreme north east corner of the Deoghar district and drains the central portion of the Santhal Parganas. It enters the Dumka district at its north western corner and follows a winding south-easternly course. In its upper course it is known as the Motihara and it is only after its junction with the Burhi in Nawada it takes the name of Mor. Another name of this stream is Morakhi or Mayurakshi. Important tributaries of the river Mor are Bhurbhuri on the east of Dudhwa Hills, the Dhabai which rises in Godda Sub-division, the Tipra, the Pusaro, the Nunbil which rise in the east of Deoghar district, the Sidh which rises in the south-east corner of Deoghar district and the Dauna.

The river Ajay rises in the Jamui district and after draining north-western corner of the district of Deoghar flows in south-easterly direction. Important tributaries of this river are the Pathro and Jainti. Both these tributaries rise in the Hazaribag Plateau. The Ajay enters the Jamtara Sub-division at Kajara and flowing southwards forms the southern boundary of the district. Most picturesque waterfall in this river System is Moti Jharna, 3 kms. south-west of the Maharajpur Rly. Station. It has two falls each 50 or 60 feet in height. There are also two small falls on the river Brahmani and Bansloi. The first is at Singhpur and another is near the village Puskira.

3. **The Subarnarekha Basin** : This river basin drains the south-eastern portion of the state. The river Subarnarekha originates south of village Nagri, 15 km. west and south-west of Ranchi town. The

river pursues a winding course to the east till it descends from the plateau forming the steep falls of Hundrughagh (320'). Then it turns southwards and forms the boundary of the district of Ranchi and then traverses the districts of Singhbhum. It contains treacherous quick sand which is dangerous to cross. The name means streak of gold and gold is found in its bed in minute quantities. The eastern part of the basin is drained by the Garra Nadi which breaks through the Dhanjori range separating the Narwa Pahar from the Rangamati Pahar. It then turns south-east and flows almost parallel to the Dhanjori range and joins the main stream, the Subarnarekha opposite Rakha mines, west of Mahulia. The river Loubhang joins the Garra Nadi where it turns south-east after flowing out of the gorge. The river Dhudh Nadi joins the Garra before the river Garra enters the gorge.

The smaller tributaries prevailing in the southern hills are rapidly cutting back through the ranges on either side of the Subarnarekha river. The Garra might have flowed west into the Kharkai across the central plain but has been captured by the headwaters of a small north-east flowing tributary of the Subarnarekha which has cut through the range. This may explain the steepness of the gorge through which the river now flows across the range. On the north side of the Subarnarekha river, tributaries like the Gurma Nand, Bagalduta Nadi and the Dimra Jhar flowing through phyllite ranges owe their present trend for similar reasons. The extreme southern part is drained by another tributary of the Subarnarekha known as Sankh which joins the main stream about 3 kms. south-east of Mosabani. The Sankh with its numerous tributaries drains the north eastern slope of the ridge whose crest line generally forms the boundary between the Purbi Singhbhum and the Mayurbhanj district of Orissa.

The plain drained by the Sankh and its tributaries is enclosed on almost sides except on the north-east along the valley of the Sankh.

The river Kharkai, principal tributary of the river Subarnarekha being formed on account of confluence of two mountain streams namely, the Terlo and the Koranjai which rise in the eastern Kolhan range of hills. The Terlo joins the Koranjai on the boundary and is known there as Kharkai. The length of the Kharkai river is about 80 kms and it is fed by several streams from the Kolhan among which may be mentioned the Iligara, the Jamiragara and the Roro. The Roro is about 54 kms. in length and is joined by the Jamira just outside Chaibasa.

The Sanjay river rises in the forest clad areas of Porahat and flows in an easterly direction forming for about 45 kms. on the boundary between the Kolhan on the south and Porahat and Kharsawan on the north and joins the river Kharkai near Lengataisai. The Sanjay receives no streams worth mentioning from the south but from the north it receives the Binjai which drains Karaikela, Sankua from Kera and several tributaries from Kharsawan.

4. **The South Koel Basin** : This basin occupies the south-western part of the state. The river South Koel rises on the northern slope of the same ridge as the Subarnarekha and flows in a north-westernly direction and collects water of the Bandora Nadi and Birgora Nadi. It flows almost parallel to the highway of Ranchi-Lohardaga and turns south at a point of 12 kms. north-east of Lohardaga and flows a southerly and south easterly course till it passes into the district of Paschimi Singhbhum. In Paschimi Singhbhum district it again changes its course westward till it joins the river Sankh in Gangpur. The two rivers (South Koel and Sankh) united form the river Brahmani which flows through the district of Cuttack into the Bay of Bengal. Important distributaries of this river are the Sankh, the Karo, the Chata and the Bonai which drain the western portion of the Khuti subdivision.

The river Sankh rises in the north-west of Ranchi district and flows through the centre of the Rajadera Plateau and then turns abruptly south and descends to the plains of Barwe by a beautiful waterfall known as Sadanighagh. This river crosses the plains of Barwe in a south-westernly direction and flows almost due south. It also forms boundary between Ranchi and Raigarh (M.P.) and after passing through the hilly areas it finally joins the river South Koel. Important tributaries of this river are the Chirra, the Lawa, the Kok, the Kodar in Chainpur Area and the Kasgor, the Utial, the Palamara, the Girma, the Saijor, the Banda, the Dangajar in Simdega and Kurdeg Area.

The Karo descends from the plateau partly by a fine waterfall known as Pherughagh with a drop of 66'. In the north west corner of Singhbhum the headwaters of the tributaries of the Karo river have cut deeply into the edge of the Ranchi Plateau and form precipitous gorges. Afterwards this river receives waters of the South Karo. The combined stream flows southward for another 18 kms. till near Manoharpur where it is joined by the Koina. After this it has turned to the west and passes into Gangpur (Orissa) where it joins the Sankh. The river South Koel thus describes almost a complete semi-circle in a course of about 54 kms. through the Paschimi Singhbhum.

5. **The Son and the North Koel Basin :** This river basin covers the north-western part of the state in the districts of Garhwa and Palamu. The river Son originates in the Amarkantak hilly areas of Maikal Range located in Madhya Pradesh and forms boundary between the states of Jharkhand and Bihar in the north-western portion especially in Garhwa and Palamu districts of Jharkhand and Bhabua and Rohtas districts of Bihar. This river receives its main tributary as river North Koel near Rohtas Plateau in the north. The north Koel rises in upper area of Gumla and enters Palamu below Netarhat near Rud after flowing nearly due west for about 30 kms. then turns north as an almost complete right angle through a gorge at Kutu and flows almost in between Palamu and Garhwa Districts till it joins the river Son, a few kms. north-west of Haidar Nagar.

The North Koel receives the water of the Amanat, the Auranga, the Gogra and the Maila. The tributaries Amanat, Sadabah and Auranga rise in the Hazaribag Plateau and join the river North Koel. All these tributaries are mainly rain-fed but during rainy season become furious. The river Amanat rises in Hazaribag and flows almost due west till it joins the Koel river 8 kms. north of Daltonganj. Its tributaries with the exception of Jin Joi, Maila and Piri all are small streams. The Auranga rises neâr Soheda in a pass, comes down from the Plateau and pursues a winding course in a north-westerly direction for a distance of about 80 kms. till it flows into the Koel near Kechki 16 kms., south of Daltonganj. Its main tributaries are the Sakri and Ghagri. The river Sadabah, an ever flowing stream, flows in the north of Daltonganj about 25 kms. away and has assumed importance due to the completion of an irrigation scheme which was completed in the year 1953. Other streams draining important areas are The Sarabdaha, The Tahleh,

The Banki which drain the area between Utari Road Station and the Danro which flow through Garhwa and the Karabar and these flow into the Son. below its junction with the Koel after draining Pargana Japala.

Besides the above mentioned river basins there are a number of important rivers which rise in Jharkhand state and drain into neighbouring states. The important among them are Punpun, Phalgu, Mohane, Sakari, Panchane, Kiul, etc.

The river Sakari drains an area of 810 sq. miles which has carved out of the northern face of the Chauparan, Kodarma, Giridih Sub-Plateau. The most important tributary is Kunda Nadi which joins Sakari from the south. The Ghalki and Chhotna are the tributaries on the west and Likhar and Ghaghra on the east.

The Mohane rises about 18 kms. from Hazaribag and drains the north-western part of the Hazaribag Plateau. Its tributaries from the west are Dhab and Garhi.

Drainage Pattern

The tabular landmass of the state accompanied with peneplanation, upheaval in phases, warping and tilting, igneous intrusions, etc. have played remarkable role in providing layout of the rivers and determine density and frequency of streams per unit area. The common drainage pattern has been dendritic, parallel, radial, superimposed, etc.

Dendritic pattern is commonly known as trellis pattern, in which rivers exhibit three parts - root, stem and branches. In Jharkhand state this pattern only contains the stem and branches and root lies in the plains of the neighbouring states. In this pattern tributaries are supposed to join the trunk stream at acute angles and the distributaries diverge at similar angles. Certainly such pattern indicates homogeneous rock or horizontal beds or low flat surface lacking marked slope. It implies a striking lack of structural control, for example, as the rock is uniform in hardness the river has no choice like the selection of a region of soft rocks and avoidance of that of hard rock[3]. It flows indifferently to the rocks. Almost important rivers of the state belong to this category of drainage pattern.

Parallel pattern is found in limited areas. In this pattern rivers flow parallel to each other. Rivers flowing towards Bihar state appear parallel to each other.

Radial pattern is found mainly around the hills like Parasnath where rivers radiate from the top of the hills in different direction. Around Netarhat, Pat Region, the rivers exhibit a different pattern which may be categorised as superimposed pattern. This pattern is also obvious near Rajrappa waterfall on the river Bhera.

Waterfalls & Springs

The highland of this province coupled with upheavals in phases, faulting, warping, tilting and other structural dynamism has given rise to several waterfalls and springs. Important waterfalls are on the river Subarnarekha, Damodar, South Koel, etc.

Hundru water fall in the district of Ranchi on the river Subaranrekha, 36 kms east of Ranchi, has been resulted by the scarp edge of the Ranchi Plateau. It has a drop of about 243 feet. The Dasam fall humbler than Hundru, 24 kms south-east of Ranchi on the river Kanchi has a precipitous drop of 130 feet. The Sadani waterfall on the river Sankh in western Ranchi is a picturesque fall of 200 feet high. The Jonha waterfall about 30 kms east of Ranchi town on the Purulia Road is another impressive waterfall. There is a waterfall of only 16 m. high, 6 kms. north of the Dhirol on Chaibasa-Maholia Road on the river Garha, a tributary of the river Subaranrekha. The waterfall of Hirni nearly 68 m. on the Ranchi - Chaibasa Road is another important waterfall in Singhbhum. The Rajrappa waterfall on the confluence of the Bhera river with the Damodar nearly 22 kms. east of Ramgarh on the southern scarp of Damodar valley is also important. There are a few waterfalls in the Rajmahal Highlands also. Moti Jharna on the Ajay river has a fall of about 50 metres. The highest waterfall of the state is Burhaghagh (465 feet) on the river Burha in Chechari Basin. Another waterfall of 140 feet is the Ghaghri, nearly 6 kms. north of Netarhat. Among the smaller falls there are two Peruaghagh one in Basia development block and the other in Kochadega development block. Some other waterfalls which deserve a mention are Hapad, 30 kms. from Vishnupur, is source of river Ghaghara, Kelaghagh, about 3 kms. south-east of Simdega as a perennial channel running through two hillocks, the Pairnaghagh is about 5 kms from Tapkara. Panchghagh about 14 kms. from Khunti has tragic story woven round its five streams. Serka fall, about 1 km. to the east of Vishnupur is in the course of river Serka. Sugakataghagh, about 23 kms. from Simdega is a fall of the river Sankh where it passes between villages Hardibera and Purna Pani.

There are a few springs on the steep slopes of hills, particularly along the numerous scarps bordering the different erosion surfaces in the state. These springs are caused by the underground watertable or saturation level within the earth surface. These springs give rise to streams and rivers. Also there are several hot springs in the state. In Santhal Parganas, there are seven hot springs. In the district of Pakur the hottest spring is known as Laulaudah near the bank of a small stream called the Boru, about 1 km. north-west of Sibpur village in Maheshpur Block. Another hot spring near Birki in the same development Block is known as Baramasia. In the district of Dumka there are six hot springs namely Jhariya Pani near Gopikandar, Tatloi on the bank of the Bhurbhuri river near Palasi, Nunbil near Kenolghuta, Tapat Pani on the left bank of the Mor near Kumrabad, Susum Pani on the opposite bank of the Mor near Baghmara and Bhumka on the right bank of the Mor river near Ranibahal.

References :

1. Singh, R.P. and Kumar, A. (1970) : Monograph of Bihar, Bharati Bhawan, Patna, P. 15.

2. Ibid.

3. Ahmed, E. (1965) : Bihar :Physical, Economic and Regional Geography, Ranchi University, Ranchi, p. 33

4

Climate

The landlocked state of Jharkhand is, climatically, a little different from the neighbouring states. Its chief climatic elements such as temperature, precipitation, pressure, wind, relative humidity, etc. reveal variations. These variations are influenced by its geographical location in tropical and sub-tropical regions, the overhead presence of tropic of cancer, considerable areas having high elevation, nearness to the Bay of Bengal, etc.

This state extends from 21^{0}58' N. Lat. to 25^{0}18' N. Lat. and 83^{0}22'E. Long. to 87^{0}57'E. Long. The tropic of cancer lies overhead Ormanjhi. Therefore, the region experiences both the tropical and sub-tropical conditions. The overhead presence of tropic of cancer should enable hot climatic condition over here but 1/3 area having an elevation of more than 400 m. and 2/3 area having elevation over 300m. have enabled the region to have temperature about 6^{0}F. less than the areas of Ganga plain. In dry season the temperature of this province remains 11^{0}F. lesser than the Ganga Plain. The impact of elevation is also realised on the pressure of wind which leads to the formation of Low Pressure and wind does not remain stagnant. As a result of this the weather conditions are congenial for human being. This state comes in the influence of both branches of monsoon which come from Arab sea and the Bay of Bengal. It is, however, the impact of the monsoon winds, the Bay of Bengal remains more effective in the region because the Bay of Bengal is hardly at a distance of 150 kms. The regional variations are, however, witnessed due to orographic variations.

Monsoon winds coming from the Bay of Bengal and the Arabian sea influence the climatic conditions of the region distinctly. The wind coming from the Bay of Bengal contains more moisture content because of its nearness than that of Arabian sea. It is, therefore, the temperature also remains relatively lower in case of Bengal monsoon. But the winds coming from Arabian sea passes over a large peninsular area and it, therefore, contains lesser moisture and relatively higher temperature but when this wind reaches the highlands of Jharkhand it is raised and meet to the monsoon of Bay of Bengal. Under such circumstances, the formation of front takes place. The wind of Bay of Bengal remains beneath the wind of Arabian sea. This front remains in the direction of North-West in South-East Region.

The place of front remains fluctuating depending upon the intensity of winds. In case the winds of the Bay of Bengal is forceful the Arabian winds exhibit retreating trend and vice-versa.

The climatic year is divided into three broad seasons : (i) The winter season lasting from November to February, (ii) The summer season from March to May and (iii) Rainy season from June to October.

(i) **The Winter Season** : During this period the state experiences anti-cyclonic conditions. The high pressure system of north western India sends an arm in the south -east which includes the region. The wind blows from the north west to south-east but the actual direction of the wind is controlled by the relief features. The temperature remains between 60^0F and 70^0F. This period is characterised by charming, clear and fine weather with pure air and azure blue sky. The days are slightly warm and the nights are cold. The months of December and January are the coldest. During this period Dumka, Ranchi and Dhanbad record the temperature of 7.5^0C, 5.2^0C and 9.5^0C respectively. Under the effect of temperate cyclones the region receives smaller amount of rainfall. The rain is, however, of immense importance for Rabi crops. In the month of January, some parts of the region record cold winds and frost. The wind throughout the season is mostly slight (3-5 kms. per hour) and is predominantly from north-west and west. These winds are fed by the downward setting of the air over north-western India, caused by low temperature and high pressure. These continental winds from north-west are not merely a surface feature, virtually, they are experienced in the upper strata.

(ii) **Summer Season** : This season begins from the very beginning of March. The temperature starts rising from the first week of this month. During this period, the sun gradually moves northwards and the duration of the day increases bringing more and more insulation to the area. The Sun comes over the tropic of Cancer on 21st June which passes through the northern portion of the state.

As a result low pressure area develops on the plateau. During April the wind blows from the west. It remains relatively hot and the temperature rises upto 160°F around Ranchi. By the end of May, the low pressure intensifies and extends towards the west. Afterwards the westerly winds begin to fade away. During this period slight precipitation takes place. A characteristic feature of the season, however, is the afternoon or evening storm burst after the heat of the day and form the main source of rain of the season. In this season eastern part of Santhal Parganas and south-east Singhbhum receive 152mm rain, Ranchi receives 106mm, Hazaribag receives 64mm and Palamu district 51mm. During the season average rainfall varies from 50 to 150mm. This season normally experiences dust-storms, thunder storms and whirl winds. The wind remains normally very hot. The north-eastern part of the state comes under the influence of Norwester which causes little rain and subsequently it reduces the temperature. The higher elevation and the vegetal cover also influence the region in recording relatively lower temperature. However, average temperature remains less than 38°C excepting the North-Koel Basin, Chaibasa plain and Industrial areas. The temperature increases from north-east to south-west direction. Average temperature recorded at Dumka, Ranchi, Hazaribag, Chaibasa, Jamshedpur,and Daltonganj is 28°C, 31°C, 31.6°C, 32.8°C, 40°C and 41°C respectively. Jamshedpur shows higher temperature mainly due to the impact of the industries. During this season relative humidity decreases and the pressure of winds also decreases. The wind becomes dry and warm.

(iii) **Rainy Season** : This season normally begins from the middle of June, when the monsoon winds bring moisture containing clouds from the Bay of Bengal. Rainfall during this period seems to be partly guided by altitude as the Ranchi plateau, the south-western dissected lands and the Dhanjori uplands receive over 1250mm. The occurrence of heavy precipitation is the most important characteristic feature of this season. The distribution of rainfall is not even throughout the rainy season. The occurrence of heavy rain during short period creates flood. The districts of Garhwa and Palamu receive relatively lesser rainfall i.e. 120cm. Chaibasa receives 1250mm. Pat region receives the highest amount of rainfall (175cm or 1750mm) because of high elevation.

Table - 4.1
Average Rainfall and Temperature
(Rainfall in MM and Temperature in °C)

Months	Dumka		Ranchi		Daltonganj		Chaibasa		Dhanbad	
	R	T	R	T	R	T	R	T	R	T
J	13.21	17.95	26.7	16.85	27.3	16.1	20.6		14.9	19.7
F	17.17	20.60	43.9	18.8	35.5	18.1	33.4		25.6	21.65
M	25.2	26.20	29.0	23.9	19.5	24.2	21.8		19.8	26.15
A	22.5	30.35	21.8	28.5	9.6	28.35	28.8		17.6	28.6
M	88.2	31.70	54.6	30.6	18.6	32.25	72.2		51.1	32.2
J	251.7	30.55	216.7	28.8	165.4	33.35	212.0		192.6	32.2
J	337.0	28.55	296.7	26.0	367.0	31.35	279.8		345.5	29.9
A	239.0	28.35	373.1	25.5	395.1	28.9	352.0		323.2	28.9
S	233.7	28.30	232.4	25.5	212.6	27.6	210		212.1	27.85
O	94.5	26.55	90.4	23.6	61.8	28.6	81.1		93.5	25.0
N	10.2	21.70	20.4	19.5	17.1	19.45	16.6		10.5	21.1
D	2.5	18.65	7.1	16.6	5.6	16.4	4.3		4.2	18.9
Annual Rainfall and Temp.	14272		1512.7	284.15	1335.1	304.65	1334.6		1310.6	312.15

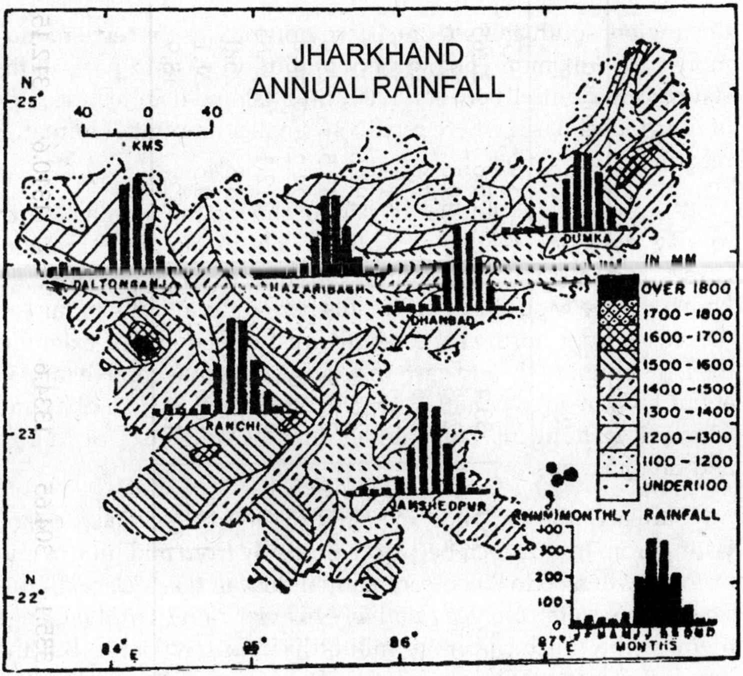

Fig. 7

The table 4.1 further provides the pattern of temperature and rainfall for some important towns of the state. Their analysis explains comparable pattern of rainfall and temperature as stated in above paragraphs.

Regional Variation in Rainfall and Temperature

Seasonal variations in climatic elements like rainfall, temperature, humidity, wind direction, etc. have been discussed in the earlier pages. In this portion emphasis has been laid on regional variation in rainfall and temperature. Variations in rainfall and temperature are influenced by the differential elevation, alignment of plateaus and ridges, hills, proximity of the Bay of Bengal, vegetal cover, etc. The average annual rainfall in the state ranges between 1250mm and 1500mm but because of variations in relief and other conditions considerable differences in the amount of precipitation are observed. The Pat region having very high elevation in the state records more than 1500mm rainfall. Besides this region, south and south-eastern portions also receive rainfall more than 1400mm. The northern and north-eastern parts of the state receive rainfall between 1200 and 1400mm. It is only the parts of north-Koel Basin where relatively smaller amount of rainfall is recorded.

As regards distribution of rainfall during the months March to May when hot weather season prevails, the amount of rainfall is received in smaller amount in the form of Mango-shower i.e. known as Pre-Monsoon. During this period the amount of rainfall decreases from north east to south west direction. For example, the Dhalbhum plain and eastern Santhal Parganas receive rainfall about 110mm but the rainfall decreases to 90mm at Hazaribag and 43mm at Daltonganj. This rainfall remains congenial for Mango and Bhadai crops.

Similarly the variation in rainfall is felt in the rainy season lasting from June to October, more precisely from mid June to mid October when 80 to 90 percent annual rainfall takes place. In this season this state receives relatively heavier rain than the Ganga plain in the north. Average rainfall is about 1200mm. But the heaviest rainfall takes place in the Pat region because of higher elevation and the south-western portion receives larger amount of rainfall because of the presence of hilly areas covered with dense forest. In this period North-Koel Basin receives rainfall ranging from 1200mm to 1250mm but the south-western portion obtains rainfall between 1350mm and 1450mm.

During the Winter season, the variation in rainfall is obviously observed. The amount of rainfall, no doubt, is meagre in this season but decreases from west to north-east and from south to north. The westward increase is understandable but the southward increase, at first sight, is perplexing. This is due to some rainfall from early spring storms in February and it is this which swells up the total. As a result of the retardation of the maximum till February, there is a danger of the rains coinciding with the flowering time of the Rabi crops in which case serious damage may be done.[1] In this season the average rainfall is above 50mm but north-western Pat region receives rainfall over 80mm.

Similar to variation in rainfall regional variation in temperature is also realised. The mean minimum temperature varies between

55^{0}F and 60^{0}F on the high plateaus of Ranchi and Hazaribag. In December the mean temperature falls down by nearly 6^{0}F to 8^{0}F. In case of Santhal Parganas the range of temperature is not recorded very high except during the hot weather months of March, April and May when the westerly winds coming from Central India cause high temperature with very low humidity. The mean temperature in January at Dumka falls to 17.95^{0}C. During summer season the highest temperature is recorded in the month of May i.e. 31.70^{0}C. At Tilaiya minimum monthly temperature in January is recorded 16^{0}C and the maximum monthly temperature in May is recorded 31.5^{0}C. Netarhat records mean minimum temperature in January much below 45^{0}F.

The mean annual temperature of this region also records spatial variation. Most areas of Ranchi plateau and highlands of Pat record 23^{0}C but the south-eastern portion records mean annual temperature above 26^{0}C . The north-western portion records mean annual temperature of above 25^{0}C. The isotherm reveals increasing trend of mean annual temperature from West Central portion to towards north-west and south-east.

As regards the degree of comfort to the people of the region, the highlands of Jharkhand even having tropical location provides sub-tropical and sub-temperate conditions over large part of its elevated region. The region remains mostly free from the hot loo winds, dust raising winds, dust storms of the plains[2]. The relative coolness of the climate during rain is a point of distinct superiority over the muggy condition of the season. The moving cool air provides pleasant and energizing conditions. The temperature during rainy season remains lower than plains. As such the area provides better and comfortable conditions than the plains, the relative absence of pestering insects which remains so common over lamps during rains in the plains is a boon to the plateau. Moreover the region also remains to a considerable extent free from dreaded epidemics. Because of undulating surface, the area hardly experiences bad effects of flood. Even this region lacks unwelcome muddy surface after the rain. Certainly little climatic discomfort is realised in the period mid-April to May when the temperature rises a little higher.

While observing the data of the rainfall and temperature of recording stations in different areas of the region and analysing the isotherms and isohyets, some areas provide clear formation of climatic zones :

1. Pat Region provides relatively more rainfall and mild temperature because of higher elevation and considerable forest cover.
2. The higher areas of Ranchi and Palamu : The higher areas of Ranchi Plateau covering parts of Lohardaga, Gumla and Southern portion of Palamu record rainfall between 1400mm and 1700mm. These higher areas also experience temperature relatively lesser than the remaining areas of Ranchi Plateau.
3. The third zone lies in the area having rainfall between 1400 and 1500mm. This area includes most parts of south-eastern portion,and most of the Hazaribag Plateau. The central portion of North Koel Basin and the Panch Pargana Plain.
4. The highlands of Rajmahal forms a separate climatic region because of the proximity of the Bay of Bengal and the rainfall occurs between 1400mm and 1500mm.

References :

1. Dayal, P. (1953) : Bihar in Maps, Kusum Prakashan, Patna, p. 20.
2. Ahmad, E. (op.cit.), p. 51.

5

Soils

Soil forming the upper layer of earth crust has been important natural resource for human habitation, agriculture, pactoral activities, vegetative growth, etc. As such this resource is being given due importance during the recent years by the scientists and scientific analysis has been accelerated in order to solve its problems. No doubt soils of any region depend upon parent materials coupled with geological formations of the area and configuration of land surface and associated erosional and depositional processes. Certainly in characterising soil properties as well as its composition the role of decomposition of plant fossils and human interference may be ignored.

Soils consist of various chemicals, physical and biological properties. The colour of soils which is seldom derived from the parent rock but acquired by subsequent changes, often indicate their organic content. Important components of soils are clay, sand particles and silt and these three mainly determine texture. Other contents like organic matters and chemicals determine the colour of the soils. High content of ferric oxide and bauxite which provide red colour. As such soils vary regionally in colour from laterite or lateritic soils on high plateaus surfaces to yellow grey loam and black and brown soil in valley beds. Important constituents are lime content, compound of silicon, aluminium and iron. Soils rich in lime content are called pedocals and while those rich in iron and aluminium and with low lime content are known as pedalfers, The distribution of these two types of soils are mainly influenced by the differential amount of rainfall. Pedocal soils are found in areas having less than 1000mm rainfall but pedalfers are mostly found in moist areas.

Soils differ in properties vertically based on dissolved substances due to percolation of water and intensity of human interference. It becomes obvious by the analysis of soil profiles. The upper most layer contains more substances like dissolved chemical properties, higher silicon content and plant fossils. Below the topmost soil there lies higher content of salt because of the precipitation of salt by percolation processes. The lower layer remains relatively unaltered soil layer.

It is true that parent materials are main determinants of soil formation but the role of denudational agents can not be ignored because they modify the original characteristics that determine the nature of soils. They bring about the decomposition of the mineral forming the rock waste. They effect the translocation of substances within the solid mass. They cause new structures. They develop the biological relationship of the soil and evolve new entity that a body of a matter quite unlike the original debris[1]Thus the main soil forming factors become the parent materials, climate with its alternating humidity and desication, natural vegetation, fauna, land terrain and human interference[2].

Since the entire area of the state has dominance of metamorphic and igneous rocks the area presents quartz and its different varieties of oxides of silica and other mineral contents which have important role. Mechanical disintegration of quartz gives rise to the hard silica particles of which sand mainly consists. In the formation of sandy soil the role of quartz become appreciable. Similarly the role of feldspar becomes important in the formation of clay because of silicate of aluminium with silicate of potassium or of lime and sodium. Feldspar undergoes quicker mechanical erosion and chemical decomposition than quartz. After the decomposition of feldspar it gives rise to sand from silica and clay from kaolin and such chemical constituents as carbonates of calcium, sodium and potassium. The composition of hornblende and augite also gives rise to similar soils which originate due to felspar because these minerals contain silicate of aluminium, calcium, magnesium and iron. The role of mica in the formation of clay becomes important because of silicate of potassium, aluminium and iron. The decomposition of calcite enriches the soil with calcium carbonate and lime content. Similarly dolomite also provides lime content.

In the formation of soils the role of different rocks is also not less remarkable. The area has vast coverage of rocks like granite, gneiss, schist, laterite, quartzite, etc. Limited areas also have sandstone, limestone, slate, etc, These rocks also influence the variation in soils. Granite consisting mostly of orthoclase feldspar, quartz, mica, etc. gives rise to partly clay and partly sandy soil. In areas having granite containing hornblende in place of mica provides soil with sufficient lime content. The soils at higher levels remain coarse and sandy but in the valleys remain clayey. Areas having sand stone particularly in humid areas poor sandy soil comes into existence. Shale gives rise to clayey soils which remain rich in potash and poor in lime and phosphoric acid. Soils derived from lime stone and dolomite are mostly clayey having rich lime content. Laterites of this region have given rise to infertile porous soil. As such laterite soils are thoroughly leached and are poor in lime, magnesia, phosphorous, potash and nitrogen. Red colour of laterite soil is derived due to iron oxide.

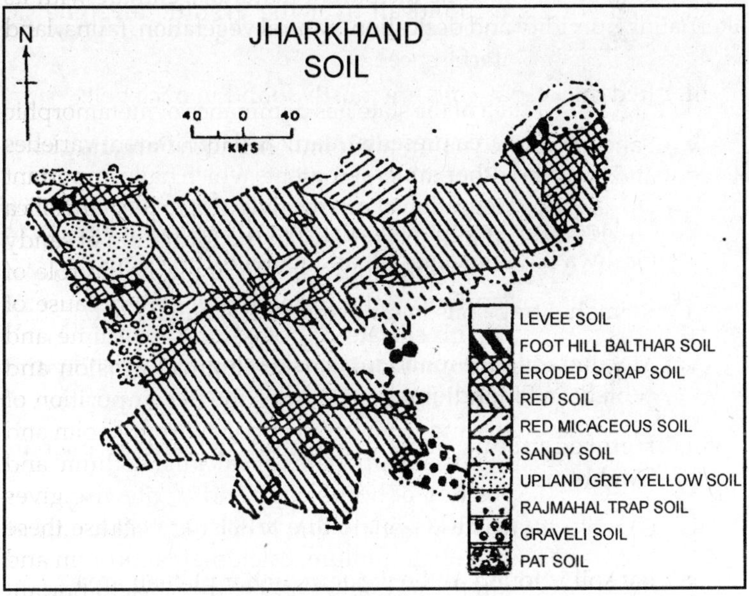

Fig. 8

The area is noted for considerable vegetative cover. The soils of this region have got largely influenced by the decomposition of plant fossils. Certainly areas having considerable forest cover have soils containing more organic content than the areas having rocks exposure. The rate of decomposition determined by climatic conditions like moisture and temperature also matters.

Soil Types :

The undulating areas made of varied geological structure having differential forest cover, human habitation and interference of erosional and decompositional processes, etc. have given rise to various types of soils. These soils differ in their physical, chemical and biological characteristics. Prof. R.P. Singh and Dr. A. Kumar have analysed soils of this region and divided into ten groups which are the following:-

i. Eroded scarp soils are found in scarp areas along the edges of the highlands.

ii. Red soils are found all-over the granitic and gneisic surfaces.

iii. Red micaceous soils are mostly found in mica belt.

iv. Sandy soils are commonly found in Panch Pargana plains and southern Santhal Parganas.

v. Upland grey yellow soils are generally found in Palamu & Garhwa having shale, silt and metamorphic rocks.

vi. Rajmahal trap soils in lava capped Rajmahal area.

vii. Metallic soils are found in areas having rocks rich in iron content in Singhbhum.

viii. Heterogeneous soils are found in central part of Singhbhum.

ix. Gravel soils occur in tertiary gravel belt of Dhalbhum and

x. Pat soil is found in the western higher plateau area.

Broadly the soils of this state can be classified into four types:-

1. Red and Yellow soil.
2. Loose sandy soil.
3. Black clayey soil.
4. Laterite soil

1. **Red and Yellow Soil**: This soil is found in most areas of the

state except in narrow trough basin of Damodar and laterite capped areas of Rajmahal highlands. These soils have been formed by the decomposition of crystalline metamorphic rocks like granite, gneiss, etc. because the gneiss and schist of this region contain mineral particles like biotite mica, hornblende and iron content. As such this soil gets red colour. Certainly the higher areas have soils with light red colour but the lower areas have relatively dark colour. Similarly the proportion & size of sand differ from region to region. It lacks nitrogen, phosphorus, acid and humus. Potash and lime content are sufficiently found.

2. **Loose Sandy Soil** : This soil is mostly confined in trough basin of Damodar. The formation of the soil has been based on Gondwana rocks. Sandstone and shales of this region have given rise to loose sandy soil and sandy loam because of their decomposition. This soil has poor humus content and is relatively lesser fertile. In colour it is yellowish.

3. **Black Clayey Soil** : This soil becomes sticky when it remains wet and becomes hard when it remains dry. It contains higher proportion of clay but it has high water retaining capacity. This soil is usually formed of basalt and has much more productive capacity than the soil formed of granite and other rocks. The higher areas contain thin layer of this soil having light colour but in lower areas this soil has very thick layer and is relatively of dark colour. In terms of fertility also this soil of lower areas has better capacity than the soils found in higher areas.

4. **Laterite Soil** : This soil has relationship with lava cappings held in early geological period. Such areas are available in the highlands of Rajmahal, Dhalbhum region and higher western Ranchi Plateau and Pat region. This soil has dark red or brown colour. It has got such colour because of iron content. The main defect of this soil is its acid reaction.

Soil Erosion & Conservation

The undulating areas of this state coupled with differential deposition and cultivation and varied resistant rocks have caused soil erosion in different proportion. Areas where people have mercilessly cut the forests have developed rocky surfaces. Such areas have experienced maximum soil erosion. Virtually softer soils of these areas have been washed away and hard bare rocks are exposed where neither forest nor cultivation can be done. At the same time people have unscientifically developed agricultural plot ignoring the slope which leads to erosion of soft soils in the process of tillage. Areas lying between two rivulets known as Tanr easily loose soft soil in process of cultivation and become less fertile. On the other hand the areas adjacent to the rivulets lying between two Tanrs known as Don land get accumulation of softer soils and become fertile.

During last four decades contractors and others have also ignored the consideration of soil erosion caused by deforestation. They have exploited the traces without considering the problems of soil for immediate gains. Even the innocent tribal inhabitants were not given other alternatives of livelihood by the government as these people continued these process of cutting the trees for maintaining their living. The govt., agencies did not take appropriate action to restrict mercilessly uprooting of trees and developing agricultural land here and there without considering the slope and problem of soil erosion.

It is only recently that people have realised the adverse impact of deforestation in terms of soil erosion, development of waste land and environmental degradation. The Govt. has developed a few major and minor schemes in order to control soil erosion, protect forest resource and wild life. The development of Damodar valley project has also played important role in afforestation, control over soil erosion and devastating impact of flood. Smaller irrigation schemes have also been developed for storing rain water, maintaining and developing forest. Koel-karo project has also been undertaking for providing multi facet development, whose impact will be also on controlling soil erosion and preserving forest resource. Still there is need of acquiring people about the problems of soil erosion and its adverse impact on agriculture, forest resources and environmental degradation.

References :

1. Wolfanger, L.A. (1929): The world soil groups, Geographical Review, Vol. XIX, P-100.

2. Ahmad, E (1965) : Op.cit, P-55.

6

Vegetation

Jharkhand, the land of forest, has got its nomenclature, primarily on the basic of wide spread dense cover of forest. The state has 23478 Sq. Km. of land underforest which is about 29.45% area of the state. Its valuable forest resources have been ruthlessly depleted by ignorant aboriginals, greedy contractors and other agencies without considering the implications of forest cover on environmental degradation, soil erosion and occurrence of flood in the lower parts of the region. It has resulted vast stretch of culturable land in homogeneous or gentle sloping flat plateau areas or in river basins or in other easily accessible areas, waste land in the hilly areas due to erosion of softer soil and deforestation in the mining and industrial areas where forest cover appear almost non-existent. However, some areas having hilly landscape, rough terrain, narrow valleys, steep scarps and other poorly accessible areas have still the presence of reasonable percentage of forest land. Spatial variation in the intensity of forest cover, plant species and growth of plants depend upon variation in land profile, elevation, drainage, amount of rainfall, typology of soils, etc. Broadly the forest of this region falls in two categories : Tropical Moist Deciduous forest and Dry Deciduous forest. However, some areas contain some plant species of cold climate like pine, chir, etc. in Pat region where cool conditions prevail in winter season. Some areas also represent mango, mahua bamboo, etc. of tropical evergreen forest. Among the forests three categories of vegetation are found: These vegetation include tall trees, dense bushes and the grasses and give a look of three layers.

A large part of the state receiving more than 120 cm. of rainfall is occupied by forest and woodland ranging from stunted open growth on slopes, bare soil on the southern drier slopes facing the sun, to fairsized dense jungle in more favourable situations, e.g. areas of thick soil, moist valleys and ravines, etc.[1] Thus the rainfall and other climatic conditions being largely uniform topography, soil, distribution of moisture and other aspects become important conditions in influencing the local nature of vegetation. Different types of rocks forming different soils encourage varied type of plants. Areas having phyllites become more suitable for bamboos, and sals grow in better way in areas having schist and ancient sediments. Most plants of this region become leafless in early summer but Sal having undergrowth remain mostly evergreen. Sal predominates in the moisture and lowlying situations of thick soil while on the drier, steeper or upper surfaces having thin soil plants of smaller size e.g. Salai, Dhak, Aonla, Hara Piar, etc. get an upper hand on Sal. In areas having thick soil protected valleys and basins dense cover of Sal is found alongwith Asan, Kend, Gambhar and Semal.

The dry deciduous forest occurs in areas having less than 120cms. rainfall. It covers the poorer forest. The common forest is a mixed stunted growth consisting of bamboos, Amaltas, Semal, Hara, Abnus, Plas, Khair, Mahua, Asan, etc. Sabai and Kus grasses are also found. This type of forest is supplanted by field crops, fruit garden and orchard and the vegetative cover looks like park land type.

In the North Koel Basin the lower portion has become almost devoid of forest cover but the upper portion and the divided terrain still have considerable forest cover. In this region about 44% area has been under forest. Notable concentration of forest is mainly confined to hilly terrains of the south where narrow valleys, steep sloping areas, high peaks, etc. are covered in the thick forest. Forests located in the divided terrains have been reduced to thorny bushes with isolated big trees. There are the government owned forest like Betla Reserve Forest, Khalha Reserve Forest, etc. Important trees in this area are Sal, Asan, Karam, Kend, Guri, Kusum, Galgal, Ginjan, Piar, Sins, etc. Bamboos thickets are found almost everywhere.

In Ranchi plateau the forest areas are mainly confined to its hilly tracts. In this area forests cover nearly 25% area. The central plateau is comparatively bare of vegetation where culturable land predominates. The largest compact forest areas occur towards the north, north-west and western higher areas. The northern scarp facing the Damodar Basin has forest cover especially where poor accessibility prevails. In the north-western portion where the elevation increases and hilly terrain provides less suitable condition for agriculture, forest cover has not been more affected by human interference. In the western portion where land is more hilly and amount of precipitation also increases, luxurious growth of plants is visualized. Another forest area occurs in the south-eastern corner in Tamar. The forest in Angara, Silli and Ormanjhi areas are very scattered and now only the upper slopes of the hills are covered with comparatively poor type of forests. Forests of this region have been depleted for cultivation by the aboriginals who started felling trees and burning them to enrich the soil on which they raise crops Sal is found every where. In the valleys especially in the sheltered situations the principal companions of Sal are the Asan, Gambhar Kend, Semal, etc. Mahua is known throughout the region but in the forests it is chiefly comfined to the hills. In the interior Sal forest on the hills are found Dhaura, Piar, Sidha, Khair, Amaltas and bamboos are also found. The palas is often gregarious in cultivated and waste lands.

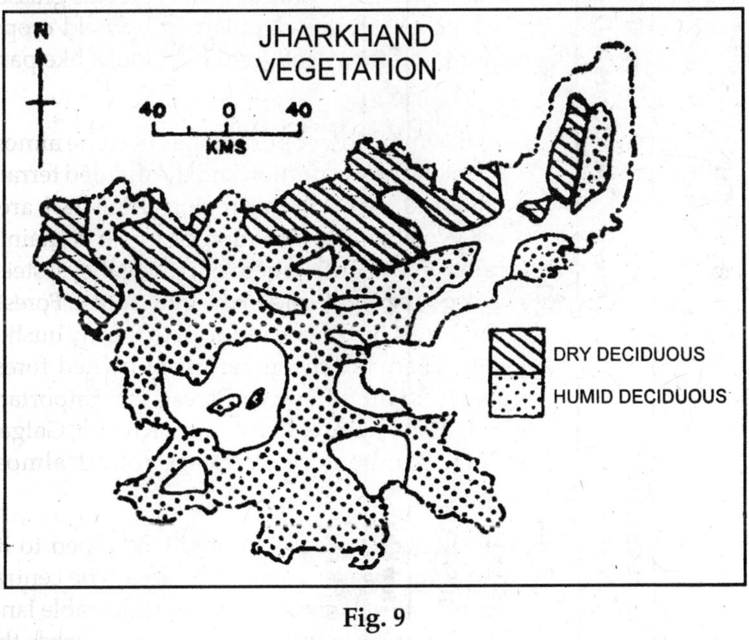

Fig. 9

The Damodar Basin has recorded remarkable depletion of forest cover because of ruthless exploitation, development of mining areas, industrial and urban growth and other developments. This area has nearly 8% land under vegetation. The

Table - 6.1

Forest Covers in Different Attitude Zones : Jharkhand

Sl. No.	Altitude Zone (Metre)	Forest Cover (Sq. Km.) Very Dense	Moderately Dense	Open Front	Total
1.	0-500	1,021	6681	8293	15,995
2.	500-1000	1508	2852	2903	7263
3.	1000-2000	59	130	31	220
	Jharkhand	**2588**	**9663**	**11227**	**23478**

Source : Based on SRTM, Digital Elevation Model, 2015.

forest of this region is mostly confined to the north and west. In the upper valleys rough terrain and less human interference have enabled the region to have some forest but in the lower portions intensive mining, urban and industrial growth have depleted the forest cover. In this region some forests have been protected. At the same time efforts have been made to develop forest to minimize the soil erosion. Important trees are Sal, Sisam, Murga, Siris, Palas, Mahua, Kend, Gambhar, etc. This region also has some fruit trees in the cultivated areas like Mango, Sajina, Imli, Jamun, Kadam, Kathal, Aura, etc.

The source region of Damodar and the hilly Pat region are covered with dense forest. The wet deciduous dominate with Sal, Gambhar, Kusum, Palas trees and tall Bamboo clusters. This type of forest also dominates over the steep face of the Pats and their piedmont section. The central part has been mostly cleared and converted into arable lands. The areas receive comparatively higher rainfall than the adjoining regions.

In case of Subarnarekha basin and adjacent areas vegetal cover has also been much influenced by the nature of land surface as well as by the human interference in the form of mining activities and industrial-cum-urban development. The lower portion of this basin has mostly homognous alluvial land in the Dhalbhum region where forest has been completely depleted and arable land has been developed. Only the adjoining areas having rough terrain provide considerable forest cover. In the upper course of this river it has comparatively undulating landsurface where forest cover is available. Important forest cover is localised in the Dalma range through which this river passes and in south-west and north-east of the basin having highlands present different conditions for vegetal cover. The Dharwarian rocks especially phyllites, schist and sedimentary deposits have favoured the healthy growth of Sal and Bamboo thickets. The Kolhan upland and Dhanjori range have considerable forest cover because of rough terrain and less human interference except iron ore and other mines. Saranda forest region is noted for the concentration of Sal and Teak trees. Virtually Saranda known for cluster of 700 hills provide ideal condition for the growth of forests and this forest area has spread over 900 sq. kms. Important trees in this region are Sal, Semal, Gambhar, Mahua, Karanj, etc. In the Singhbhum plains and the Kharkai basin

Table - 6.2
District Wise Forest Cover : Jharkhand

Sl. No.	District	Geographical Area	Very Dense Forest	Moderate Dense Forest	Open Forest	Total	< of GA	Scrub
1.	Bokaro	1929	61	231	273	565	29.29	39
2.	Chatra	3732	250	869	662	1781	47.72	19
3.	Deoghar	2470	0	14	190	204	8.23	10
4.	Dhanbad	2996	0	44	158	202	6.74	19
5.	Dumka	6212	0	277	383	660	10.62	58
6.	Garhwa	4092	124	409	859	1392	34.02	67
7.	Giridih	4963	76	337	465	878	17.69	22
8.	Godda	2110	14	269	129	412	19.53	20
9.	Gumla	9077	323	921	1417	2661	29.32	34
10.	Hazaribah	5998	271	632	1190	2093	34.89	40
11.	Koderma	1435	68	321	209	598	41.67	2
12.	Lohardaga	1491	173	217	109	499	33.47	10
13.	Pakur	157	3	173	111	287	18.27	20
14.	Palamu	8657	533	1824	1237	3594	41.52	106
15.	W. Singhbhum	9907	483	1558	1871	3912	39.49	77
16.	E. Singhbhum	3533	52	589	429	1070	30.29	22
17.	Ranchi	7698	138	712	1226	2076	26.97	51
18.	Sahibganj	1834	19	266	309	594	32.39	69
	Jharkhand	**79714**	**2588**	**9663**	**11227**	**23478**	**29.45**	**685**

Source : India State of Forest Report, 2015.

forest cover has considerably been depleted because these areas are comparatively less undulating and favourable for habitation and other cultural activities.

The Rajmahal highland having rough terrain, relatively more rainfall because of the proximity of the Bay of Bengal and humid conditions in the neighbouring areas of West Bengal forest cover is of considerable importance. The Damini koh area is noted for forest resource. Certainly the adjoining plains, the gentle sloping land and the river valleys have become devoid of forest because of the development of culturable land. Forests have remained confined to hills, steeper ravines and slopes. In Sahibganj, Pakur and Godda districts the jungle has not been spared even on such hilly sites for Maler or Sauria Paharias settlements who have developed patches of arable land even on the steep slopes. There are a few reserve forests situated in Dumka, Godda and Pakur districts. The best and heavily wooded portion consists of the old reserve in Dumka district which extends over about 50 sq. kms and is situated in hilly region having an elevation of 200 metres to 250 metres. In the remaining forest the growth is poor but some portions are well wooded and contain trees of considerable value, among which Sal predominates. The forests, nearly all, occupy hilly region, consisting of hill ranges with gentle slopes and broad rounded crests and of intervening valleys, which are usually broad. In all the forests, Sal is found in greater or less abundance, usually accompanied by Kend, Bamboos, Mahua, Semal, etc. In the valleys and along the lower slopes, poles of Sal alongwith other trees are mostly found. Next to Sal are Asan, Murga, Kusum, Bamboos, Sabai grass, etc.

The uneven distribution of forest cover also becomes obvious by the analysis of the table 6.2. Higher percentage of land engaged in forest cover has been in the districts of Chatra, Koderma & Palamu. These districts have more than 40% land under forest because in these districts plateaus are comparatively more rugged and also have inaccessible areas. Even the steep scarps and narrow valleys of this region are inhabited and forests of these areas are less denudated. Similarly the districts of W. Singhbhum & Hazaribagh together have more than 34% land under forest. Next in importance is Garhwa where 34.02% land is under forest cover. In this district also undulating areas and scarps are still densely forested. Certainly mineralised areas have recorded depletion of forest. The district of Lohardaga having higher Ranchi Plateau region has 33.47% land under forest. In this area hilly landscape, relatively higher precipitation and less population and less population pressure have favoured presence of forest cover.

Table - 6.3
Forest Cover : Jharkhand

Sl. No.	Type of Forest Cover	Canopy Density (%)	Area (Sq. Km.)	Area (%)
1.	Very Dense Forest	>70	2588	11.02
2.	Moderately Dense Forest	40-70	9663	41.16
3.	Open Forest	10-40	11227	47.82
	Jharkhand	-	**23478**	**100.00**

Source : State of Forest Report, 2015, FSI

The districts of Pashchimi and Purbi Singhbhum relatively have a higher % of land under vegetation. In these two districts there are hilly areas like Porahat-Dalma range, Dhanjori upland, Kolhan upland, Saranda hills where rough terrain, favourable soil and comparatively suitable climatic conditions have contributed to vegetative cover. Certainly industrialization, urbanisation and mining operation have been affecting the forest cover of this region. The district of Gumla has 29.32% land under vegetation. In this portion Simdega Plateau region having rugged land surface , the western undulated areas have considerable forest cover. Certainly plants of these areas are of medium to smaller size and bushes are more plentiful. The district of Ranchi has 26.47% land under vegetation. In this district the forest areas are confined to only south-eastern portions and in northern scarp areas. The vast central portion has been arable.

The districts of Dhanbad, Deoghar, Dumka, Giridih, Pakur have vegetation cover between 6% and 18%. The districts of Dhanbad has 6.74% area under forest which is due to mining industrial & population pressure. The district of Deoghar has only 8.23% because of more agricultural development. The district of Dumka has a poor vegetal cover of just 10.62%. Districts having less vegetal covered is mainly due to intensive mining, dense net work of transport, clustering of urban industrial centres and agricultural development.

The forest resource of this state needs due attention towards its shrinking size, felling of valuable trees, associated soil erosion, awareness towards afforestation and preserving environmental conditions. The main problems of the forests of the region concern human interference in the form of clearing bushes for food crops for increasing population, greed of the contractors, lack of alternative livelihood for the aboriginals who still depend upon the forest and unplanned development of industrial, urban growth, etc. These problems have caused depletion of forest in mineralised, industrialised, urbanised and densely populated areas of the region. At the same time unscientific felling of immatured trees is converting the forest into bushes. The problem of soil erosion is prevalent. Waste land area is increasing the deforested areas and rocky surfaces are getting exposed. As such there is need to conserve the forest by planting trees in the deforested areas, preventing unscientific felling of trees, declaring protected forest areas and developing social forestry in the inhabited areas. It is true that people have realised the importance of vegetation in terms of environment, wild life, economy and cultural development. The forests provide us valuable woods for building material, furnitures, railway sleeper, etc. Bamboo, grasses and other vegetations are also used for paper industries which have become important for educational and official purposes. It may be noted that the need of paper is increasing day by day because of awareness of people towards education. Several other forest products enable people to develop industries like lac industry, biri industry, honey manufacturing, medicine related to herbal plants. During recent years special attention has been paid to maintain environmental conditions because forest influences precipitation, evaporation, moisture content, soil fertility, etc.

National Park & Wildlife Sanctuary

The state of Jharkhand has a vast geographical extent which contains a vide variety of flora & fauna. This diversity is a natural heritage for us and the preservation & conservation of this natural heritage is also necessary for the ecosystem of our earth. Although, the establishment of National Park & Wild life sanctuary is to protect the biodiversity of the area, but they also deliver other invaluable economic, social, cultural & health benefits. These

National parks & Wild life sanctuaries are home to various endangered species of wild animals. It is a place where these animals are safe from hunting, predation or competition. To protect these wild animals and their habitat. The government has established National Park and wild life sanctuaries across the state (Table 6.4)

Table - 6.4
National Park and Wildlife Sanctuaries : Jharkhand

Sl. No.	Name	District	Legal Status	Year of Notification	Area (Sq. Km.)
1.	Palamu Tiger Reserve Forest	Latehar	NP	1976	1026.00
2.	Hazaribagh Wildlife Sanctuary	Hazaribagh	WLS	1976	186.25
3.	Mahuadanr Wolf Sanctuary	Latehar	WLS	1976	63.25
4.	Dalma Wildlife Sanctuary	E. Singhbhum	WLS	1976	193.22
5.	Topchanchi Wildlife Sanctuary	Dhanbad	WLS	1978	8.75
6.	Lawalong Wildlife Sanctuary	Chatra	WLS	1978	207.00
7.	Koderma Wildlife Sanctuary	Koderma	WLS	1985	105.62
8.	Parasnath Wildlife Sanctuary	Giridih	WLS	1981	49.33
9.	Palkot Wildlife Sanctuary	Gumla	WLS	1990	183.18
10.	Udhwa lake Birds Sanctuary	Sahebganj	WLS	1991	5.65
11.	Gautam Budha Wildlife Sanctuary	Koderma	WLS	1976	121.14
12.	Palamu Wildlife Sanctuary	Palamu	WLS	1976	794.33

National Park is an area which is strictly reservd for the betterment of the Wildlife & biodiversity and where any activities like developmental, forestry, poaching, hunting, grazing and cultivation are not permitted. In these parks, even private ownership rights are not allowed. Their boundaries are well marked and circumscribed. In the national parks, the emphasis is on the preservation of a single floral or faunal emphasis. Whereas Sanctuary is declared for the purpose of protecting, propagating or developing wildlife or its environment. Certain rights of people living inside the Sanctuary could be permitted like collecting minor forest products and grazing as long as they do not interfere with well being of the animals.

There wildlife sanctuaries and national parks have been become happening tourist destination for nature and wildlife lovers. They are visited by millions of domestic & foreign tourists each year. These destinations are not only preserving the rich wildlife of the state but also gives a economic boost by the visiting tourists to the state economy.

Reference :

1. E. Ahmed: (Op.cit.), page no. 72.

7

Land Use and Agriculture

Land use and agriculture being associated with mankind from the beginning of human civilization have been fundamental assets all over the world. Their importance still persists in developing parts where the agriculture is the chief source of livelihood. As the state of Jharkhand forms a part of the developing world the proper use of land becomes significant. In this region all the land has not been properly used. During the British regime a proper utilization of land was neither realised nor it was deemed necessary to be realised. The Britishers did not take much initiative in this region because this area was poorly accessible. Even the people of this region were not much inclined towards using the land in proper way and developing agriculture because they were partly unaware of agricultural practices and they were easily getting source of livelihood from forest gatherings.

In course of time the people of this area witnessed substantial changes in life style due to inter-action with the people of the plain and social reformers of other areas. At the same time the development of transport network, agricultural infrastructure, restriction of ruthless depletion of forest, educational development have motivated people to realise the need of proper use of land and agriculture. Despite the above facts the people of this region are not using the land properly. It is, therefore, necessary to know the various uses to which the land is put and to note its potential uses.

Land Use :

There are nine landuse categories in Jharkhand. Annual season and crop report for the agricultural years 2008-09 for the Jharkhand state have been used for analysis and interpretation.

Table - 7.1
Land Use Pattern : Jharkhand (2008-09)

Sl.No.	Classification	Area in '000 Hectares	% of total area
1.	Net Sown Area	1504	18.87
2.	Forest	2239	28.09
3.	Unculturable Waste Land	336	4.22
4.	Non-Agricultural Use	1332	16.71
5.	Land under Misc. Tree Crops & Grooves	93	1.17
6.	Permanent Pasture & other Grazing Land	110	1.38
7.	Culturable Land other than Current Fallow	336	4.22
8.	Fallow land other than Current Fallow	962	12.07
9.	Current Fallow	1394	17.49
		7970	**100%**

***Reporting Area for Land Utilisation 7970 ha instead of 7972 ha.**
Source : Land use statistics, Ministry of Agri., GOI, 2008-09.

The above table no. 7.1 indicates broad utilization of land in the state. Each classification is defined below :-

Net Sown Area : This represents the actual land sown with crops during the particular year.

Forest : This represents all forest lands or administered as forest under any legal enactment dealing with forests whether state-owned or private.

Unculturable Waste Land : It is also called barren and unculturable land. This represents all barren and unculturable lands like bare rocks, usar lands, etc.

Non-Agricultural Use : Such lands concern uses like settlement, transport, playground, industries, etc.

Land under Miscellaneous Tree Crops & Grooves : This land use represents areas under grooves, fruit trees, etc.

Permanent Pastures & other Grazing Land : Such lands include grazing lands.

Culturable Waste Land : It is also called cultivable waste land. Cultivable waste is more significant than culturable waste. This represents all land available for cultivation, whether not taken for cultivation or abandoned after a few years for one reason or the other.

Fallow Land other than Current Fallow : These include cropped lands which are kept fallow for a period of between one and five years.

Current Fallow : These include cropped lands which are kept fallow for a period of upto one year.

A glance at table 7.1 reveals considerable coverage of vegetation despite depletion of forest caused by expansion of culturable land, non-agricultural uses of land, etc. The state has only 18.87% N.S.A.areas against 28.09% forest covered areas. Current and other fallows consist of 29.56% area because of infertile character of soil in most areas, lack of insured irrigation facilities, etc. The area has 4.22% unculturable land and 16.71% under non-agricultural use. It may be noted that the area under settlement is comparatively less than the plain land located regions because of comparatively less population. Orchards and pastures have engaged 1.17% and 1.38% land respectively. 4.22% area is devoted to culturable waste land.

The pattern of land use is determined by two sets of factors : 1. Physical factors like topography, climate and soils which broadly determine the capabilities of the land and 2. Human factors like the length of occupance, density of population, economic and social status coupled with accessibility to markets, transport and communications control the extent to which the resources of land are utilized. Due to regional variations in the above mentioned conditions the land is not of the same quality. Consequently particular land of particular region remains suitable for particular purpose in the service of man.

As regards dominance of forest covered land in the state of Jharkhand, the role of rough terrain, drainage pattern, poor accessibility, suitable climatic conditions can be pointed out. There are areas consisting of steep scarps, very rugged terrains, inaccessible narrow valleys, etc. where human habitation becomes difficult and forest has been less exploited. Even today the native people to a considerable extent depend on these forest for their livelihood. More over there is protective measures of the government for the conservation of forests resource in the state partly for protecting the valuable woods and also for maintaining ecosystem, controlling soil erosion and protecting thin soil covered rocky land. During recent years depletion of forest has been considerably controlled and afforestation schemes have been accelerated.

Agricultural use of land is only next to forest because problem of population pressure is not so acute. Agricultural lands have been developed on the gentle sloping lands and in the valleys only. Other areas are comparatively difficult for the extention of agriculture. More over the prevalence of other sources of livelihood like forest products, hunting, job opportunities in mines and industries, etc. has motivated people of this region not to reclaim more agricultural land.

The development of mining centres, industries, transport network, settlements, etc. have brought considerable land under non-agricultural use.

The large percentage of land under fallows depends upon the leaching of soil and loss of fertility. It may be noted that the soils on sloping land and terraced fields are easily drained down in case of heavy rains and in few years of regular ploughing the cultivable lands are left fallow to regain the fertility . Certainly people have been using fertiliser and compost for this purpose.

Orchards have developed close to the settlements almost everywhere. Orchard provide seasonal fruits and fuels. It is easy to develop orchards due to presence of suitable climatic conditions.

Here, the land use pattern data at the district level is available for one year 2004-05, so analysis & interpretation at the district level has been done accordingly. Upto 2004-05, only 22 districts were functional, so data were available for these 22 districts in the state of Jharkhand.

In terms of net sown areas the state of Jharkhand has 22.11% land but there is lack of uniformity in the development of Net

Table - 7.2
Land use Pattern : Jharkhand (2004-05)

Sl. No.	District	Geographical Area	Net Sown Area	Forest	Unculturable Waste Land	Non-Agricultural Use	Land under Misc. Trees Crops & Groove	Permanent Pasture & other Grazing Land	Culturable Waste Land	Fallow land other than current Fallow	Current Fallow	Area Sown more than once
1.	Bokaro	288.97	15.51 (537%)	72.23 (24.99%)	25.11 (8.69%)	45.92 (15.89%)	2.12 (0.73%)	2.22 (0.77%)	10.29 (3.56%)	34.40 (11.90%)	81.17 (28.09%)	10.43
2.	Chatra	375.52	45.07 (12.00%)	226.82 (60.40%)	20.25 (5.39%)	10.15 (2.70%)	5.69 (1.51%)	1.62 (0.43%)	3.95 (1.05%)	28.66 (7.63%)	33.31 (8.95%)	5.92
3.	Deoghar	248.15	63.68 (25.66%)	34.63 (13.96%)	13.45 (5.42%)	19.26 (7.76%)	3.61 (1.45%)	10.03 (4.15%)	14.91 (6.01%)	34.71 (13.99%)	53.87 (21.71%)	4.04
4.	Dhanbad	204.16	34.60 (16.15%)	18.93 (9.27%)	32.59 (15.96%)	43.15 (21.14%)	3.30 (1.62%)	0.57 (0.28%)	11.30 (5.53%)	20.50 (10.04%)	39.22 (19.21%)	5.40
5.	Dumka	379.03	19.10 (31.42%)	42.71 (11.27%)	24.00 (6.33%)	41.99 (11.08%)	8.17 (2.16%)	18.59 (4.90%)	22.72 (5.99%)	42.43 (11.19%)	59.32 (15.65%)	6.05
6.	Garhwa	428.82	64.77 (15.10%)	191.15 (44.58%)	24.75 (5.77%)	15.10 (3.52%)	2.36 (0.55%)	2.80 (0.65%)	6.55 (1.53%)	39.00 (9.09%)	82.34 (19.2%)	24.34
7.	Giridih	493.22	77.59 (15.73%)	158.42 (132.12%)	30.42 (6.17%)	33.62 (6.82%)	13.47 (2.73%)	9.96 (2.02%)	17.98 (3.65%)	54.97 (11.45%)	96.79 (19.62%)	16.60
8.	Godda	231.84	78.21 (33.73%)	31.32 (13.51%)	8.71 (3.76%)	18.08 (7.80%)	3.34 (1.44%)	6.33 (2.73%)	6.37 (2.75%)	35.16 (15.17%)	44.32 (19.11%)	6.31
9.	Gumla	538.92	161.75 (30.01%)	81.14 (15.06%)	44.75 (8.30%)	29.99 (5.56%)	9.23 (1.71%)	0.30 (0.06%)	27.10 (5.02%)	63.00 (11.69%)	21.66 (22.57%)	7.57
10.	Hazaribagh	604.63	106.14 (17.55%)	255.66 (42.45%)	54.10 (8.95%)	37.67 (6.23%)	779 (1.29%)	3.93 (0.65%)	9.16 (1.51%)	56.00 (9.26%)	73.18 (12.10%)	12.17

11.	Jamtara	179.17	55.34 (30.89%)	30.20 (16.86%)	9.87 (5.51%)	20.32 (11.34%)	3.92 (2.19%)	8.79 (4.91%)	9.69 (5.41%)	16.06 (8.96%)	24.98 (13.93%)	2.82
12.	Koderma	130.20	17.04 (13.09%)	55.23 (42.42%)	14.71 (11.30%)	9.24 (7.10%)	1.90 (1.46%)	1.44 (1.11%)	2.98 (2.29%)	8.16 (6.27%)	19.50 (14.48%)	4.50
13.	Latehar	319.19	59.77 (18.73%)	137.97 (43.23%)	19.31 (6.05%)	12.22 (3.83%)	4.34 (1.31%)	1.14 (0.36%)	6.05 (1.90%)	29.30 (9.18%)	49.09 (15.38%)	12.75
14.	Lohardaga	153.62	39.64 (25.80%)	44.36 (22.88%)	9.40 (6.12%)	9.48 (6.17%)	1.39 (0.90%)	0.06 (0.04%)	9.81 (6.39%)	16.74 (10.90%)	22.74 (14.80%)	5.83
15.	Pakur	181.70	57.84 (31.83%)	20.79 (11.44%)	12.51 (6.88%)	14.66 (8.07%)	5.34 (2.94%)	5.67 (3.12%)	7.46 (4.11%)	20.50 (11.28%)	36.93 (20.92%)	5.21
16.	Palamu	524.69	97.75 (18.63%)	226.85 (43.25%)	31.69 (6.04%)	19.91 (3.79%)	7.53 (1.44%)	1.87 (0.36%)	9.96 (1.90%)	48.04 (9.16%)	81.09 (15.45%)	20.85
17.	W. Singhbhum	562.70	145.99 (25.94%)	227.54 (40.44%)	50.01 (8.89%)	29.53 (5.24%)	7.53 (1.34%)	3.15 (0.56%)	24.42 (4.34%)	32.58 (5.79%)	41.95 (7.45%)	14.20
18.	E. Singhbhum	556.69	83.28 (14.96%)	122.82 (22.06%)	42.15 (7.58%)	155.45 (27.42%)	9.16 (1.64%)	2.68 (0.48%)	21.49 (3.86%)	20.55 (3.71%)	99.01 (17.79%)	63.60
19.	Ranchi	758.25	255.85 (33.73%)	159.14 (20.99%)	39.50 (5.21%)	74.50 (9.83%)	10.70 (1.40%)	2.03 (0.27%)	26.32 (3.47%)	66.20 (8.73%)	124.01 (16.35%)	16.80
20.	Sahebganj	201.75	41.46 ((20.55%)	42.74 (21.18%)	15.68 (7.77%)	14.43 (7.15%)	3.93 (1.95%)	2.82 (1.40%)	5.56 (2.76%)	28.71 (14.23%)	46.42 (23.01%)	7.50
21.	Saraikela Kharsawan	237.23	60.42 (25.47%)	95.96 (40.45%)	22.10 (9.32%)	13.03 (5.49%)	3.05 (1.28%)	1.33 (0.56%)	10.90 (4.63%)	12.74 (5.37%)	17.70 (7.46%)	4.86
22.	Simdega	371.63	81.67 (21.98%)	55.94 (15.05%)	30.72 (8.27%)	20.57 (5.54%)	6.40 (1.72%)	0.13 (0.03%)	18.65 (5.02%)	43.10 (11.60%)	114.45 (30.80%)	5.29
	Jharkhand	**7970.08**	**1762.47 (22.11%)**	**2333.47 (29.28%)**	**575.78 (7.22%)**	**688.27 (8.64%)**	**124.27 (1.56%)**	**87.46 (1.10%)**	**283.62 (3.56%)**	**751.51 (9.43%)**	**1363.05 (17.10%)**	**263.04**

Source : Directorate of Statistics & Evaluation, Jharkhand.

Area Sown. Areas having considerable homogeneous land in the form of flat plateaus, broad valleys or very gentle sloping land have higher percentage of land under cultivation than the areas having comparatively rough terrain, dense forest, sparse population, etc. The highest percentage of Net Sown Area has been recorded in the district of Godda (33.74) because north-western portion of this district has homogeneous land where agriculture has developed which is clear vide table 7.2 . Other districts having more than 30% Net Sown Area include Ranchi (33.73%), Pakur (31.83%), Dumka (31.42%), Jamtara (30.89%) & Gumla (30.01%). It may be noted that most areas in these districts are suitable for agricultural use of land. Districts having Net Sown Areas less than the state average of 22.11% include Dhanbad (16.95%) because of most mining developments, Giridih (15.73%) including Kodarma (13.89%) also because of mines and rough terrain, Hazaribag (17.55%) including Chatra (12.00%) due to rough terrain, Palamu (18.63%) including Garhwa (15.1%) because of divided rough terrains. In the north-eastern portion of the state, the district of Sahibganj has partly adjoining plains of the river Ganga in the north and the east where more than 90% land is under cultivation but the south-western Damin-i-koh area where rough terrains prevail, percentage of culturable land sharply decreases leading to the average of only 32.81%. In the district of Ranchi the flat plateaus having gentle sloping land has given rise to good terraced Tanr and Don lands where cultivation has considerably developed. The Don areas are more fertile and cultivators grow more than one crops. In the less fertile Tanr land areas cultivators also grow coarse grains and pulses. In the districts of Palamu and Garhwa where culturable land is more dominant in the north have wide spread distribution of cultivated land but the southern hilly areas, the south-western water divide and the south-eastern divided terrains have restricted expansion of culturable land.

It has been noted earlier that the state of Jharkhand has considerable forest cover in most areas but the distribution of forest covered areas is also not uniform. The state has 29.28% areas under forest. The highest percentage of forest covered areas lie in the district of Chatra (60.40%). In this district the Upper Hazaribag plateau consisting of steep scarps, rugged terrains, narrow valleys, etc. have only isolated patches of agricultural land and vast areas are beyond the reach of cultivable use. As such vast areas still have forest. Moreover there are reserved forests also. The second position in terms of forest cover goes to the district of Garhwa where 44.58% land is under forest. It may be noted that the west-southern areas of district of Garhwa where lie the hilly areas, scarps are beyond the access of cultivators. There are gorgeous valleys where lie natural protected of forest because of extreme paucity of transport. The divided terrains of the south-eastern of Palamu district and the water divided areas lying in the south-western portion of Garhwa district also provide suitable conditions for natural growth of forest. It is only the northern valleys of the Koel where forest has been almost depleted for agriculture. The position of Palamu (43.25%), Latehar (43.23%), Hazaribagh (42.45%), Koderma (42.42%), Saraikela-Kharsawan (40.45%), W. Singhbhum (40.44%) are very appreciable in terms of forest cover. In these districts forest areas are confined to only hilly areas, steep scarp areas and reserved areas. Homogeneous areas of these districts have lost the wood land due to expansion of settlement and agriculture. The areas having relatively poor share of forest include Dhanbad (9.27%), Dumka (11.27%), Pakur (11.44%), Godda (13.52%), Deoghar (13.96%), Simdega (15.05%) etc. In all these districts only the hilly areas have forests and the remaining areas have become devoid of this resource. Even forests have lost tall trees and only bushy cover provide look of forest scenery. The district of Dhanbad in the region has the lowest forest, due to extensive mining activities, industrial growth and development of urban agglomeration.

The barren & unutilised land consists of rocky, bare, rugged, barren land prevailing in the hilly areas or dissected plateau areas. In such areas the development of agricultural land becomes very difficult. Such land has been 7.22% in the state of Jharkhand. The highest percentage has been in the district of Dhanbad including Bokaro (15.96%). Industrial refuse has converted considerable areas of this region into unculturable waste land. Other districts having substantial uncultural waste land include Kodarma (11.30%), Saraikela-Kharsawan (9.32%), Hazaribagh (8.95%), W. Singhbhum (8.89%) etc. In the districts of Purbi and Pashchimi Singhbhum mining activities exposed rocky areas, etc. have caused considerable culturable waste land. In the districts of Hazaribag also this category of land lies in the areas having rough terrain where deforestation has led to soil erosion. Comparatively less unculturable waste land exists in the district of Godda (3.76%), Ranchi (5.21%), etc. where land is comparatively homogeneous.

In case of non-agricultural uses land under settlement, communication, water bodies, industries, etc. are included. These uses of land share 8.64% land of the state. The percentage of this category of land is more marked in the areas having human agglomerations and infrastructure development. The highest percentage of non-agriculture use of land has been recorded in the district of Purbi Singhbhum (27.92%) where industrial growth, urbanization and dense population are more marked. Considerable areas have been under industrial use, mines, settlement, transport network, etc. The second position in this respect goes to Dhanbad district where lies very high density of population due to the presence of agglomeration of urban centres which have originated on account of mines and industries and are growing at a faster rate. Moreover, widespread mines and human settlements and infrastructural facilities have engaged 21.14% area of the district. Other districts having land under non-agricultural use more than the state average (8.64%) include Bokaro (15.84%), Jamtara (11.34%), Dumka (11.08%) & Ranchi (9.83%). In these districts the population is also of reasonable size and non-agricultural functions have more developed. The remaining districts have non-agricultural land less than the state average. These districts include Chatra (2.70%), Garhwa (3.52%), Palamu (3.79%), Latehar (3.83%).In these areas population is relatively smaller and infrastructure is comparatively less developed.

Land under Miscellaneous tree crops and grooves cover 1.56% of the total land of the state. This class of land although occupies a fraction of the land but plays important role in fruit markets. During recent years people have developed interest in planting fruit trees and developing grooves near settlements. The highest percentage of such land has been recorded in the district of Pakur (2.94%) where mango orchards and other fruit trees are developed partly because of salubrious climate. Other districts having considerable orchards are Giridih (2.73%), Jamtara (2.19%) & Dumka (2.16%). The district of Deoghar has also good coverage of mango orchards and fruit trees because of demand in the markets. The religious significance of Deoghar town provides good demand of fruits. In Sahibganj district mango orchard is also well developed. Districts having comparatively less land under grooves are Garhwa (0.55%), Bokaro (0.73%), Lohardaga (0.90%), Saraikela-Kharsawan (1.28%) etc.

The state also has some areas under permanent pasture and grazing land. These uses have involved 1.10% land of the state. Pasture land is considered essential for the domestic animals because agriculture depends on cattle but pasture lands have been mostly converted into other uses. Even then some areas of the state have significant percentage of pasture land. The highest percentage has been in the district of Jamtara (4.91%) being followed by Dumka (4.90%), Pakur (3.12%) & Godda (2.73%). In these areas cattle rearing is remarkable. Areas having negligible percentage of pasture include Simdega (0.03%), Latehar (0.04%), Gumla (0.06%), Palamu (0.36%) & Latehar (0.36%).

Culturable waste land classed under this category are not cultivated for more than five years. Such land has been 3.56% in the state. These are marginal land and can be used for cultivation if efforts are made to improve it by managing irrigational facilities. The highest percentage of culturable waste land has been in the district of Lohardaga (6.39%) where problem of leaching of soil is more marked. The cultivators abandon the cultivable land for regaining fertility. The second position goes to Deoghar where 6.01% land is under culturable waste. In this district also conditions like Dumka prevail. Other districts having considerable areas under this category are Deoghar, Dumka, Dhanbad & Jamtara etc. where shifting cultivation and abandoned fallows explain this phenomenon. In the mining areas, mining activities have caused similar damage to agricultural land in most areas of Dhanbad and Pashchimi Singhbhum.

Current and other fallows include these lands which remain uncultivated during the year concerned and those lands which were under cultivation for a period of not less than one year and not more than five years respectively. Current fallows are caused due to unfavourable weather conditions but other fallow lands are caused due to loss of fertility or other causes. These categories of land use have been more pronounced in almost all districts because of hilly terrain where permanent cultivation is limited to the fertile plains only. Abandoning of cultivable fields is common feature because of leaching of soil, lack of irrigation, etc. The highest percentage has been recorded in the district of Simdega (42.40%) because this district has considerable areas having rugged topography where cultivators frequently abandon the cultivable fields. The second position is held by the district of Bokaro (39.99%). In this district also hilly areas provide similar condition like in Gumla. Other districts having higher percentage of current and other fallows, more than the state average (26.53%), are Sahibganj, Deoghar & Godda etc. These districts have also considerable undulating areas where fallows are bound to develop. The remaining districts have comparatively lower percentage of land under this category. These districts are Saraikela-Kharsawan, W. Singhbhum, Chatra etc.

There has been considerable changes in the land use of the state because of the development of non-agricultural activities. Certainly some areas have recorded even increase in culturable land. The comparative analysis of the land use data of the years 1990-91 & 2008-09 in the table 7.3 reveal decrease in Net Sown area and increase in non-agricultural use and fallow land significantly. In the year 1990-91 the net sown area of the state has 25.07% of land under this category which has decreased to 18.87% in the year 2008-09. The area under non-agricultural use in 1990-91 was 8.27% which has almost doubled in 16.71% in the year 2008-09 because of the urban, industrial & other developments.

Table - 7.3

Comparative Land Use Pattern : Jharkhand

Sl. No.	Classification	1990-91 '000 Hectares	2008-09 'Hectares
1.	Net Sown Area	1996 (25.07%)	1504 (18.87%)
2.	Forest	2332 (29.29%)	2239 (28.09%)
3.	Unculturable Waste Land	573 (7.20%)	336 (4.22%)
4.	Non-Agricultural use	658 (8.27%)	1332 (16.71%)
5.	Land under Misc. Trees Crops & Groves	87 (1.09%)	93 (1.13%)
6.	Permanent Pasture & other grazing land	100 (1.26%)	110 (1.38%)
7.	Culturable Waste land	311 (3.90%)	336 (4.22%)
8.	Fallow Land	1904 (23.92%	2356 (29.56%)
	Total	**7961 (100.00%)**	**7970 (100.00%)**

Source : 1. Statistical Office, Bihar (1990-91)

2. Land Use Statistics, Ministry of Agri., GOI, 2008-09.

The unculturable waste land has significantly decreased from 7.20% in 1990-91 to 4.22% in 2008-09. Even in land under miscellaneous trees, crops & grooves, there has been very little increase from 1.09% in 1990-91 to 1.13% in 2008-09 due to awareness towards plantation of fruit trees like mango, guava etc. There has been increase also in permanent pasture & grazing land and culturable waste land between the two periods. The Fallow land which includes current & other fallow land have increased from 23.92% in 1990-91 to 29.56% in 2008-09, which is very surprising. The area under forest has also declined from 29.27% in 1990-91 to 28.09% in 2008-09, despite taking many efforts at the government level.

Agriculture :

Most areas of the state of Jharkhand have remained aloof of agricultural activities for long time because of hilly terrain, forest cover and less involvement of native population in agriculture. Agricultural activities have remained more confined to smaller plains, homogeneous plateau areas and other accessible areas. These activities increased in the plateau areas only after the immigration of people from the Bihar plain or other areas who either took shelter in this forest clad region because of fear of then rulers or for the lust of business and developing financial improvement. As such, in this region, there are concentration of native population which have comparatively lesser agricultural developments than the areas dominated by the non-tribals. These two areas differ in terms of the intensity of cropping pattern, yield of crops, etc. At the same time these two areas also reveal contrast in development of agricultural infrastructures and adoption of modern agricultural developments.

It is important to note that the agriculture of this region is comparatively less developed because of the dominance of tribal ecology, lesser inclination towards improved agricultural techniques because of traditional orthodox belief, paucity of fund and awareness towards agricultural developments held elsewhere.

Even today in most areas agriculture is mostly based on natural precipitation and only such crops are grown which are easily produced based on natural conditions. Certainly during recent years development of minor irrigation schemes and river valley projects have enabled cultivators to grow some other crops which are grown in seasons having negligible rainfall. Virtually people of some areas of this state have developed interest in vegetables, seasonal fruits for supply in urban centres of distant places. As such agricultural activities have not remained confined to rainy season. Virtually all the year round some crops are grown somewhere depending upon natural precipitation, use of ground and surface water.

Agro-Climatic Regions :

The Planning Commission now 'Niti Ayog' Government of India has included this state under zone-VII known as Eastern Plateau and Hill region. The state has been further sub-divided by the ICAR into 3 sub-zones (MAP-1) which are following:

Agro Climatic Sub-Zones

	Agro-climatic sub zones	Districts covered
i.	Central & North Eastern Plateau (Zone-IV)	1. Dumka 2. Deoghar 3. Godda 4. Pakur 5. Sahebganj 6. Jamtara 7. Hazaribagh 8. Koderma 9. Chatra 10. Ramgarh 11. Giridih 12. Dhanbad 13. Bokaro
ii.	Western Plateau (Zone-V)	14. Ranchi 15. Khunti 16. Lohardaga 17. Gumla 18. Palamu 19. Garhwa 20. Simdega 21. Latehar
iii.	South-Eastern Plateau (Zone-VI)	22. East Singhbhum 23. West Singhbhum 24. Saraikela-Kharsawan

Harvest Pattern :

For the convenience of analysis the agricultural harvests have been categorised as Bhadai, Aghani, Rabi and Garma. The table 7.4 reveals agricultural harvests in the state of Jharkhand in the year 1997-98. It contains area involved in various harvests and their percentage. The table portrays the fact that even today there is dominance of Aghani harvest having 69.75% land engaged in it. Bhadai crops get the second position having 20.26% culturable area and these two harvests are followed by Rabi and Garma harvests having 9.26% and 0.73% area respectively.

As regards spatial distribution of different agricultural harvests there has been contrasting variation based on nature of terrain and population characteristics. Bhadai harvest is still important in the tribal region because poor farmers and landless labourers, lacking foodgrains for the whole year also depend upon early Bhadai harvest like Marua, Bhadai paddy, maize, bajra, etc. Maize is a common product of this region. The production of Marua has gone down because of lesser food value. But the production of maize has increased. In this season Bhadai pulses, vegetables, til and other crops are also grown. The state of Jharkhand involves 20.26% area under Bhadai harvest but there are areas having more important Bhadai crops more than the regional average include the district of Gumla (41.28%), Ranchi (33.02%), Garhwa (30.52%), Palamu (29.77%) and Lohardaga (29.84%). It may be due to high concentration of the native population and rough terrain. Areas having moderate concentration of Bhadai crops ranging between 10 to 20% included Bokaro (19.72), Chatra (17.85), Hazaribag (14.71), Dumka (14.53), Godda (13.18), Kodarma (12.26) and Giridih (11.81). These areas have also considerable concentration of tribal population and less development of irrigation facilities. The remaining districts have less than 10% area under Bhadai harvest. These districts include Pakur (9.23), Deoghar (9.16), Sahibganj (8.44), Dhanbad (7.75), Pashchimi Singhbhum (7.17) and pubi Singhbhum (2.51). These districts are either better industrially developed or have more concentration of non-tribals.

It has been noted earlier that even today Aghani harvest has dominance over other harvests. The state of Jharkhand has 69.75% area under Aghani harvest. Districts having Aghani harvest more than the regional average include Purbi Singhbhum (69.29%), Pashchimi Singhbhum (90.46%), Dhanbad (87.97%), Deoghar (82.10%), Pakur (81.03%), Giridih (81.09%), Bokaro (75.79%), Sahibganj (75.25%) and Hazaribag (73.74%). Most of these districts have location in the eastern portion having comparatively better rainfall and in these areas production of Aghani crops involve less irrigation. Districts having moderate Aghani harvest include Godda (63.31%), Chatra (63.08%), Lohardaga (59.26%), Ranchi (59.88%), etc. In these areas land is comparatively more rough except the district of Ranchi. Areas having lower percentage of Aghani harvest include Gumla (54.00%), Palamu (43.38%),

Garhwa (37.67%), etc. In these areas there is comparatively lesser rainfall and rough terrain.

Table - 7.4
Harvestwise Area in 1997-98 in Jharkhand
(in Hectares and Percentage)

Sl. No.	District	Bhadai	Agahani	Rabi	Garma	Total
1.	Dumka	24053	131557	9445	571	165626
		14.53	79.43	5.70	0.34	100.00
2.	Godda	10720	51500	18860	272	81352
		13.18	63.31	23.18	0.33	100.00
3.	Deoghar6786	60816	6123	353	74078	
		9.16	82.10	8.27	0.48	100.00
4.	Sahibganj	4461	39795	8263	365	52884
		8.44	75.25	15.62	0.69	100.00
5.	Pakur	6194	54386	6428	112	67120
		9.23	81.03	9.58	0.17	100.00
6.	Hazaribag	15647	78455	11790	508	106400
		14.71	73.74	11.08	0.48	100.00
7.	Kodarma	2883	17716	2873	45	23517
		12.26	75.33	12.22	0.19	100.00
8.	Chatra	8821	31172	9192	232	49417
		17.85	63.08	18.60	0.47	100.00
9.	Giridih	13460	92442	7795	302	113999
		11.81	81.09	6.84	0.26	100.00
10.	Bokaro	9702	37293	1997	213	499205
		19.72	75.79	4.06	0.43	100.00
11.	Dhanbad	3294	37409	1328	493	42524
		7.75	87.97	3.12	1.16	100.00
12.	Ranchi	86663	157158	14375	4250	262446
		33.02	59.88	5.48	1.62	100.00
13.	Lohardaga	14778	29348	4100	1297	49523
		29.84	59.26	8.28	2.62	100.00
14.	Gumla	112190	146753	9191	3641	271775
		41.28	54.00	3.38	1.34	100.00
15.	Purbi. Singhbhum	4301	164762	1284	752	171099
		2.51	96.29	0.76	0.44	100.00

16. Paschimi Singhbhum	16087 7.17	203033 90.46	5084 2.27	228 0.10	224432 100.00
17. Palamu	52269 29.77	76158 43.38	46119 26.27	1029 0.58	175575 100.00
18. Garhwa 26704	32966 30.52	27347 37.67	488 31.25	87505 0.56	100.00
Total	419013 20.26	1442719 69.75	191594 9.26	15151 0.73	2068477 100.00

From the point view of Rabi crops this state is not so popular because of lack of homogeneous fertile alluvial land and developed irrigation facilities. Certainly, interest towards the production of Rabi crops has increased because of changing food habit and increasing popularity of wheat and other pulses. Areas having comparatively homogeneous land have higher percentage of Rabi cropped land. Even today, this state has only 9.20% area under Rabi harvest. Districts having high percentage of Rabi cropped area include Garhwa (31.25), Palamu (26.27), Godda (23.18), Chatra (18.60), Sahibganj (15.62), etc. In these areas the plain land of two Koel Basins and other river basins have motivated farmers to grow Rabi crops. Districts having comparatively lesser concentration of Rabi crops include Kodarma (12.22%), Hazaribag (11.08%), Lohardaga (8.28%), Deoghar (8.27%), etc. In these areas rough terrain has prevented the production of rabi harvest. Districts having low concentration of Rabi crops include Dumka, Giridih, Dhanbad, Gumla, Purbi and Paschimi Singhbhum, Bokaro, etc.

No doubt Garma harvest is still less developed in this state having only 0.73% area. Certainly some districts have developed more interest in growing vegetables and seasonal fruits. These districts include Lohardaga (2.62%), Ranchi (1.62%), Gumla (1.34%) and Dhanbad (1.16%). In these districts demand of vegetables and seasonal fruits have motivated cultivators to develop Garma crops. Districts having moderate concentration include Palamu (0.58%), Garhwa (0.56%), Hazaribag (0.48%), Sahibganj (0.69%), Deoghar (0.48%), Chatra (0.47%), etc. Districts having negligible development of Garma crops include Pakur, Kodarma, Giridih, Paschimi Singhbhum, Godda, Dumka, etc.

While considering the role of irrigation in the production of different harvests Rabi crops get the dominating position because cultivators grow Rabi crops or Garma crops only where irrigation facilities are available and Aghani crops need irrigation only when monsoon rainfall does not occur. Rabi crops involves 48.10% irrigation being followed by Aghani crops (44.51%), Garma - (5.96%) and Bhadai - (1.43%). The table 7.5 reveals the statement of seasonwise irrigated area in Jharkhand for the year 1997-98. The higher percentage of Rabi irrigated area was noted in the district of Deoghar (80.77) being followed by Dumka (79.02), Bokaro (78.42), Giridih (77.69), Godda (75.16), Kodarma (73.42), etc. In these areas Rabi crops are grown based on irrigation only. Districts having lower percentage of Rabi cropped irrigated area less than regional average include Garhwa, Palamu, Paschimi Singhbhum, Purbi Singhbhum, Ranchi, etc. Aghani crops involve 44.51% irrigated area of the state. The highest percentage of Aghani irrigated area has been recorded in the district of Pashchimi Singhbhum (87.12%) being followed by Palamu (65.62), Garhwa (63.86), Sahibganj (48.99), Ranchi (43.97), Hazaribag (42.72), Gumla (35.67), etc. Districts having lower percentage of irrigated area include Dumka, Godda, Deoghar, Pakur, Kodarma, Bokaro, Dhanbad, Giridih, Purbi Singhbhum, etc.

It has been also noted earlier that Bhadai crops are grown not depending on irrigation. In the beginning of Pre-Monsoon rainfall cultivators grow maize, marua and other Bhadai crops and these crops get ready with the help of monsoon rainfall. It is, however, in some districts, negligible Bhadai crops are irrigated. These crops involve only 1.43% irrigation. The highest percentage is only 8.51% in Chatra district being followed by Kodarma (6.65,) Purbi Singhbhum (5.46,) Hazaribag (3.79,) Palamu (2.22,) Garhwa (0.63), etc. Garma crops are mostly grown with the help of irrigation. Certainly cultivators grow these crops only on limited land where irrigation are available. These crops involved only 5.96% area. The highest percentage has been in the district of Purbi Singhbhum (39.45) being followed by Dhanbad (25.04,) Ranchi (16.67,) Gumla (14.65,) Lohardaga (12.22,) Bokaro (8.98), etc. Areas having major urban centres have motivated poor farmers to grow vegetables and fruits for urban areas. Districts having poor percentage of Garma crops irrigated area include Dumka, Godda, Pakur, Hazaribag, Kodarma, Chatra, Garhwa, Palamu, Paschimi Singhbhum, etc.

Harvestwise Irrigation :

It has been noted earlier that previously cultivators used to grow crops mainly based on natural precipitation during rainy season and there was negligible production of such crops which depended entirely upon irrigation. Even today there is dominance of Aghani harvest having close relationship with natural precipitation but development of irrigation has enabled remarkable increase in production of Rabi and Garma crops. The table 7.5 reveals seasonwise irrigated area of the state of Jharkhand in the year 1997-98. The table further reveals the fact that Bhadai crops are produced even today not much depending upon irrigation. It may be due to the fact that the cultivators prepare seedlings of Ragi and Bhadai paddy in the beginning of monsoon rainfall and also grow maize in the beginning of rainfall itself and these crops get ready within two to three months during rainy season as such Bhadai crops hardly need irrigation. It may be due to the fact that the cultivators prepare seedlings of Ragi and Bhadai paddy in the beginning of monsoon rainfall and also grow maize in the beginning itself and these crops get ready within two to three months during rainy season as such Bhadai crops hardly need irrigation.

The state of Jharkhand has irrigated area of 192850 hectares, out of which 48.10% area is occupied by Rabi crops, 44.51% area by Aghani crops, 5.96% area by Garma crops and only 1.43% area is occupied by Bhadai crops. Virtually out of 18 districts of this state, 8 districts of Godda, Deoghar, Bokaro, Dhanbad, Ranchi, Lohardaga, Gumla and Pashchimi Singhbhum did not record irrigated area during Bhadai season. Bhadai harvest consisted of the highest percentage in the district of Chatra(8.51) being followed by Kodarma (6.65,) Purbi Singhbhum (5.46,) Hazaribag (3.79,) Palamu (2.22,) Garhwa (0.63,) Sahibganj (0.19,) Dumka (0.17,) Giridih (0.16) and Pakur (0.08).

Aghani crops partly depend upon irrigation especially at the time of failure of monsoon. More over use of fertilizer and improved seeds compel cultivators to irrigate standing paddy crops more timely. The state of Jharkhand having more undulating land surface and lacking wide spread clayey soil hardly permit accumulation of water for paddy fields for required time. As such irrigation of paddy fields becomes necessary. Aghani paddy remains more dependent upon irrigation in areas having comparatively less rainfall like districts of Palamu, Garhwa and more undulating areas of Paschimi Singhbhum. The highest percentage of Aghani harvest depending upon irrigation has been recorded in the district of Pashchimi Singhbhum (87.12%) being followed by Palamu (65.62,) Garhwa (66.86,) Sahibganj (48.99,) Ranchi (43.97,) Hazaribag (42.72,) Gumla (35.67), etc. Aghani harvest using irrigation less than 30% included the districts of Kodarma, Chatra, Giridih, Bokaro, Dhanbad, Purbi Singhbhum, Dumka, Deoghar, etc.

The use of irrigation is very common during Rabi season. The production of wheat needs hardly 5 to 6 times irrigation. These days cultivators have developed interest towards growing wheat and other Rabi crops. The highest percentage of irrigation has been noted in the district of Deoghar (80.77). Other districts having production of Rabi crops based on irrigation more than the regional average (48.10%) included Dumka, Godda, Pakur, Kodarma, Giridih, Bokaro, Lohardaga, Dhanbad, etc. Districts having production of Rabi crops based on irrigation less than 48% included Ranchi, Purbi Singhbhum, Pashchimi Singhbhum, Palamu and Garhwa.

Table - 7.5

Statement of Seasonwise Irrigated land in Jharkhand, 1997-98 (in Hectares and Percentage)

Sl. No.	District	Bhadai	Agahani	Rabi	Garma	Total Irrigated
1.	Dumka	18	1642	8400	570	10630
		0.17	15.45	79.02	5.36	100.00
2.	Godda	—	2556	8539	266	11361
			22.50	75.16	2.34	100.00
3.	Deoghar—	1006	5656	341	7003	
			14.37	80.77	4.86	100.00
4.	Sahibganj	14	3703	3518	323	7558
		0.19	48.99	46.55	4.27	100.00
5.	Pakur	2	648	1816	86	2552
		0.08	25.39	71.16	3.37	100.00
6.	Hazaribag	674	7589	9008	494	17765
		3.79	72.72	50.71	2.78	100.00
7.	Kodarma	196	545	2164	42	2946

	6.65	18.50	73.42	1.43	100.00
8. Chatra	670	1839	5133	231	7873
	8.51	23.36	65.20	2.93	100.00
9. Giridih	15	1837	7495	300	9647
	0.16	19.04	77.69	3.11	100.00
10. Bokaro	—	299	1861	213	2373
		12.60	78.42	8.98	100.00
11. Dhanbad	—	173	1255	477	1905
		9.08	65.87	25.04	100.00
12. Ranchi	—	9576	8572	3631	21779
		43.97	39.36	16.67	100.00
13. Lohardaga	—	976	3100	604	4680
		20.85	66.23	12.92	100.00
14. Gumla	—	3887	5413	1596	10896
		35.67	49.68	14.65	100.00
15. Purbi. Singhbhum	104	289	761	752	1906
	5.46	15.16	39.93	39.45	100.00
16. Pashchimi Singhbhum	—	7304	888	192	8384
		87.12	10.59	2.29	100.00
17. Palamu	9.9	26896	12253	931	40989
	2.22	65.62	29.89	2.27	100.00
18. Garhwa 149	15090	7952	438	23627	
	0.63	63.86	33.66	1.85	100.00
Total	2760	85863	92762	11486	192850
	1.43	44.51	48.10	5.96	100.00

Source : Statistical Office, Bihar.

It has been also mentioned that production of Garma crops especially vegetables has become important in some areas of the state. Cultivators have started growing vegetables for marketing in industrial and urban centres which enable them to earn more money. This is why the production of seasonal fruits, green vegetables has increased. The highest percentage has been noted in the district of Purbi Singhbhum (39.45) and this district has been followed by Dhanbad (25.04,) Ranchi (16.67,) Gumla (14.65,) Lohardaga (12.92,) Bokaro (8.98), etc. Districts having Garma crops irrigated area less than 5% included Godda, Pakur, Kodarma, Giridih, Garhwa, etc.

Table 7.6
Harvest wise Percentage of Irrigated land in Jharkhand (1997-98)

Sl. No.	District	Bhadai	Agahani	Rabi	Garma
1.	Dumka	0.07	1.25	88.00	99.89
2.	Godda	0.00	4.96	45.28	97.79
3.	Deoghar	0.00	1.65	92.37	96.60
4.	Sahibganj	0.31	9.31	42.58	88.49
5.	Pakur	0.03	1.19	28.25	76.78
6.	Hazaribag	4.31	9.67	76.40	97.24
7.	Kodarma	6.80	3.08	75.32	93.33
8.	Chatra	7.60	5.90	55.84	99.57
9.	Giridih	0.11	1.99	96.15	99.34
10.	Bokaro	0.00	0.80	93.19	100.00
11.	Dhanbad	0.00	0.46	94.50	96.75
12.	Ranchi	0.00	6.09	59.63	85.44
13.	Lohardaga	0.00	3.37	75.61	46.57
14.	Gumla	0.00	2.65	58.89	43.83
15.	Purbi Singhbhum	2.42	0.17	59.27	100.00
16.	Pashchimi Singhbhum	0.00	3.59	17.47	84.21
17.	Palamu	1.74	35.32	26.57	90.48
18.	Garhwa	0.56	45.77	29.08	89.75

Source : Statistical Office, Bihar.

The analysis of the table 7.6 reveals the percentage of irrigated area harvestwise in the state of Jharkhand. It shows the dominance of irrigation during Rabi and Garma harvest. In the district of Dumka only 0.07 percent Bhadai crops, 1.25% Aghani crops, 88.0% Rabi crops and 99.89% Garma crops depended upon irrigation. In the district of Godda, Bhadai crops did not depend upon irrigation. Only 4.96% Aghani, 45.28% Rabi and 97.79% Garma crops required irrigation. Similarly in the district of Deoghar 96.60% Garma crops 92.37% Rabi crops 1.65% Aghani crops depended upon irrigation. The district of Sahibganj consisted of 88.49% Garma, 42.58% Rabi, 9.31% Aghani and 0.31% Bhadai crops. The production of Bhadai crops based on some irrigation was found in the districts of Chatra, Kodarma, Hazaribag, Purbi Singhbhum, Palamu, etc. In case of irrigation based Aghani crops, a few districts got due share. These districts included Garhwa, Palamu, Ranchi, Hazaribag, Chatra, etc. Rabi crops based on irrigation remained dominant in almost all districts. Among these Giridih, Deoghar, Dumka, Bokaro are important. Similarly Garma crops also depended upon irrigation in all districts. Districts having more than 90% irrigation based production of Garma crops included Dumka, Godda, Deoghar, Hazaribagh, Chatra, Giridih, Purbi Singhbhum, Palamu, etc. Districts having comparatively less irrigation based production of Garma crops included Gumla, Lohardaga, Pashchimi Singhbhum, etc.

Crops :

It has been analysed that there are crops which are grown without irrigation and with irrigation with seasons in different areas. There are certain crops which are grown in more than one season like paddy, maize, vegetables, etc. Crops grown in Bhadai season include Bhadai paddy, maize, jute, marua, jowar, bajara etc. Out of these Paddy and Maize are also grown in Aghani season. Virtually Aghani paddy becomes the dominant crop. In this season people harvest sugarcane, Til, etc. Rabi crops are mainly grown in only one season except a few vegetables. Rabi crops include wheat, gram, masoor, peas and several vegetables. Certainly these crops are confined to limited areas especially grown in areas having irrigation facilities. Similarly production of Garma crops also depend upon the development of irrigation facilities and is becoming popular day by day because of increasing demand of vegetables and seasonal fruits. The production of Garma maize and rice has not gained momentum because of heavy cost involvement in irrigation. The table 7.7 reveals cropwise area production & yield in the year 2007-08..

Table 7.7

Normal Area, Production & Yield of Major Crops in Jharkhand (5 Years Average from 2003-2004 to 2007-08)

Sl. No.	Crops/Group of Crops	Area ('000 hect.)	Production ('000 tons)	Yield (Kg/hectare)
1.	Rice	1446.60	2432.60	1682
2.	Maize	200.19	283.13	1414
3.	Wheat	73.54	115.75	1574
4.	Ragi (Marua)	14.54	9.99	687
5.	Other Millets	8.92	5.64	632
	Total Cereals	**1743.79**	**2487.11**	**1633**

6.	Gram	50.29	46.23	919
7.	Arhar	94.50	72.39	768
8.	Urad	83.47	45.39	544
9.	Other Pulses	99.88	54.17	542
	Total Pulses	**328.14**	**218.18**	**665**
	Total Foodgrain	**2071.93**	**3065.29**	**1479**
10.	Mustard & Pareseed	52.16	28.62	549
11.	Linseed	22.05	8.44	383
12.	Groundnut	19.94	18.21	913
13.	Other Oilseeds	21.19	9.51	449
	Total Oilseeds	**115.34**	**64.78**	**562**
14.	Fruits (Ton/hec)	33.76	370.46	10.97
15.	Vegetables (Ton/hec)	204.18	3005.60	14.72
	Total Horticulture	**237.94**	**3376.06**	**14.19**

Source : Directorate of Statistics & Evaluation, Govt. of Jharkhand.

Table 7.8
Food Grain Production : Jharkhand
(In Thousand Tons)

Sl. No.	Food Grain/Year	2012-13	2013-14	2014-15	2015-16
1.	Rice	3164.9	2810.6	4324.4	2669.8
2.	Wheat	319.5	370.4	330.8	287.3
3.	Coarse Cereals	463.8	526.1	487.7	395.2
4.	Pulses	686.2	578.6	597.1	495.1
5.	Oilseeds	197.2	182.9	174.5	183.4
6.	Sugarcane	461.9	462.8	469.8	463.82

Source : Directorate of Agriculture, Govt. of Jharkhand.

Maize :

Maize is an important staple crop in this region. The poor people of this region also prefer growing maize because it needs less care and provides better yield. Moreover, the undulating land and porous soils provide better conditions for growing Bhadai maize, when more people grow over large area. Certainly Garma and Rabi maize remain confined to limited areas. Bhadai maize covers 90.8 thousand hectares area while Garma and Rabi maize together cover only 1.63 thousand hectares (2009-10). This crop is also not grown everywhere in same proportion. Districts having larger area under maize include Palamu, Garhwa, Dumka, Giridih, Sahebganj etc. In these districts more than eight thousand hectares got involved under maize. Districts having less than 3000 hectares area included Ranchi, Khunti, Dhanbad, Ramgarh, Bokaro, Koderma etc.

From the point view of irrigation Bhadai maize involves only 9 thousand hectares area but Garma and Rabi maize involved irrigation over 1459 hectares area.

Production of maize remained more important in Dumka, Palamu, Garhwa, Giridih, Simdega which is clear from the table 7.9. Production remained less important in Ranchi, Khunti, Dhanbad, Gumla etc. The Yield is highest in Simdega (2254 Kg/ha) which is followed by Dumka (1823) & Lohardaga (1638).

Table - 7.9
District Wise maize Area, Production & Yield : Jharkhand (2009-10)

Sl. No.	District	Area (In Hectare)	Production (In Metric Ton)	Yield (Kg/Ha)
1.	Dumka	314,244	25,969	1823
2.	Palamu	122,002	20,512	932
3.	Garhwa	218,706	16,746	895
4.	Giridih	411,005	11,541	1049
5.	Simdega	4517	10,184	2254
6.	E. Singhbhum	7003	9,872	1409
7.	Sahebganj	58682	9,180	1057
8.	Lohardaga	5,397	8,842	1638
9.	Hazaribagh	68,155	8710	1068
10.	Pakur	4543	7390	1627
11.	Deoghar	7385	6536	885
12.	Chatra	6422	6437	1002
13.	Saraikela-Kharsawan	4849	6040	1245
14.	Latehar	4231	6034	1426
15.	Godda	3352	4692	1400
16.	W. Singhbhum	3991	4014	1006
17.	Bokaro	2321	2993	1290
18.	Jamtara	3141	2873	915
19.	Ramgarh	2491	2755	1106
20.	Koderma	2561	2283	891
21.	Gumla	3050	2094	687
22.	Dhanbad	1822	1650	906

23.	Khunti	1471	1096	745
24.	Ranchi	921	313	340
	Jharkhand	**1,63,240**	**1,90,699**	**1168**

Source : Directorate of Agriculture, Govt. of Jharkhand, 2009-10.

Paddy :

Paddy is a dominating crop in the state. The land of this region is more favourable for the production of paddy and other crops than wheat. As such cultivators give more preference to the production of Aghani paddy and some importance to Bhadai paddy. Among paddy occupied area Aghani paddy engages 86.52% while Bhadai and Garma paddy engage 13.20% and 0.28% respectively. Bhadai paddy is grown only near the settlement while Aghani paddy is grown over almost all areas suitable for paddy cultivation. Certainly Garma paddy remains confined to the areas where irrigation facilities are available and this variety of paddy is not gaining importance because of the involvement of more production cost.

As regards Bhadai paddy, poor farmers of this region give due importance to the production of different varieties of Bhadai paddy because this crop enables poor farmers to meet the shortage of food before Aghani harvest. This crop is also preferred because of negligible irrigation involvement and almost insured production as failure of monsoon hardly affect this crop. This crop is more important in the districts of Gumla, Palamu, Pashchimi Singhbhum, Lohardaga, Ranchi, etc. may be because of more liking of the tribal people. Areas where Bhadai paddy is not so popular include Sahibganj, Kodarma, Dhanbad, Purbi Singhbhum, Bokaro, Chatra, etc. Garma Paddy is important in the districts of Ranchi, Gumla, Purbi Singhbhum, Sahibganj, Dumka, etc.

Aghani Paddy is almost important everywhere. It is, however, more significant in the districts of Pashchimi Singhbhum, Purbi Singhbhum, Ranchi, Dumka, Gumla, etc. These districts have more Aghani involved areas because of comparatively more suitable land and better rainfall. These days fine and coarse Aghani paddy are grown depending upon the nature of land. Cultivators have started using high yielding varieties of paddy and chemical fertilisers.

It is important to note that cultivators irrigate Aghani paddy in case of failure of monsoon especially where irrigation facilities are available. Areas where more Aghani paddy is irrigated include the districts of Palamu, Garhwa, Pashchimi Singhbhum, Ranchi, Sahibganj, Godda, etc. Garma paddy is more irrigated in the districts of Ranchi, Purbi Singhbhum, Gumla, Sahibganj, Dumka, Lohardaga, etc.

The maximum area of paddy cultivation is in W. Singhbhum, which is closely followed by Gumla, E. Singhbhum, Ranchi, Saraikela-Kharsawan & Simdega which is clear from the given table 7.10. However, Simdega district comes first in the paddy production in the whole of the state, which is followed by Ranchi, W. Singhbhum, E. Singhbhum, Dumka. So far as the Yield is concerned Simdega district tops also in the state having a Yield of 3022 Kg./hectare. This is due to better irrigational facilities, use of fertilizers & HYV Seeds. The 2nd position is held by Jamtara which has a paddy Yield of 1975 Kg./hectare.

Table - 7.10

District Wise Paddy Area, Production & Yield : Jharkhand (2009-10)

Sl. No.	District	Area (In hectare)	Production (In Metric Tons)	Yield (Kg./Hectare)
1.	Simdega	74,8506	2,26,218	3022
2.	Ranchi	82,6504	1,18,513	1434
3.	W. Singhbhum	1,49,5961	1,04,458	1039
4.	E. Singhbhum	82,9763	96,086	1158
5.	Dumka	61,351	83,971	1369
6.	Gumla	98,3972	76,846	781
7.	Saraikela-Kharsawan	75,4365	69,552	922
8.	Jamtara	34,359	67,859	1975
9.	Godda	31,793	61,170	1924
10.	Deoghar	32,348	54,895	1697
11.	Sahebganj	32,591	51,656	1585
12.	Pakur	37,720	43,160	1156
13.	Giridih	27,775	39,512	1412
14.	Khunti	37,746	31,329	830
15.	Dhanbad	21,211	27,002	1273
16.	Chatra	14,732	19,184	1302
17.	Bokaro	14,906	18,831	1236
18.	Garhwa	14,840	18,698	1260

19.	Lohardaga	11,125	16,097	1447
20.	Hazaribagh	15,299	14,212	929
21.	Palamu	7,104	11,565	1628
22.	Ramgarh	14,693	11,284	763
23.	Koderma	7,200	4,282	595
24.	Latehar	995	1,534	1541
	Jharkhand	**9,81,693**	**15,18,364**	**1546**

Source : Directorate of Agriculture, Govt. of Jharkhand 2009-10.

Wheat :

Wheat is an important crop in the region. It was cultivated on 99,960 hectares in the year 2009-10. It is largely cultivated in the districts of Palamu, Sahebganj, Hazaribagh. It has got moderate size of area under cultivation in the districts of Godda, Giridih, Lohardaga, Garhwa & Deoghar etc. Rest districts comprise small area under cultivation of wheat.

Table - 7.11

District Wise Wheat Area, Production & yield : Jharkhand (2009-10)

Sl. No.	District	Area (In hectare)	Production (In Metric Tons)	Yield (Kg./Hectare)
1.	Palamu	72,701	13,187	1,830
2.	Pakur	4,797	11,932	2,521
3.	Sahebganj	61,172	11,331	1,872
4.	Chatra	4,656	10,701	2,330
5.	Latehar	4,397	10,060	2,321
6.	Hazaribagh	61,173	9,784	1,616
7.	Lohardaga	52,376	9,456	1,828
8.	Deoghar	51,648	9,180	1,800
9.	Garhwa	52,207	9,023	1,750
10.	Giridih	53,665	8,963	1,691
11.	Godda	57,894	8,381	1,872
12.	Dumka	50,569	7,817	1,566
13.	Jamtara	3,644	7,339	2,056
14.	Koderma	4,389	7,295	1,687
15.	Simdega	2,690	6,985	2,660
16.	E. Singhbhum	4,203	5,766	1,393
17.	Ranchi	2,631	5,228	2,036
18.	Dhanbad	2,518	3,786	1,542
19.	Gumla	2,687	3,613	1,377
20.	W. Singhbhum	3,156	3,517	1,138

21. Bokaro	3,046	3,166	1,052
22. Ramgarh	2,014	3,081	1,581
23. Saraikela-Kharsawan	2,054	2,525	1,296
24. Khunti	1,443	1,085	787
Jharkhand	**99,660**	**17,312**	**738**

Source : Directorate of Agriculture, Govt. of Jharkhand, 2009-10.

In the districts of Palamu, Sahebganj & Hazaribagh this crop is relatively more important because of more homogeneous area and better facilities of irrigation. Even in remaining districts of the state people are trying to increase wheat covered area because of changing food habit. This crop is mostly based on irrigation.

From the point of view of production of wheat the highest position is held by the district of Palamu being followed by Pakur, Sahebganj, Chatra & Latehar etc. It may be noted that production of wheat is also increasing because of the introduction of high yielding varieties and use of chemical fertilizer and compost.

Gram :

Gram is another important crop in the state. This crop is also important in the districts having comparatively more homogeneous areas. The highest place in terms of area under gram is held by Pakur being followed by Simdega, Garhwa, E. Singhbhum & Latehar. Districts having less area under gram include Bokaro, Ramgarh, Khunti, Deoghar etc. In terms of production the highest place is also held by the district of Pakur, followed by Gumla, Garhwa, Lohardaga, Simdega, E. Singhbhum etc.

Table - 7.12

District Wise Gram Area, Production & yield : Jharkhand (2009-10)

Sl. No.	District	Area (In hectare)	Production (In Metric Tons)	Yield (Kg./Hectare)
1.	Pakur	58,871	8,242	1,400
2.	Gumla	30,577	4,233	1,385
3.	Garhwa	36,503	4,161	1,140
4.	Lohardaga	2,640	3,300	1,250
5.	Simdega	42,302	3,300	780
6.	E. Singhbhum	32,244	3,240	1,005
7.	Chatra	2,636	3,033	1,150

8.	Dumka	2,366	2,756	1,165
9.	Koderma	2,645	2,647	1,001
10.	Sahebganj	2,908	2,472	850
11.	Hazaribagh	2,208	2,462	1,115
12.	W. Singhbhum	31,296	2,457	785
13.	Latehar	31,855	2,293	720
14.	Saraikela-Kharsawan	2,615	2,105	805
15.	Palamu	3,013	2,049	680
16.	Deoghar	1,870	1,761	945
17.	Godda	2,468	1,407	570
18.	Jamtara	2,410	1,325	550
19.	Giridih	2,580	1,162	450
20.	Dhanbad	2,190	961	438
21.	Ramgarh	6,702	762	1,137
22.	Ranchi	2,123	686	323
23.	Bokaro	4,901	485	980
24.	Khunti	8,173	214	262
	Jharkhand	**63,011**	**57,590**	**914**

Source : Directorate of Agriculture, Govt. of Jharkhand

Other Pulses :

Other pulses include Kulthi, Mung, Khesari, Masoor, Arhar, etc. These crops are also more important in the districts of Palamu, Garhwa, Godda, Pakur, Sahibganj, Ranchi, etc. These crops are less important in Dumka, Deoghar, Giridih, Bokaro, Dhanbad, Purbi Singhbhum, etc. Production of these crops is also important in the districts of Palamu, Garhwa, Godda, Pakur, Sahibganj, Ranchi, etc.

Fruits :

The state of Jharkhand is also famous for the production of a vide variety of fruits. The total production of fruits in the state was 943277.56 tons which was grown in 89486.74 hectares (2015-16). The highest area under fruit is occupied by Mango, which is followed by Jackfruit, Banana, Guava, Limes & Lemon etc. (Table 7.13). Whereas Pomegrante has the least covergae area followed by Ber, Bil, Papaya, Aonia, Litchi etc.

Table - 7.13
Area & Production of Fruits : Jharkhand
(2015-16)

Sl. No.	Fruits	Area (In hectare)	Production (In Tons)
1.	Mango	44,317.92	3,90,131.07
2.	Jackfruit	11,631.58	1,07,204.94
3.	Papaya	1,615.55	1,06,226.60
4.	Guava	8,070.26	79,758.66
5.	Limes & Lemon	4,767.36	47,652.43
6.	Litchi	3,451.33	40,009.40
7.	Beal	654.03	35,475.11
8.	Banana	10,236.27	30,889.78
9.	Aonia	2,482.43	3,480.04
10.	Ber	475.00	2,827.00
11.	Pomegranate	5.01	7.53
12.	Other Fruits	1,780.00	99,535.00
	Total	**89,486.74**	**9,43,277.56**

Mango is the highest produced fruit in the state which is followed by Jackfruit, Papaya, Guava, Limes & Lemon etc. The fruits are produced almost in all the districts of the state. The tropical fruits consisting of Mango, Jackfruit, Papaya & Guava are very important in terms of production & distribution because of the suitability of the climate. The least produced fruits in the state are pomegranate.

Vegetables :

The state of Jharkhand is also noted for the production of a vide variety of vegetables. The total production of vegetables in the state is 27,80,763.00 tons which is grown in about 2,05,550.12 hectares (2015-16). Tomato has the maximum area coverage among vegetables followed by Potato, Cabbage and Brinjal in the state (Table 7.14) Capsicum has the least area coverage followed by Pumpkin, Cucumber, Carrot, Radish etc.

Table - 7.14
Area & Production of Vegetables : Jharkhand
(2015-16)

Sl. No.	Vegetables	Area (In hectare)	Production (In Tons)
1.	Potato	44,738.29	6,24,169.58
2.	Cabbage	22,816.34	2,92,424.76
3.	Onion	15,557.08	2,63,452.12
4.	Cauliflower	19,414.44	2,46,656.75
5.	Tomato	1,69,779.39	2,29,598.69
6.	Brinjal	20,195.63	2,16,874.19
7.	Pea	13,548.25	1,92,051.79
8.	Beans	10,375.80	1,54,553.86
9.	Green Chilli	1,185.00	1,43,242.84
10.	Okra	3,155.26	63,423.54
11.	Pumpkin	325.00	19,405.00
12.	Carrot	892.00	15,195.60
13.	Bottle Gourd	3,209.37	15,538.49
14.	Bittler Gourd	3,209.37	11,773.38
15.	Radish	1,011.57	11,326.35
16.	Cucumber	821.03	6,998.46
17.	Capsicum	276.00	3,720.00
18.	Other Vegetables	19,002.60	2,72,360.60
	Total	**2,05,550.12**	**27,80,763.00**

Source : Dept. of Agriculture, Co-operation & Farmers Welfare.

Potato is the highest produced vegetable in the state, which is followed by Cabbage, Onion, Cauliflower & Tomato. The area under potato is incresing day by day. In certain areas even two crops of potato one grown in a year because this region provides ideal conditions for early varieties of potato which enable cultivators to earn more money especially around urban centres. As such, the districts of Ranchi, Ramgarh, Hazaribagh, Palamu, Garhwa, Godda, Dumka, Chatra, Giridih, Gumla etc. have become very important as potato & other vegetables growing areas. The vegetables are grown especially around urban centres having flat terrain & good irrigation facilities.

Milk, Egg, Meat & Wool Production :

The production of milk, meat, egg and wool production in

the state has shown significant increase between 2008-09 to 2015-16 (Table 7.15). During this period, the production of milk has increased from 1466.35, thousand MT in 2008-09 to 1812.38 thousand MT in 2015-16, registering about 23.60% increase in milk production.

Table - 7.15
Milk, Egg, Meat & Wool Production : Jharkhand

Sl. No.	Item/Year	Unit	2008-09	2015-16
1.	Milk	Thousand MT	1,466.35	1,812.38
2.	Egg	Lakh No.	3,994.77	4,832.84
3.	Meat	Thousand Tons	44.17	50.71
4.	Wool	Thousand Kg	149.00	105.82

Source : Dept. of Animal Husbandary & Fisheries, Govt. of Jharkhand.

The total no. of livestock population in the state is 1,75,70,555 (2012 Census) which constitutes about 3.52% of the country. Here, we see a large no. of the cattle population which constituted about 4.57% share of the country. Similarly, the production of eggs, meat & wool has also witnessed remarkable increase over the two periods in the state. The demand of these items are growing day by day in the state and as a result commercialisation & industrial farming has contributed a lot to meet these demands. Majority of these livestock and cattle are reared in rural areas around urban centres.

8

Irrigation and River Valley Projects

Uneven and irregular distribution of rainfall during agricultural seasons make the necessity of irrigation. These days production of various crops during different agricultural harvests has further accelerated the requirements of irrigation. It may be noted that previously people used to grow a few crops depending upon the natural precipitation and seasons lacking rainfall used to be without farming activities. Presently the production of crops even during seasons having negligible rain has compelled farmers to develop irrigation means. Virtually production of Rabi and Garma crops and a lot of vegetables to a considerable extent depend upon irrigation and these crops have gained significance because of changing food habits and consumption. More over the dominance of undulating terrains in the state hardly allow rainfall to stagnate and drain off quickly. The region also lacks perennial rivers. Most of the rivers and rivulets are rainfed. Even when rivers and rivulets containing water hardly help in irrigating the fields away from the river course because of higher terraced fields on both sides.

Based on above conditions use of irrigation water becomes necessary wherever agricultural activities have intensified. Surface and underground water have enabled cultivators to adopt different methods of irrigation. Because of rough terrain natural water bodies are negligible. Certainly the region provides ideal conditions for the development of artificial reservoirs or tanks, etc. It may be noted that people block the flow of the rivulet or river valley and store rain water for irrigation. In homogeneous areas Nala, Ahar, Pyne or smaller water bodies also provide the facility of surface water. In case of underground water the region has some limitations. Because of rocky structure the development of tube well becomes difficult. Even wells dry up during summer season and hardly facilitate irrigation on large size of land. A few rivers have also enabled the facility of canal irrigation.

In the absense of sufficient rainfall, the agriculture in

Jharkhand state is primarily depended on irrigation. Since, this state is a water starved and deficient state, so the managing and utilising the available water resource is the prime concern of the state. As per the water Resource Department, the availability of water resource in the state is only 2,87,810 lakh m^3 out of which 237890 lakh m^3 is from surface water and the remaining 49,920 lakh m^3 is from ground water. The total utilisation of surface and ground water for irrigation purposes in the state so far is only 47,360 lakh m^3 of which 39,640 lakh m^3 is from surface water and 7,720 lakh m^3 is from ground water.

Table - 8.1
Source of Irrigation : Jharkhand
(2012-13)

Sl. No.	Sources	% Irrigation
1.	Canal	17.53
2.	Pond	19.07
3.	Tubewell	08.25
4.	Well	29.38
5.	Others	25.77
		100.00

Source : Irrigation Department, Govt. of Jharkhand

The Jharkhand state has a total agricultural land of 29.74 lakh ha. Against the total cultivable area of 29.74 lakh hectare, irrigation potential of 24.25 lakh hectare has been identified so far. About, 1276 thousand hectare irrigation potential has to be created through major & medium irrigation projects and 1184 thousand hectare irrigation potential has to be provided by minor irrigation schemes and other sources.

Table - 8.2
Irrigation Fact Sheet : Jharkhand

Sl. No.	Item	Jharkhand
1.	Irrigation Coverage	31.98%
2.	The cultivable are to total area	47.67%
3.	NSA to total area	29.63%

Sources : DES & Economic Survey, 2011-12.

Table - 8.3
Ultimate Irrigation Potential : Jharkhand
('000 hectare)

State	Major & Medium Potential	Minor Potential			Total
		Surface	Ground	Sub-Total	
Jharkhand	1276	354	830	1184	2460

Source : Water Resource Department, Govt. of Jharkhand

The analysis of table 8.4 reveals sourcewise net irrigated area in the year 1997-98. In terms of canal irrigation six districts namely Deoghar, Sahibganj, Pakur, Kodarma, Giridih and Bokaro are devoid of irrigation by canal. Even the districts of Godda and Dumka have even less than 1% irrigated area by canal. Districts having net canal irrigated area less than 10% and more than 1% consisted of Hazaribag, Chatra, Dhanbad, Gumla and Purbi Singhbhum. The remaining districts like Ranchi, Lohardaga, Pashchimi Singhbhum, Palamu and Garhwa consisted of net irrigated area by canal more than 10%.

Tank irrigation is relatively in better position in this state. It irrigates 15.75% of the Net Irrigated Area. The highest percentage of tank irrigated area has been in the district of Pakur (55.30%) being followed by Godda (48.40%), Deoghar (36.63%), Lohardaga (35.26%), Purbi Singhbhum (29.27%), Dumka (23.59%), Bokaro (17.77%), Chatra (17.52%), etc. Districts having tank irrigated area less than the average of the state included Garhwa, Pashchimi Singhbhum, Gumla, Ranchi, Giridih, Kodarma, etc.

It has been mentioned earlier that the structure of the area is not favourable for the development of tube well. It is however, developed in some districts. Districts having tube well irrigation even less than 1% included Dumka, Godda, Deoghar, Pakur, Dhanbad, Lohardaga and Gumla. The state of Jharkhand has 13.40% tube well irrigated area. The districts having more than 30% irrigated area by tubewell included Bokaro (40.67), Giridih (40.13) and Kodarma (34.87). Districts having tube well irrigated area between the regional average (13.40%) and 30% included Chatra (26.91), Ranchi (21.22) and Hazaribag (21.20). Only a few districts consisted of tube well irrigated area more than 1% and less than the regional average. These districts comprised of Palamu, Garhwa, Pashchimi Singhbhum, Purbi Singhbhum, Sahibganj, etc.

Well irrigation is comparatively better developed as it irrigates 31.39% of net irrigated area. All districts have some areas under well irrigation. The highest percentage has been in the district of Dumka (58.11), being followed by Giridih (50.83), Gumla (49.21), Lohardaga (45.20), Dhanbad (43.34), Godda (41.70), Bokaro (40.53), Ranchi (35.89) Hazaribag (34.80), etc. Districts having well irrigated area less than the regional average included Garhwa, Pashchimi and Purbi Singhbhum, Sahibganj, Pakur, etc. Even in terms of other sources of irrigation the highest percentage has been recorded in the district of Sahibganj (80.38%). This district is being followed by Purbi Singhbhum (57.22%), Gumla (46.65%), Dhanbad (39.26%), Deoghar (31.55%), Paschimi Singhbhum (30.23%), etc. Other sources of irrigation irrigating less area included Bokaro, Lohardaga, Palamu, Giridih, Chatra, Hazaribag, etc.

The percentage of Net Irrigated Area to Total Net Sown Area remained less than 10 per cent in the districts of Dumka, Deoghar, Pakur, Bokaro, Dhanbd, Lohardaga, Gumla, Purbi Singhbhum and Pashchimi Singhbhum. Districts having moderate Net Sown Irrigated Area included Garhwa, Palamu, Kodarma, Chatra, Sahibganj, Ranchi, etc.

As regards source wise gross irrigated area in the year 2012-13 remarkable spatial variation has been noted. The table 8.1 reveals gross irrigated area by different sources of irrigation in percentage. The state of Jharkhand has 29.38% well irrigated area and 25.77% gross irrigated area by other sources. Canal, pond and tube well irrigated area consisted of 17.53%, 19.07% and 8.25% respectively. Canal gross irrigated area is non-existent in the districts of Deoghar, Sahibganj, Pakur, Kodarma and Bokaro. The percentage of gross irrigated area by tank less than 5% included Dumka, Godda, Gumla and Dhanbad (Table 8.4). Districts having canal irrigated area between 5% and 14.02% (regional average) consisted of Hazaribag and Purbi Singbhum. The remaining districts namely Pashchimi Singhbhum, Palamu, Garhwa, Lohardaga, Ranchi and Chatra consisted of higher percentage more than the regional average.

Tank irrigated area remained important in more districts. The highest percentage has been noted in the districts of Pakur (58.76) (Table 8.4) being followed by Godda (57.99), Lohardaga (34.68),

Deoghar (32.94), Dumka (23.73), Purbi Singhbhum (25.71), Chatra (22.27) etc. Districts having lower percentage included Giridih, Ranchi, Gumla, Pashchimi Singhbhum, Garhwa, etc. Tube well Gross Irrigated Area remained non-existent in the districts of Dumka, Deoghar, Lohardaga and Gumla. This irrigation remained less important in the districts of Godda, Pakur, Sahibganj, Purbi Singhbhum, Pashchimi Singhbhum, Dhanbad, etc. Only a few districts consisting of Hazaribag, Kodarma, Chatra, Giridih, Ranchi recorded relatively better position.

Gross irrigated area by well remained almost prevalent in all districts. The state of Jharkhand consisted of 35.38% gross irrigated area by well. The highest percentage has been in the district of Dumka (57.40) being followed by Giridih (55.51), Gumla (48.46), Bokaro (46.44), Dhanbad (45.66), Ranchi (44.59), Lohardaga (43.91), Hazaribag (39.71), Godda (38.69), etc. The remaining districts consisted of gross irrigated area by well less than the regional average (35.38%). These districts included Palamu, Garhwa, Purbi, Singhbhum, Sahibganj, Pakur, Chatra, etc.

In case of other sources of irrigation all districts of the state recorded some areas. This state consisted of 24.08% gross irrigated area by other sources. The highest percentage has been in the district of Sahibganj (82.51) being followed by Purbi Singhbhum (59.60), Gumla (48.11), Dhanbad (38.48), Garhwa (36.05), etc. Districts having lower percentage of gross irrigated area by other sources included Godda, Hazaribag, Giridih, Bokaro, Lohardaga, etc.

From the point of view of the percentage of irrigated area to total cropped land it is only 9.32% in the state of Jharkhand. Certainly the distribution is uneven. The highest percentage has been in the district of Garhwa (27.0) percent. It is followed by Palamu (23.34%), Hazaribag (16.69%), Chatra (15.93%), Godda (13.96%), Kodarma (12.53%), etc. Districts having moderate irrigated area included Dumka, Giridih, Ranchi, Bokaro, Pakur, etc. Districts having lower percentage of irrigated land included Purbi and Pashchimi Singhbhum, Pakur, Gumla, etc.

The Damodar Valley Project :

The Damodar Valley Project was developed on the pattern of Tennessee Valley Project of U.S.A. in the year 1948. It was designed for an integrated economic development in the valley by the

Table - 8.4
Sourcewise Net Irrigated Area, 1997-98 (in hectares)

Sl. No.	Name of District	Canal	Tank	Tubewell	Well	Other source	Total	% of Irrigated NSA
1.	Dumka	74	1995	-	4914	1474	8457	
		0.88	23.59		58.11	17.42	100.00	5.14
2.	Godda	57	4628	4	3988	885	9562	
		0.60	48.40	0.04	41.70	9.26	100.00	12.38
3.	Deoghar	-	2307	-	2004	1987	6298	
		-	36.63	-	31.82	31.55	100.00	8.67
4.	Sahebganj	-	561	175	251	4043	5030	
			11.15	3.48	4.99	80.38	100.00	10.91
5.	Pakur	-	1096	3	353	50	1982	
			55.30	0.15	17.81	26.74	100.00	3.23
6.	Hazaribag	1062	2272	2825	4639	2526	13324	
		7.97	17.08	21.20	34.80	18.95	100.00	13.61
7.	Kodarma	-	310	860	749	547	2456	
			12.57	34.87	30.37	22.19	100.00	13.93
8.	Chatra	445	1008	1548	1992	760	5753	
		7.74	17.52	26.91	34.63	13.20	100.00	12.21
9.	Giridah	-	38.1	3472	4398	401	8652	
			4.40	40.13	50.83	4.67	100.00	10.07

10.	Bokaro	-	372	852	849	22	2095	
			17.77	40.67	40.53	1.07	100.00	5.77
11.	Dhanbad	16	246	7	670	607	1546	
		1.04	15.91	0.45	43.34	39.26	100.00	4.10
12.	Ranchi	4975	192	3582	6058	2072	16879	
		29.47	1.14	21.22	35.89	12.28	100.00	66.98
13.	Lohardaga	513	1242	-	1592	175	3522	
		14.57	35.26		45.20	4.97	100.00	7.62
14.	Gumla	142	172	-	3737	3542	7593	
		1.87	2.27		49.21	46.65	100.00	2.87
15.	Purbi Singhbum	108	420	73	99	735	1435	
		7.53	29.27	5.08	6.90.	51.22	100.00	1.57
16	Paschimi Singhbum	4874	574	154	33	2442	8077	
		60.34	7.11	1.91	0.41	30.23	100.00	4.02
17.	Palamu	9062	5164	4965	8991	6810	34992	
		25.89	14.76	14.19	25.69	19.47	100.00	23.88
18.	Garha	4322	1772	2511	3930	6501	19036	
		22.70	9.31	13.19	20.65	34.15	100.00	30.17
Total		Jharkhand	25650	24712	21031	49247	36259	156899
		16.35	15.75	13.40	31.39	23.11	100.00	

Source : Statistical Office, Bihar.

Table - 8.5
Sourcewise Gross Irrigated Area, 1997-98 (In hectares)

Sl. No.	Name of District	Canal	Tank	Tubewell	Well	Other source	Total	% of Irrigated area & total cultivated land
1.	Dumka	80	2523	-	6102	1925	10630	
		0.75	23.73		57.40	18.12	100.00	6.42
2.	Godda	57	5907	4	4395	998	11361	
		0.5	51.99	0.04	38.69	8.78	100.00	13.96
3.	Deoghar	-	2307	-	2398	2298	7003	
			32.94		34.24	32.82	100.00	9.45
4.	Sahebganj	-	668	175	304	5411	6558	
			10.19	2.67	4.63	82.51	100.00	12.40
5.	Pakur		1422	3	353	772	2550	
			55.76	0.13	13.84	30.27	100.00	3.80
6.	Hazaribag	1071	2827	3698	7054	3115	17765	
		5.86	15.91	20.82	39.71	17.70	100.00	16.69
7.	Kodarma	-	464	951	893	638	2946	
			15.75	32.28	30.31	21.66	100.00	12.53
8.	Chatra	1445	1753	1780	1761	1134	7873	
		18.35	22.27	22.61	22.37	14.40	100.00	15.93
9.	Giridah	-	342	3532	5355	418	9647	

			3.55	36.61	55.51	4.33	100.00	8.46
10.	Bokaro	-	397	852	1102	22	2373	
			16.73	35.90	46.44	0.93	100.00	4.82
11.	Ranchi	5039	192	3914	9711	2923	21779	
		23.14	0.88	17.97	44.59	13.42	100.00	8.30
12.	Lohardaga	715	1623	-	2055	287	4680	
		15.28	34.68		43.91	6.13	100.00	9.45
13.	Gumla	159	213	-	5280	5238	10896	
		1.46	1.95		48.46	48.11	100.00	4.01
14.	Purbi Singhbhum	108	490	73	99	1136	1906	
		5.67	25.71	3.83	5.19	59.60	100.00	1.11
15.	Paschimi Singhbhum	4912	765	154	33	2498	8362	
		58.74	9.15	1.85	0.39	29.87	100.00	3.72
16.	Palamu	9279	5809	5175	12351	8375	40989	
		22.64	14.17	12.63	30.13	20.43	100.00	23.34
17.	Garhwa	4322	2102	3058	5629	8516	23627	
		18.29	8.90	12.94	23.82	36.05	100.00	27.00
18.	Dhanbad	16	279	7	870	733	1905	
		0.84	14.65	0.37	45.66	38.48	100.00	4.50
	Jharkhand	27043	27776	23356	68238	46437	192850	
		14.02	14.40	12.12	35.38	24.08	100.00	9.32

Source : Statistical Office, Bihar.

Damodar Valley Corporation. Now it is managed by the combined efforts of the Jharkhand and West Bengal Governments. This project includes parts of Jharkhand and West Bengal and comprises an area of about 8500 sq. miles. It aimed to construct multipurpose storage dams with hydro-electric plants to irrigate lands by developing irrigation channels with distributaries measuring 2280 km. including 136 km. long irrigation - navigation canal and to establish thermal power stations along the transmission lines.

The project consists essentially of the following —

(i) Eight dams on the Damodar and its tributaries with a total reservoir capacity of 4.7 million acres feet, protecting the valley from a "design" flood of one million cusecs and hydro electric stations.
(ii) Thermal power stations, using low grade coal to even out the seasonal fluctuations of hydro-electric power and to provide in combination with the hydro-electric system.
(iii) A power transmission grid covering 470 miles of 132 KV primary lines with necessary length of 66KV and 33 KV secondary lines.
(iv) To provide irrigation.
(v) Cheap transport facility through the canal.
(vi) Efficient drainage of about 1120 sq. kms. of water logged land in the lower basin.
(vii) Land reclamation and improvement.
(viii) Anti-Malarial measures and health improvement.
(ix) Development of Pisciculture.
(x) Forest based industries.
(xi) Recreation centres and tourist places.
(xii) To control flood and conservation.
(xiii) Soil and afforestation in the upper basin.

There are eight major dams at Tilaiya, Bal Pahari, Maithon, Panchet, Durgapur, Aiyar, Konar and Bokaro upon which hydel plants and reservoirs for irrigation have been developed. In addition Tenughat Dam has been constructed to provide water to Bokaro steel plant through canal and to facilitate other advantages. These dams have been constructed in two phases. In the first phase Tilaiya and Maithon Dams on the river Barakar (1957), Panchet

Dam on the river Damodar (1959) and Konar Dam on the river Konar were completed. At the same time the facility of hydel power production was developed at Maithon and Panchet. A barrage was constructed at Durgapur in the year 1960 for developing irrigation channels. The Tilaiya Dam consisting of 340 metres length and 30 metres height produces 4000 kw. hydel power. Konar Dam is 3960 metres in length and 50 metres in height and was completed in the year 1955. Maithon Dam and Panchet Dams were completed in 1957 and 1959 respectively. Bokaro Thermal Power Station was installed on the river Konar in 1953 with a production capacity of 1.5 lakh kw. but its fourth unit was installed in 1960-61 and its production increased 2.25 lakh kwt. This is the largest power station in India. A township was built at the worksite and was provided with the modern facilities. The capacity of Maithon dam is 60 thousand kwh and Panchet Dam has a capacity to produce 40 thousand kw. Thermal Power Station at Chandrapura was installed to produce 250 lakh kw. In addition to above mentioned main dams, canals and thermal power station there are 20 additional head water storage schemes of small size, some of which have been completed and the dams and reservoirs for drinking water projects constructed at Ranchi and Hazaribag and the irrigation dam at Gonda fall under this group. The Tilaiya High Level Irrigation Scheme has the capacity to irrigate 15000 acres. Durgapur Barrage provides the facility of irrigation for 8 lakh acres, much of the irrigated area lies on the left of the river in Burdwan and Hooghly districts of West Bengal. Only 1.5 lakh acres of canal irrigated area lie in Bankura district. This irrigation has helped in accelerating rice acreage and yield.

This project helped in conserving soils in the upper portion of the valley. It is estimated that about a million acres have been rendered waste by soil erosion resulting from deforestation. Out of this 40 thousand acres have been reclaimed as a first measures which are scattered over a wide area of the valley. About 5 thousand acres of land have been afforested. Those displaced from the Tilaiya, Bokaro, Konar, Panchet, Maithon projects have been rehabilitated in villages having road, drinking water, community centre, schools, etc. Power produced in the region has mainly consumed in the major industrial, mining and urban areas. Main consumers of power are the iron and steel industries at Bokaro, Burnpur, Durgapur, Asansol, Railways electrification from

Kolkata to Mughal Sarai, aluminium and mining industries, Kolkata Electric Supply Corporation, Indian Copper Corporation, etc., the Locomotive works at Chitranjan, the Hindustan Cables of Kolkata, the Electric Supply Corporation, the Copper mines at Ghatshila, the Coal fields at Jharia and Ranignaj, the State Electricity Board of Bihar, Jharkhand and West Bengal.

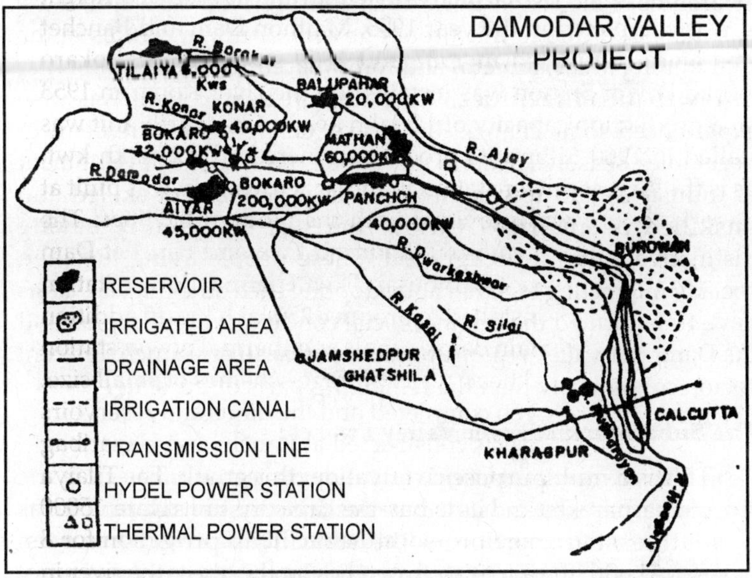

Fig. 10

The development of the Damodar Valley Project in Jharkhand State and West Bengal has not only facilitated in controlling soil erosion having flood in the lower portion, loss of life and property but has also enabled the production of hydel power in bulk for industrial, urban and other uses. As a matter of fact this project has changed the physical environment, industrial landscape, urban habitat and socio-economic characteristics. It is bringing a new order in its geography by not only illuminating the valley and its adjoining areas but by accelerating the region in the realm of agriculture, mining, manufacturing, urban, rural, transport network, social ecology, etc. The green belt of the valley has been punctuated with plants and their colonies which somewhere have attained a greater magnitude particularly in Dhanbad complexes and growing complexes of the upper Damodar valley in Hazaribag[1]. Land use pattern has also, to some extent, changed. Further depletion of forest has been stopped. Afforestation has helped in improving the share of forest cover. Culturable waste land has been reclaimed due to irrigation facilities and the awareness of the people towards agriculture. Non-agricultural use has increased due to mushroom growth of urban centres in the coal mining areas, mineral and forest based industries and other infrastructural facilities. Agricultural use of land has also experienced some changes because of changing requirements of vegetables, seasonal fruits in the market centres. The development of industries and towns attracted large number of skilled and unskilled people from various parts of the country having distinctive cultural background. These people living together, socially interacting together and adopting each others social habits have transformed the original social conditions of the region and the Damodar valley region appears comparatively much different socio-economically than the remaining parts of the state.

The Subarnarekha River Valley Projcct :

This is a multipurpose river valley project. It is based on the river Subarnarekha and its tributaries. It covers considerable areas of south-eastern portion of the state. This project has been developed with the intention to control flood in the lower portions, minimize soil erosion, irrigation of crops and water supply to the industrial concerns and urban centres. The advantages of this project are not only confined to this state but it also provides some benefits to neighbouring state of Orissa and West Bengal. The river Subarnarekha originates in the south-eastern portion of Ranchi Plateau and drains the eastern portion of the Ranchi Plateau. Its tributaries Kharkai and Kanchi enable the river to serve the area in terms of water supply and irrigation.

Considering the advantages of water supply, irrigation, soil erosion and generation of hydel power there are proposals to construct dams near Chandil,& Galudih near Ghatshila. These two dams would enable the formation of large water reservoir from which irrigation channels and generation of power would be possible. Similar to these two dams there are proposals of two more dams on the river Kharkai, the main tributary of the river

Subarnarekha. These two dams would be constructed at Ichabag near Chaibasa and Gajia near Jamshedpur. These dams and reservoirs would enable the irrigation of about 2.4 lakh hectares area of the state.

Koel Karo Project :

This project emerged in the form of proposal as early as 1973-74 but it remained without concrete frame due to local political constraints. This project is, however, attracting the attention of the government for the generation of power and development of area by irrigation facilities and other advantages like control over soil erosion, afforestation, etc. This project is based on the rivers Koel and Karo. The river south Koel rises in the district of Ranchi and joins the river Sankh in Orissa and then it is called as Brahamani. Karo river is the main tributary of the river South Koel. These two rivers form the basin of about 7 thousand square km. and the entire area is dominated by the tribal people. At the same time this area has also considerable forest cover. There is proposal to construct 44 metres dam across the river Koel and 55 metres dam across the river karo. These two dams would enable the formation of about 5 km. canal. These works would enable generation of 710 MW hydel power and irrigation of a large area. In this project there is likelihood of huge expenditure of about 600 crores rupees, acquisition of 50 thousand acres of land, re-allocation of 50 villages having more than 8 thousand families.

Mayurakshi Project :

The project is based on the river Mor or Morakhi in Santhal Pargana. This project would provide advantages to the state of Jharkhand and West Bengal. In the upper portion of this river Canada dam has been constructed for irrigation and generation of power. The dam has been constructed at Masanjor in the district of Dumka. This dam is 640 metres long and 47 metres high. This dam controls the drainage of upper Mayurakhshi and enables the irrigation of 10 thousand hectares of land. Down to this dam there is a barrage at a distance of 23 kms. at Tilpara. This barrage is 310 metres long and both sides of this barrage canals have been developed which irrigate about 3 lakh hectares of land. This project also enables the production of hydel power of 4 thousand kw. which is being used in the districts of Santhal Parganas and West Bengal.

North Koel Project :

This project is based on North Koel river, a main tributary of the river Son. This project has been designed to control flood, soil erosion, to generate hydel power, to develop irrigation channels and to have other advantages. This area is noted for having flood in rainy season and comparatively dry during summer seasons. As such, there is a proposal to construct a dam of 66 feet high at Kutku which would enable generation of 90 MW power and development of 112 km. main irrigation canal and 240 km. branch irrigation canal.

Reference :

1. Singh, R.P. & Kumar A. (1970) : op. cit., p. 66.

9

Minerals

The state of Jharkhand having geological formations of different geological periods and varied landscapes is not only rich in forest resource but is rather more important in terms of minerals. The state is not only notable for minerals in the country but also gets position on the world map. This region is considered as store house of minerals of the country accounting for nearly 7% of its mineral production in value terms (2016-17). This state has huge reserve of good quality coal (80% of India) and produces about 18.95% coal of the country. The area has 32.82% reserve of iron ore of the country and produces 12.24% iron ore of the country (2015-16). It also produces 53% mica, 7.5% bauxite, 6.84% copper, 21% fireclay, 26.95% graphite, 64% kyanite, Pyroxenite & Flint stone 100% each etc. In addition to these minerals there are considerable amount of other metallic, non-metallic and atomic minerals.

Huge reserve of different minerals in the state has close relationship with different geological periods. The Dharwarian sedimentation enabled the state to have good reserve of superior quality iron ore in south-eastern portions of the state. The Gondwana deposits in the trough basin of the Damodar valley facilitated the formation of thick coal seams of bituminus and other types of coal. The formations of Vindhyan deposits in the north-western and south-eastern margins have provided the occurrence of limestone. Lava cappings during cretaceous period enriched this region with laterites containing bauxite. Besides these important minerals containing formations of igneous intrusions, erruptions, etc. have also played important role in the occurrence of a number of minerals. The following table reveals the relationship between the geological periods, rock formations and mineral associates.

During recent years accelerated production of minerals and its diverse uses has played pivotal role in improving economic structure and multi-faceted developments. The region has become the focal points of migrant attractions towards the industrial pockets, urban agglomerations and related activities. Consequently mineralized and industrialised areas of the state have experienced phenomenal economic development and social transformations better than other areas.

Reserves :

Jharkhand state is one of the leading mineral bearing state of India and is the store house of more than 30 types of minerals, hence aptly called as "Museum of Minerals". No region in the world is gifted with such a vast mineralisation in such a close proximity as the Jharkhand enjoys. It has reserves of more than 30 types of minerals which includes energy, ferrous, non-ferrous, fertilizer, industrial, refractory, atomic, strategic, precious and semi-precious stone (Table 9.1 & 9.2)

Table - 9.1
Mineral Availability at a Glance : Jharkhand

1.	Mineral Resources in Plenty	China Clay, Fire Clay, Magnetite, Graphite, Kyanite, Copper, Bauxite, Feldspar, Mica & Decorative Stones.
2.	Mineral Resources Sufficient	Cola, Iron-Ore, Limestone, Quartz, Quartzite, Dolonite, Pyroxenite.
3.	Mineral Resources Insufficient	Andalusite, Manganese, Chromite, Beryl, Talc, Gold, Bentonite.

Table - 9.2
Status of important Mineral Reserves : Jharkhand (2013-14)

Sl. No.	Mineral	Mineral Reserver (In MT)	% Share to India's Reserver	Rank in India	Available Grade
1.	Coal	71864.00	29.24%	Ist	All Grade
2.	Haematite	3750.00	32.82%	2nd	>58% to 67% fe
3.	Cobalt	9.00	20.04%	2nd	Low grade
4.	Apatite	7.27	29.56%	2nd	9.60% As P_2O_3
5.	Quartzite Silica Sand	170.44	5.85%	2nd	SiO_2 96%
6.	Chromite	736.00 (Th. Tonne)	0.41%	3rd	All Grades

7.	Asbestos	0.15	0.81%	3rd	Amphibole Variety
8.	Copper Ore	178.82	13.36%	3rd	1.00% to 1.85% Cu
9.	Mica	1494430 Kg	2.49%	3rd	All Grade
10.	Graphite	9.22	5.78%	3rd	3.74-25% Carbon
11.	Kyanite	4.68	4.56%	3rd	Various Grade
12.	Bentonite	1.01	0.19%	4th	All Quality
13.	Fire Clay	68.11	9.81%	4th	Good Quality
14.	Vermiculitc	30004800 (Tonne)	11.44%	5th	Good Quality
15	China Clay	210.18	9.14%	6th	Superior Quality
16.	Lime Stone	662.47	0.38%	7th	Various Grade
17.	bauxite	177.35	4.01%	7th	All Grade
18.	Felspar	1.32	1.88%	7th	Ceramic Grade
19.	Manganese Ore	29.95 (Th. Tonne)	0.010%	7th	High Magnetic Content
20.	Magnetite	7.54	0.07%	8th	High Magnetic Content
21.	Barytes	0.03	0.37%	10th	High Quality
22.	Gold	0.008 (Tonne)	0.004%	111th	1.5-3% Au
23.	Dolomit	57.18	0.80%	12th	All qualities
24.	Magnesite	2.30	-	-	Various grade
25.	Pyroxenite	0.83	-	-	Various grade

Source : http://www.india.gov.in

Table - 9.3

Geological Associates Period	Character	Mineral
Archean	Sedimentary rocks (Later on metamorphesed)	Iron ore, manganese ore, limestone, dolomite, kyanite, refractory
quartzite,		Steatite, chert, quartzite, slate, china clay, ochres
Archean	Basic Ingeous Rocks	Chromite, steatite, vanadium bearing iron-ore, asbestos magnesite magnetite
	Granite	China clay, building stone, road metal
	Pegmatite veins	Mica, felspar, beryl, rose quartz, apatite cassiterite, tourmaline
	Other veins in Archeans	Copper, apatite-magnetite, gold, barytes, quartz, lead

	Kolhan series Sedimentary forming unconformity	Building materials iron- ore, manganese ore, limestone
	Newer dolerite	Road metal, railway ballast
Cambrian	Vindhyan system	Limestone, sandstone shale, glass-sand, pyrite, stone, sandstone, glass-sand, etc.
Gondwana	Deposits	Coal, fireclay, china-clay sandstone, glass-sand, etc.
Cretaceous	Traps	Road metal, building stone, agate, atmethyst, opal
Tertiary		gravel , ballast for railways and concrete
	Laterite	Building stone, bauxite
	Recent Alluvium	Brick and pottery clays, sand kankar, earths for cement manufacture

Mineral, of this region can be categorised into three groups:-1. Metallic minerals containing iron-ore, copper, bauxite manganese, chromite, Tungstun, gold, silver, etc., 2. Non-metallic minerals consisting of mica, limestone, dolomite, asbestos, fire clay, china clay felspar, etc. and 3. Atomic and power generating minerals comprising of uranium, thorium, graphite, vanadium, coal, etc.

Production :

Jharkhand was the leading producer of Kyanite, Pyroxenite and granite, the 2nd largest producer of Quartzite, coal and fire clay and 3rd largest producer of Copper Ore, Copper concentrates and quartz in the country during 2011-12. The other principal minerals produced in the state were bauxite, dolonite, kaolin, felspar and manganese ore. The value of mineral production in Jharkhand during 2015-16 stood at 20710 crore. The state accounted for 7% of total value of mineral production in the country during 2015-16 and claimed 5th position in the country. Coal, the principal mineral produced in the state contributed 87% of the total value of mineral production in the state followed by iron-ore 12%. The number of reporting mines in Jharkhand during 2015-16 was 211 as against 297 in the 2010-11.

Metallic Minerals:

Iron-Ore: The state of Jharkhand occupies important place in occurrence and production of iron ore. It has 4th position in term of production. This state has 29.2% reserve of iron-ore of the country and produces only 15% iron-ore of the country. Iron-ore mines are concentrated in the south-eastern portions of the state where the Iron-ore series of Dharwarian formations are found. The Iron-ore series composed primarily of banded hematite-quartzite, phyllite, tuffs, lavas and Cherts form a horse shoe closed on the south in Orissa and open in the north in Singhbhum with ends about 13 kms apart.[1] This region has considerable reserve of haematite iron-ores containing about 5.5% to 50% iron. Important mines are Noamundi, Panrsia Buru, Buda Buru, Gua, Jamda, Jhiling Buru, Lukud Buru, Notu Buru, etc. Noamundi is surrounded by Itarbaljuri, Kumirta, Utsill, Surbil, Kumirha, Gundijore, Baljori and Bara Baljori. Gua is surrounded by Budha hill, Bauki, Ghatkuri and Jamda is surrounded by Bijai, Bamiaburu, Baria Buru, Taiba, Baljori, Mohudi, Dijrapahar, Mergada Lupunga, Karamapada, etc. Most of these iron ore mines are located in Kolhan region. This belt provides iron ore to the steel plants of Jamshedpur, Bokaro, Rourkela, Durgapur, Bhilai, etc. Low grade iron-ore is also found in some areas of Damodar Basin as iron-stone shales in Dhanbad specially in Raniganj field. The ore occurs in the form of limonite or siderite. Small amount of magnetite iron-ore is found near 11 kms. south-west of Daltonganj. Iron-ore is also found in considerable quantities in the basaltic trap and trappean beds of Rajmahal, as well as, in the sand stones and are worked by iron smelters locally called Kols. In the Damodar basin iron-ore occurs both in the Jharia Coal field and in Raniganj coal field but are more abundant in later coal field which lies in West Bengal. In addition to main ore deposits of the Kolhan region, there are several small deposits. At Poralıat much of the hematite occurs as a fine powder. The deposits are scattered and impersistent. The hills near Lukud Buru contain bands of Hematite quartz schist interbedded with phyllites. Some of the former are almost iron-ores. Hematite schist and hematite quartzites cropped out in the ranges which form the boundary between Dhalbhum and Manbhum, north of Kundun. Another belt of ferruginous schist varying to banded hematite quartzite and iron extends eastward from the Kharkai river at Samran in Seraikela. In a more or less continuous bed of banded hematite quartzite which grades at intervals to ferruginous phylitic-schist, east and west of Jaikan, high grade specimens of iron-ore may be obtained. An extensive banded hematite-quartzite crops out on some hill tops overlying serpentine and hornblende-schist, south-east from Bhitardari. Near Daltonganj important deposits of iron-ore occur at Gore, Datum, Biwabathan and Sua. At Gore magnetite occurs associated with trimolite schists on a small hillock. The quality of ore is variable.

The magnetite schist grading into massive magnetite on the one hand and poor disseminated ore with quartz on the other. The out crops of richer ore may be about 4 lakh tons. At Datum magnetite occurs in bands, lenses and small pockets in the tremolite-schists. Some of the bands are about 3 to 4 feet thick and ore traceable for 100 to 150 feet. Several other small occurrences of ore are seen between Datum and Barajharia, but neither the quality is good nor the quantity is large. At Biwabathan there is a small deposit of magnetite associated with micaceous rock. According to survey upto 1988 this region has estimated reserve of hematite of 3083 million tons, out of which 1391 million tons reserves of hematite is known. In addition there is probability of 1692 million tons of hematite iron-ore. The reserve of magnetite iron-ore is about 4.2 million tons.

Copper : In terms of copper the state of Jharkhand has the leading position. The copper belt occurs in Purbi Singhbhum. It extends in a linear belt of 129 kms. From Duarparam on Bamini river traverses in an easterly direction through Kharsawan, Seraikela and Dhalbhum to terminate in its south-eastern part at Baharagora. The copper ores occur in veins in the granite and in the adjoining mica-schist, quartz-schist and epidiorites. These veins are considerably developed in the region where important copper mines are found. There are important regions having copper mines like Rakha mines, Mosabani and Dhobani mines. Copper mines of some importance are in Astakoali, Badia, Chapri, Baharagora, Charakmara, Dhadkidih, Churiapahar, Duarparam, Gohala, Jaypur, Hitku, Kanas, Kendadih, Laukesna, Khejurdari, Munda Devata, Mosabani, Nandup, Rajdah, Roladih, Sidheshwar, Surda, Thakurdih, etc. Besides there are copper mines at Pathergora and Kenduadih. There is possibility of the occurrence of copper in the Rajmahal highland and Hazaribag plateau. In the district of

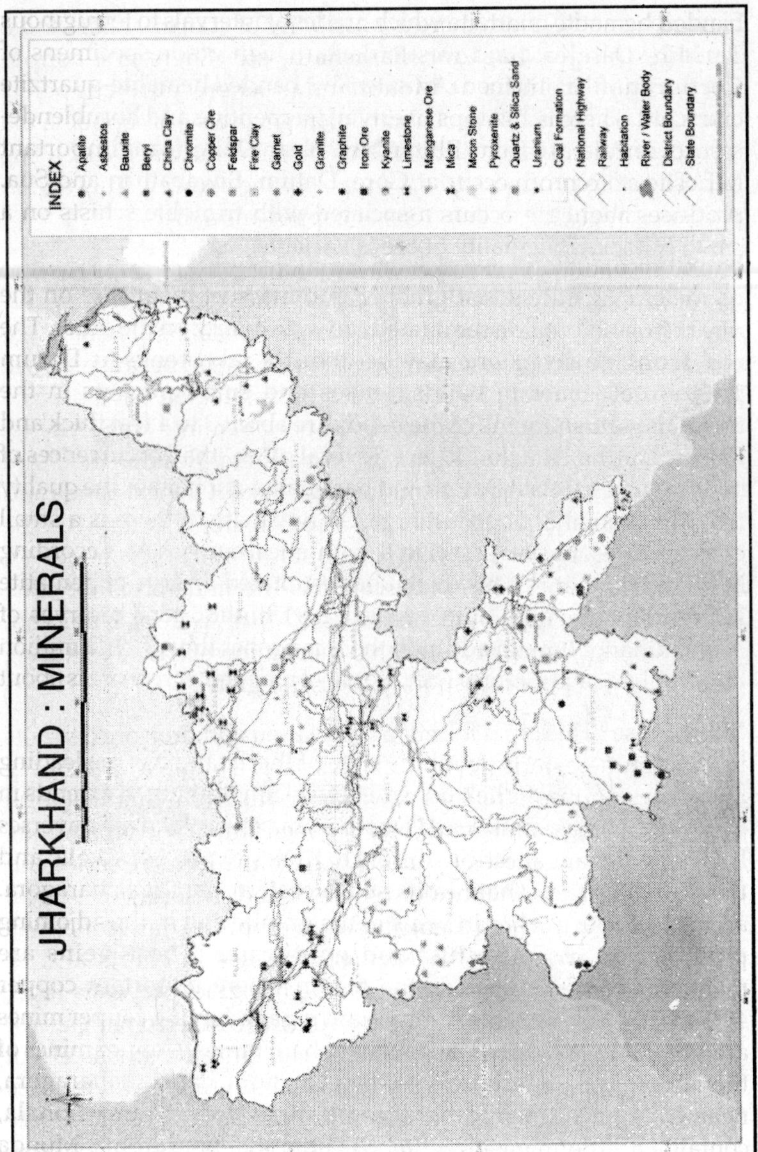

Source : Jharkhand Space Application Centre, Dept. of IT.

Fig. 11

Hazaribag and Giridih there is possibility of copper at Barganda, Taridih, Dandlo, Bagadar, Parasnath, etc. There is a copper Corporation of India at Moubhandar where pure copper is extracted. There is also probability of huge reserve of copper in the district of Palamu and Garhwa. This state has the reserve of 17882 MT. The state produces 6.84% copper of the country. This region produces about 1.3 million tons of copper annually.

Bauxite : Bauxite is found in association with laterites. The Pat region having thick deposits of laterites is the store of this minerals. Apart from Pat region the adjoining higher plateaus of Lohardaga and Gumla also have extensive laterite cover because of the presence of lava cappings. Laterites of this region have bauxite minerals containing 51-55 alumina. Important mines are Serendag, Bagru, Dudha, Katcha, Khamar, Banjari, etc. Besides there are bauxite mines at Pakhar, Chapuduadhia, Pakri, Oranga, Banda and Bar. The Rajmahal highlands have also bauxite. This state has the estimated reserve of 177.35 MT. Annual production is about 1.6 MT which is about 10.53% of the country. This mineral of the region is important for aluminium industries at Muri, Asansol, Hirakund, Alwe, Rihand. It may be noted that the mining of bauxite is facing the problem of poor transport in the Pat region.

Manganese : This ore is found in association with iron ore deposits in the south-eastern portion of the state. The area containing manganese mines stretches towards south from Gua to Limtu between Chaibasa and Jamda and between Jamda and Nuamundi. Deposits of manganese ore occurs both in the Iron ore series and the Kolhan series. The former type is found in southern Kolhan and at Ledaburu where manganese ore occurs as thin lenticles in phyllites, as irregular replacement masses in cherts and as manganiferous surface laterite. The higher grade deposits (over 48% manganese suitable for the manufacture of ferromanganese) are generally associated with cherts. The largest deposits are of the laterite type which may be also of high grade. At one place near Gua, an outlier of the basal Kolhan series conglomerate contains a little manganese. In northern Singhbhum there is an occurrence of manganiferous laterite at Langi and Basadera in north-eastern Singhbhum.

Deposits associated with the Kolhan series occur close to Chaibasa. The ores have replaced both the basal sandstone and limestone. They occur as thin lenticles parallel to the bedding of the rocks, or as lateritic material at the surface; the latter is commonly high in iron. The ores are concretionary and consist of psilomelame and pyrolusite. The manganese has been segregated by solutions which obtained their manganese content probably from the underlying magniferous rocks of the Iron-ore series. This region has the estimated reserve of 295063.00 Th. Tons. The annual production is about 3 thousand tons. This mineral has become important because of its use in the production of special type of steel, dry battery, chemical industries, glass industry, match industry, etc. It may be noted that this region has inferior quality of manganese ore as well as the amount of manganese is also meagre but its importance is due to availability in the iron-ore belt.

Chromite : It is main source of chromium and it is also used in iron and steel industry for special type of steel, chemical industries, etc. This mineral is also found in the south-eastern parts of the state. Important mines are located at Jojahatu and Keraikela near Saraikela in Singhbhum. Occurrence of this mineral at other places is under geological investigation. The chromite of this region is of superior quality as the ore contains 48% chromium. It has an estimated reserve of 178420.00 Th. Tons, out of which 2250 tons are known and the remaining is probable. The annual production of this mineral is about 170 tons.

Tungsten : It is obtained from ore of wolframite mineral. It is used in the production of high tempered steel and for manufacturing electrical goods. According to geological survey this mineral is expected to be found in parts of Hazaribag. Its exact reserve is not known and its use has not emerged on commercial scale.

Vanadium : This ore also occurs alongwith magnetite ores in Dulpara Dubalabeta and adjoining areas in the district of Singhbhum. It is a white and silvery metal. It is used in the production of special type of steel which can sustain heavy weight. Its estimated reserve is more than 25 lakh tons.

Gold : Gold is found in different parts of state with special emphasis in the Subarnarekha valley. It is also found in valley of Son, in the extreme north west of the state and in the valley of Damodar. But its significant availability is only confined to the Subarnarekha valley where sands contain few smaller particles. The source of gold appears to be the numerous small veins of quartz which occurs in both the phyllite and epidiorites. Quartz veins of Kendarkocha in Dhalbhum has gold but its production is uneconomic.

Silver : This mineral is found in association with sulphur, lead, copper etc. It is found in Hazaribag, Chatra, Palamu, Ranchi, Singhbhum, etc. but its occurrence is not on commercial scale every where. It is obtained in the processing of other minerals like copper, lead, etc. The Indian Copper Corporation is activity considering to have plan for the production of Silver. This state produces about 11 thousands kgs, nearly 31% of the country.

In addition to above mentioned metallic minerals, other minerals like lead, tin, nickel, etc. are found in negligible amount. Lead is found in scattered manner in Hazaribag, Palamu, Ranchi, Singhbhum district, etc. Tin being obtained from cassiterite ore is found in the district of Hazaribag, Giridih, Ranchi, Lohardaga, etc.

Non-Metallic Minerals :

Mica : In terms of reserve and production of mica Jharkhand state holds leading position in the world. Mica is of two types-muscovite and biotite. This region has dominance in muscovite mica. Its main area lies in the northern portion of the state. Virtually there is a mica belt stretching 145 kms. long and 26 kms. wide and this belt mainly lies in Giridih and Hazaribag but part of it also occurs in Gaya, Munger, Nawada districts of Bihar. This mica belt produces about 80 % of the world's supply of better quality sheet mica. Most of the mica of this region is found in pegmatite veins where quartz, orthoclase, tourmaline containing mica blocks are found. The blocks of mica are 15 .cms to 30 cms. in width and length and 7 to 10 cms. in thickness. Sometimes the size of mica blocks also remain 5 cms. to 7 cms. The main mica producing areas lie in Kodarma reserved forest area, from where 1/3 mica of the state is produced. Other competing areas lie near Gawan north-east of the reserved forest and north-east of Parsabad station. The state has a reserve of 59890050.00 MT of Mica.

The mica of this region has great reputation in the world market because of its Ruby colour. The mica area of Kodarma reserve forest covers an area of 92 sq. kms. The primary centres are Kodarma reserve forest, Domchanch, Dharhakola, Masnodih, Dhab, Gawan and Tisri. In addition to these mica mines, some mica is also found in Singhbhum, Palamu and Hazaribag. A palegreen mica has been prospected at Purnadihi, Bengaria and Laubera in Singhbhum. A minor occurrence of mica has been at Sikriadpur near Simdega. Negligible amount of mica is also found near Daltonganj, Lesliganj, Kini, Khorhi and Bandua in the Koel Basin.

Mica belt of this state deserves special status because it has huge occurrence of Ruby mica which fetches good money in the world market. At the same time occurrence of mica is in a bed of more than 100 feet thick which facilitates in easy and cheaper mining. There are about 600 mines out of which 100 mines are mechanised. These mines exist upto the depth of 700 feet. Kodarma and Jhumri Tilaiya are main mica industrial centres where different types of mica are processed for markets. Its importance is increasing day by day because of its varied uses in growing industrialization and electrification. Besides mica is also used in chemical industries, paper industry, rubber industry and others. About 90 % of its production goes to world markets. The average annual production is about 2527 metric tons which is 53% of the country.

Dolomite and Limestone : The state of Jharkhand is not so rich in terms of limestone because only limited areas of Palamu and Garhwa, Damodar Basin, Santhal Parganas and Singhbhum have limestone. These areas have small reserve of limestone. Crystalline dolomite and dolomitic limestone occur at a number of places in Palamu and Garhwa district. The important deposits are at Bakoria, Sadhwadih, Chauparia, Banpaher, Harhipahar, Kothila, Balumath, Pandwa, Majhaulia, etc. Some of these deposits form prominent hillocks. The dolomite is usually white in colour. In Ranchi plateau limestone occurs in Hosir, Lachaura and Dunduray. This deposit extends in part through the Hazaribag district also. It is used for manufacture of lime. In Babhani-Hoyer-Khelari area limestone bands are available. All the requirements for the cement works at Khelari are drawn for these lime stone deposits. The nodular limestone called Kankar or ghuting are found in many places in the Rajmahal hills and adjoining areas. Considerable deposits being found at Sakarigali, where qualities of lime have been manufactured for Calcutta and elsewhere. Limestones found in South-eastern portion of the state are generally impure and not suitable for industrial purposes such as flux in the steel industry, chemical and cement industry. There is a zone of limestone immediately overlying the basal sandstone in the Kolhan series where outcrops extend from Chaibasa to Jaganathpur, a distance of about 45 kms. It is of variable thickness thinning out completely in places attaining its greatest thickness of about 15 metres near Rajanka. The Associated Cement company limited are working the Kolhan limestone at Jhinkpani where lies good reserve of limestone. It is a pink, grey, greenish limestone often containing thin lamellae of phyllite, shale or chert. Dolomite is also produced in Koel Basin near Daltonganj and Chaibasa. The reserve of dolomite is about 7,084.20 MT. The annual production of dolomite is 1.66 lakh tons.

Asbestos : It is found in the crystalline Dharwarian rocks. This silky fibrous mineral is found in districts of Ranchi and Singhbhum. In Ranchi districts asbestos is found near Itki. In south-eastern portion of the state asbestos occurs at several places like Seraikela and Dhalbhum associated with serpentinised ultra basic igneous rocks. In Seraikela the ultra basic rocks, whose hydro-thermal alteration has produced the asbestos veins, are considered as inclusions in the Singhbhum granite. The largest deposits are at Barabana and Rajnagar in Seraikela. In Dhalbhum asbestos veins are associated with hydro thermally altered ultra basic rocks which appeal to be intrusion into the granite and dioritic rocks of the area. The principal occurrences are at Manpur and Gobradih. Small deposits have been found at several places such as Matku, Patkocha and Digarsai. It has a reserve of 18.45 MT tons.

Kyanite : Kyanite of high quality is found in abundance at Lapsaburu near Rajkharsawan and is worked by the Indian Copper Corporation. Huge deposits of poor grade material are lying dormant near about Ghatshila. It is used for refractory material.

Fire Clay : It occurs in the Damodar Basin, Ranchi, Koel Basin, Singhbhum, etc. Fire clay occurs plentifully on the western side of Rajmahal hills. It is mainly found alongwith pegmatite and quartz and it is chief source of china clay. The chief producing area is the Damodar basin at Jharia and Raniganj. It is mined for the manufacture of fire bricks which are used as refractory materials as well as lining of furnaces in steel plants. It is also found in Hazaribag, Rajmahal highland, Purbi Singhbhum, etc. Its annual production is about 261 tons.

China Clay : It is a white powdery clay derived from the decomposition of felspar in gneisses, schists, pegmatite, granite, etc. It occurs in almost areas of the state especially in Ranchi, Singhbhum, Palamu, etc. It is used for bleaching material and textile industry, paper industry, insulator factory, etc. Its reserve is about 2302.42 MT. The annual production is 59 thousand tons. China clay occurs in three ways: (1) As the decomposition product of Felspar in the fundamental gneisses and schists.(2) In the white Damodar sandstone where its presence is due to the decomposition of felspar originally present in the sandstone and (3) As beds of white China clay interbedded in the white Damodar sandstone. The first form is seen in some quantity at Katangi near Baskia, at Karanpura and at Dodhani. The second form is seen at Mangalhat where it is extracted by the Calcutta pottery company for the manufacture for China clay and Porcelain. And it is also present throughout the Hura coal field, in the northern and eastern boundaries of the Dhamini Coal field, and in parts of the Chaparbhita Coal field chiefly near Alubaru and Amjatri. The third form occurs in the Hura coal field as a bed from 4 to 5 feet at Soka Khanodih, Kui, Ajlatna etc. Graphite thick, about a quarter of a mile west of Piyaram.

Felspar : This mineral is found in areas having either igneous intrusioin or in areas having crystalline rocks of early geological periods. This mineral is found almost everywhere. Pegmatite veins or coarse grained granite provide good amount of felspar which is used for china clay. But this mineral is under use especially in areas having good transport. Important mines exist in the districts of Hazaribag, Giridih, Santhal Parganas, etc. Its annual production is about 261 tons.

Sulphur : It is found in association with copper ore, galena, pyrite, etc. It is found in south-eastern parts of the state.

In addition to above mentioned non-metallic minerals, some other minerals like soap stone, apatite, beryl, etc. are also found. Soapstone is bound in Purbi Singhbhum, Paschimi Singhbhum and Hazaribag district especially in the Archaen and Dharwanian rocks. It is used for making powder, pencil, etc. Apatite lies in Purbi and Paschimi Singhbhum. It is used in chemical fertilizer. Benjal is found in mica belt of northern region. Barytes is obtained from Kolpothe of Singhbhum district.

Atomic and Power Producing Minerals : The state of Jharkhand is also important from the point of view of atomic minerals which has become very important because these minerals are chief source of atomic power. All resourceful countries are running after atomic power for their supremacy in world politics. This region enables the country to have atomic minerals found in the region are uranium, thorium, berylium, monazite, zirconi, hleminite, antimony, etc.

Uranium : It is found in association with sulphide copper and oxidised iron ore. It occurs mainly in slate and phyllite. Uranium of this region are of two types-pegmatite pitchblende and uranium compound. Important mine is located in Dhalbhum near Sungri. It is also available at Narwa, Jurmadih, Bagjat, Kanyaluka and Keruadungri, etc.

Graphite : This mineral is also known Plumbgo or Bleach lead. It is a form of carbon. It is found in north Koel Basin especially in the surrounding areas of Daltonganj having pegmatite veins. The quality is generally poor due to contamination with other rocks. Important mines located are either amorphous or flaky and occur in thin veins traversing the graphite schists.

Ileminite : This mineral is found in Ranchi Plateau. Its estimated reserve is 0.74 million tons. It occurs in the pegmatite veins. It is raw material of titanium. It is used in aeroplane, rocket, and submarine, etc.

Thorium : It is found in the rugged and dissected upper surface of Ranchi Plateau and Panch Pargana. This area comprises estimated reserve of 2 lakh tons of thorium.

In addition to above mentioned atomic minerals berylium is found in Hazaribag, monazite and zircon are found in Ranchi and Hazaribag districts respectively. According to Nuclear Power Commission this region has the potentiality of generating power equivalent to power generated by 59 million tons of coal.

Coal : As regards coal resource the state of Jharkhand gets dominating position on the mineral map of the country. The Damodar Basin of this state is worth mentioning in terms of huge reserve of superior quality of coal as well as production. Coal of this region provides 88 % of the price gained from the production of minerals in the state. This state produces about 18.95% coal of the country. This region produces 95.5% coking coal of the country and remaining 4.5% coking coal comes from Madhya Pradesh and West Bengal. This state has 29.24% coal reserve of the country and almost superior quality of coal is reserved here. This state has an estimated reserve of 245693.00 MT of coal especially of coal seams more than 1.6 metres thick and is economically viable.

It is to be noted that the trough basin of Damodar contains coal of superior quality as well as huge reserve. Chief characteristics of the coal seams of this area are that most of the seams are very thick, sometimes, more than 25 metres thick which provides convenient mining of coal. In some cases coal seams are exposed on the surface of the river bed or lie very close to the upper surface of the earth which enable surface quarrying only. Coals of this region are of bituminus and sub-bituminus type of Gondwana period which are of individual use. Coal of this region contains 50 to 60% carbon but coal found in the Rajmahal hilly area is of inferior quality. Coal mines of the Damodar Basin lie in east-west direction while the coal beds of Rajmahal exist in north-south direction. The cost of production is reasonably less in this region because of availability of coal near the surface, huge reserve in limited area where infrastructures are developed. Certainly the problem of fire in the mine is prevalent in some areas. According to the Geological Survey of India this region has 3 to 6 coal seam having the thickness ranging from 8.3 metres to 62.06 metres coal seams in Brahamani Coal region near Murdangal. Bihar State Mining Development Corporation has estimated a reserve of 8 million tons Coal in the forest areas of Palamu near Duga region. Central Mining Planning and Designing Institute has conducted drilling of 15340 metres in Dhanbad district near Nirsa, Lakhima, Kapasara, etc. and the district of Dumka at Hura and Chuparibita and at Chitra in Godda district. They have estimated a reserve of 164.12 million tons in western Bokaro Coal field at Loiyo and Gose, 30 million tons at Medanirai of Huter Coal field and 362.45 million tons at Maurya and Lapanga of southern Karanpura Coal field. Mineral Exploration Corporation Limited has estimated 1029.55 million tons in northern Dhaku and Ajanta region which are part of C.C.L.B.C.C.L. has estimated 643.78 million tons of coal at Dharmaband and Angarapatra.

The coal mines of this state is divided into five major regions :

1. The Damodar valley Coal region:- Raniganj, Jharia, Chandrapura, Bokaro, Ramgarh, South Karanpura, North Karanpura, etc.

2. Hazaribag and Giridih Coal region:- Itkhori, Chop, Giridih, etc.

3. Ajay Basin Coal Region :- Jayanti, Sahjori, Kundit Karaia etc.

4. Rajmahal Hilly Coal region:- Brahamani, Pachwara, Chaparbita, Jilbari, Hara, etc.

5. North Koel valley Coal region:- Daltonganj, Hutar and Auranga.

Table - 9.4
Coal Reserves : Jharkhand
(As on 1.4.2.14) (In Million Tonnes)

Sl. No.	Coal Field	Proved	Indicated	Inferred	Total
1.	Jharia	15127.97	4302.09	-	19430.06
2.	North Karanpura	9499.42	6914.61	1864.96	18278.99
3.	Rajmahal	3211.18	11219.06	1691.82	16122.06
4.	East Bokaro	3385.77	3903.71	863.32	8152.80
5.	South Karanpura	3230.09	1867.66	1480.22	6577.97
6.	West Bokaro	3720.89	1308.71	33.66	5063.26
7.	Aurangabad	352.05	2141.65	503.41	2997.11
8.	Raniganj	1538.19	466.56	31.55	2036.30
9.	Ramgarh	710.59	495.30	58.05	1263.94
10.	Deograh	326.24	73.60	-	399.84
11.	Hutar	190.79	26.55	48	249.82
12.	Daltonganj	83.86	60.10	-	143.96
	Jharkhand	**41377.04**	**32779.60**	**6559.47**	**80716.11**

Source : Coal Directory of India, 2013-14.

1. The Jharia Coal Region : This is the most important coal field of the State. This region lies in the districts of Dhanbad and Bokaro, about 48 kms. west of Raniganj Coal field. It covers an area of about 280 Sq. kms. In this region coal seams is exposed on the surface in the beds of Damodar and its tributaries at several places. This coal field has special importance because of considerable reserve of superior quality coking coal. It has a reserve of 24.07% coal of the state which is available upto 900 metre depth. This region has 5470.19 million tons superior quality of coking coal which is 10% of the total coal reserve of the country. This mine has the problem of fire in a pocket having about 100 million tons of prime coking coal. This region has 18 to 20 coal seams worth mining. This coal area has adversely been affected by the problems associated with administrative and technical problems of B.C.C.L. and Coal India Ltd. However, B.C.C.L. has increased its production 29 million tons. It is true that coal seams of this region are near the surface or in less depth; but the problems of faults and dykes existing in mining areas are realised. At the same time dense population and habitation in the region create development of mines. Fire affected mines are also causing heavy loss for protecting other mines. This region has three main mining plans at Munidih, Putaki and Balihari. An another new plan at Mukanda with the help of Russian Government is being developed. B.C.C.L. has aimed to produce 900 million tons coal through open mining and 2700 million tons through underground mining by the first quarter of 21st Century.

2. **The Bokaro Coal Region :** This coal region is only 5 kms. west of Jharia Coal field. It lies in Bokaro district and covers an area of 352 sq. kms. Through this region there flows the river Bokaro and divides the coal field into Eastern and Western Coal fields. In Eastern Bokaro Coal field it is produced from Kargali coal seams having the thickness of about 28 metres. Besides this there are 18 more coal seams. The eastern portion of this coal field is more developed because the coal seams are comparatively more thick and are of better quality. The occurrence of coal is even upto 900 metres depth. This mine is under railways. As such coal produced in this region is consumed in Railways. Coal of this region contains more ash, however, coal is of coking variety more useful for steel plant.

The western coal field has 23 coal seams having the thickness of 1 to 3 metres. Coal is available upto depth of 600 metres and has an estimated reserve of 1270 million tons of superior quality while the Eastern coal field has estimated reserve of 500 million tons of superior quality coal. These two regions together have reserve of 13216.06 million tons i.e. a little more than 16.37% of the state. This

Coal region has given rise to several coal washeries. The steel plant of Rourkela depends for metallurgical coal of this region.

3. **The Karanpura Coal Region :** This coal region is also bifurcated into Northern and Southern Coal fields. These two Coal fields exist west of Bokaro Coal field. The Northern Karanpura Coal field covers an area of 760 Sq. kms. and the Southern Karanpura Coal field covers an area of 120 Sq. kms. But the northern Karanpura region is comparatively more developed. The northern Karanpura Coal field has the reserve of 18278.99 million tons of all types of coal which is upto 600 metre depth. The Southern Karanpura Coal field has the reserve of 6577.97 million tons which lie upto 900 metres depth. These two regions together have 30.79 % coal reserve of the state. Coal of this region is of A and B grade and most of the coal is of coking variety.

4. **The Ramgarh Coal Region :** This coal region lies in the west of Karanpura Coal field. It has a reserve 1263.94 million tons. This region consists of 22 coal seams worth mining which are available upto the depth of 900 metres.

5. **Auranga, Daltonganj and Hutar Coal Areas :** These Coal regions are located in North Koel Basin of Palamu and Garhwa district. These are not of much importance. One of these Auranga Coal region has an area of 161 Sq. kms. and has a reserve of 2977.11 million tons that is 3.68% of the state. Coal of this region contains moisture and ash. The Daltonganj Coal field has an area of 322 Sq. kms. and has a reserve of 144 million tons which is only 0.18% of the state's reserve. The Hutar Coal field has an area of 129 Sq. kms. and has a reserve of 250 million tons i.e. 0.31% of the reserve of the state. Auranga and Hutar Coal fields have coal upto the depth of 600 metres while Daltonganj has coal upto 300 metres.

6. **The Rajmahal Coal Region :** In Rajmahal hilly areas specially along the western border there lie several coal fields which may be enumerated from north to south as (1) The Hura Coal field, a trap about 15 miles long from north to south commencing about 13 miles south-east of Colgang.(2) The Chaparbhita Coal field about 10 miles further south in the valley of the Gumani, (3) The Pachwara Coal field in the Bansloi valley and (4) The Brahamani Coal field in the valley of the river from it derives its name. The south-western portion of the Rajmahal highland contains the small Deoghar Coal

fields and the northern edge of the Raniganj Coal field. The Coal in Deoghar fields is neither plentiful nor of good quality. Most of other Coal fields contain inferior quality of coal. Altogether the Rajmahal Coal region has 16122.06 million tons of coal reserve which represents 20.06% of the total coal reserve of the state.

Coal resource still plays dominant role in power sectors, steel plants and other industries. There is need to conserve coal for these industrial uses because it is an exhaustible resource.

10

Industries

The state of Jharkhand having abundance of minerals, power potential, forest products, etc. coupled with locational advantages like solid rock base, transport network and cheap native labour has emerged as one of the most important industrial regions of the country. Virtually this region has potentiality to develop as the most important industrial region of the country, provided the proper planning for the development of industries is made. The state being the store house of minerals has the capacity to give rise to various metallic and non-metallic minerals based industries. Immense forest resources have capability to boost up forest based industries. Even agricultural products of this region can accelerate agro-based industries in agricultural areas. For the development of above mentioned industries this region has adequate power resources in the form of thermal, hydel and atomic power. This state is worth mentioning for huge reserve of coking coal and other types of coal for industrial use. As a matter of fact even the industries of other regions of the country considerably depend upon the coal of this region. The hydel power potential is also not less remarkable. The only developed hydel power system of the Damodar Basin has been meeting the requirements of this region as well as the adjoining provinces. Reserve of metallic minerals like iron ore, manganese, copper, bauxite, etc. can give rise to a number of metallic industries. Non-metallic minerals are also not less important in encouraging non-metallic minerals based industries. Mica is the leading non-metallic based industry in the world. This region has also capability to establish a number of forest based industries like Saw mills, Lac industry, Paper industry, Biri making industry, etc.

Besides, the concentration of minerals, power potential and other industrial infrastructure in close proximity to each other facilitate industrial development by reducing transport cost and minimizing the problems of industrial location. This region has also cheap labour force to be engaged in various industries.

The industries in this region can broadly be classified under three heads . Firstly there is heavy and organised industries carried on large scale and run by the help of power, viz. (1) Iron and Steel (2) Engineering (3) Food Industries (4) Chemical Industries (5) Copper Industry (6) Cement Industry (7) Glass Industry (8) Hume Pipe Industry (9) Wood Working Industry, etc. In the second category, small scale industries carried on usually without the use of power may be included such as Tusser rearing industry, Biri making industry, Trunk and Bucket manufacturing, Ice and Ice candy manufacturing, Small Wood working factory, Handloom weaving, Lac industry and other small cottage industries such as basket making, pottery, rope making, etc. In the third category, the mining and mineral processing industries occupy prominent position.

Despite the above facts the region having vast industrial resources and favourable conditions for industrial development only limited areas have registered phenomenal industrial development. Important areas have developed in the Damodar and the Subarnarekha basins. Main concentration has developed in the coal mining areas of Dhanbad and Bokaro, Jamshedpur, Ranchi, Ghatshila, etc. In coal mining areas of Dhanbad coal washeries, iron and steel and related industries developed. In areas around Ranchi H.E.C. (Heavy Engineering Corporation) at Ranchi, Ball bearing at Ratu, Aluminium factory at Muri, High Tension Insulator factory at Namkum, Electrical Appliances at Tati Silwai have developed. Jamshedpur region has encouraged the development of Tata Iron and Steel industry at Jamshedpur, Glass and Ceramic industries at Kandra and industries like electrical cable, Railway Wagon engineering and Agricultural implements Moubhandar near Ghatshila is noted for Copper Industry.

Iron and Steel and Associated Industries : It is a raw material oriented industry. In terms of iron and steel and associated industries this state holds an important place. There are two main steel plants at Jamshedpur and at Bokaro. The steel plant of Jamshedpur was established in private sector in the year 1907 but the production started from the year 1911. This plant was established by Jamshedji Tata at Sakchi on the confluence of the two rivers- Subarnarekha and Kharkai. The locational advantages of the steel plant included the availability of superior quality of iron ore from the iron ore mines of Noamundi, Joda east, Gurumahisani, Badam Pahar, Sulaipat mines which are owned by Tata. There are several other iron ore mines having good quality hematite iron ore. Manganese is made available from Gua and Manoharpur mines, Chromite from Chaibasa mines. Good quality of coal comes from the Jharia coal field. Virtually it lies in between the Damodar valley coking coal and the chief iron ore belt of India. Even limestone and dolomite are made available from Panposh and other locations of Sundergarh are not far off. Almost all raw materials and coal are found within a radius of 100 kms. Moreover all the mines are duly linked with electric and railways and steel plants is itself on the main rail route connecting Kolkata and Mumbai. Jamshedpur is only 240 km from Kolkata, a great distributing port and centre of general engineering, trade furnishing, good market for the products of Tata Iron and Steel Company. This place also has the suitable site for the expansion of associated industries. Cheap labour is available from the neighbouring areas and skilled labourer from different parts of the country. Facility of empty railway wagons are available in carrying raw materials. In addition to above favourable factors the water is easily available from the river of Subarnarekha. The plant initially had capacity of producing 1.21 M. Tonnes of Pig Iron and 1.1 M. Tonnes of Steel per annum. This capacity has been enhanced to 3.9 MT of Pig Iron, 2 MT of Ingot Steel and 3. MT of Saleable steel. Currently, it produces about 3 MT of Saleable steel.

Recent Production : There are several associated industries manufacturing agricultural implements and machinery, the tin plate factory, the wire factory, Tata Locomotive and Engineering Works etc. It also produces iron sheet, rail, fish plate, wheel, sleeper chasis etc.

The Bokaro Steel Plant : The Bokaro Steel Plant is the largest steel plant of the country and is one of the important steel plants of Asia. This public sector industry is located near Marafari in the coal belt of the Damodar Basin. It was established in the year 1966 with the help of Russian Government. The localization of this plant is not only marked by its location in the superior quality coking coal region but there are other advantages also. There are iron ore mines of Jamda and Saranda at a distance of only 32 km but iron ore for this plant comes from the iron ore belt of Singhbhum. It avails the facility of empty railway wagons. Wagons carrying coal from this region for the steel plants of Jamshedpur, Rourkela, while in return brings iron ore, manganese, limestone for the steel plant of Bokaro. Although the limestone of Kaimur hill is nearer but it comes from Birmitrapur. This plant is facilitated by adequate water for cooling the furnaces and supply of water in the township by a canal which is linked with the dam of Tenughat on the river Damodar. This region also has the advantage of hydel power, thermal power, proximity of Kolkata market and port, cheaper labour from the surrounding areas as well as neighbouring states.

Table - 10.1
Iron & Steel Production : Jharkhand
('000 Tonnes)

Plant	Annual Installed Capacity		Production			
	Hot Metal	Crude/ Liquid Steel	Hot Metals		Crude Liquid Steel	
			2013-14	2014-15	2013-14	2014-15
Bokaro Steel Plant	4,585	4,360	4,100	4,253	3,776	3,831
Jamshedpur	NA	9,700	9,898	10,164	9,155	9,331

Source : Indian Minerals Year Book, 2015.

Aluminium Industry : The state of Jharkhand is also suitable for the development of aluminium industry because its raw material is available in bulk in this region. Aluminium is derived from bauxite mines at Lohardaga with an estimated reserve of 6.4 million tons. The Indian Aluminium Company Ltd. is located at Muri on the Subarnarekha river. It obtains bauxite from Bagru Pat by electric ropeway upto Lohardaga where from metre gauge rail brings bauxite to Ranchi where it is unloaded and reloaded in broad gauge wagan for Muri factory. The Company processes 10 thousand tons of bauxite ore at Muri and reduces alumina by electrolytic processes into aluminium at Alupuram, Alwaye in Kerala and rolls seven thousand tons of aluminium into sheets and cards at Belur near Kolkata. It prepares aluminium powder at Kalwa near Mumbai to produce 200 tons of paste, light weight of this product is a factor of economy in transport to far distant places. Since it is a weight losing industry its location becomes advantageous near the raw material site. To produce one ton of alumina it requires 2 tons of bauxite, 0.2 ton coal and 0.1 ton caustic soda. At the same time its location needs cheaper and adequate power supply because to produce one ton of aluminium it needs 2 tons alumina and about 25 tons kw power. It is therefore, located preferably near hydel power. For the localisation of this industry the Ranchi Plateau region provides ideal conditions because of abundant bauxite in the Pat region, coal in the Damodar, hydel power form the Damodar basin, better transport system and proximity of Kolkata market. The Muri plant uses about ten thousand tons of bauxite. The Indian Aluminium Company has also started manufacturing of aluminium at Hirakund entirely based upon the alumina of Muri. There are aluminium plants at Asansol in West Bengal and Rihand in Punjab which receive bauxite from Lohardaga region. This industry is expected to expand with the development of cheap hydel power to be developed in other river basins of this region.

Copper Industry : For the development of copper industry, the south eastern portion of the state provides ideal conditions. The copper mines of the region provide raw materials for this industry. There is a belt containing geological formations of Dharwarian rocks stretching 120 km with good quality of copper ore. The Indian Copper Corporation established copper industry at Ghatshila in the year 1924 with an investment of about 2 lakhs. In the 1930 to produce brass sheet and yellow metal, rolling mill was established. This industry obtains raw materials from the copper mines of Mosabani and Badra. First of all raw material is brought to Moubhandar by rope way where mineral is processed and sulphur is extracted from the copper ore. Copper sheet, copper wire and other goods are manufactured at Ghatshila. Here nearly 5 thousand tons of copper is processed per day. Brass is manufactured by mixing zinc with copper and tin. Ghatshila has become the largest copper industry where 1/3 copper of the country is produced. In 1988 nearly 1348 thousand tons of copper ore was produced near Ghatshila.

Engineering Industry : This industry includes different types of industries but goods manufactured by iron and steel are primarily taken into consideration. Industries producing machines, tools and other appliances concerning railways, motor cars, ships, aeroplanes etc. are given due importance. During post independence years this industry has witnessed remarkable growth and development because of growing industrialization. Hatia is the centre of numerous engineering industries like heavy machine building plant, machine and tool plant, foundry forge plant, etc. High tension Insulator Factory is located at Namkum as where Sri Ram Ball Bearing Co. Ltd. is located at Ratu near Ranchi. The heavy machine building plant was developed with Russian assistance with a capacity to produce 45 thousand tons of machinery in the first stage and 80 thousand tons in the second stage. The foundry forge plant was developed with Czech assistance. It has a capacity to smelt 1.4 lakh tons of steel and concerned metals. In heavy machine building plant, large and heavy machines and tools concerning iron and steel industry and other industries are manufactured. This plant has the capacity to produce 55 thousand tons of heavy machineries and tools.

The above mentioned centres of engineering industries have locational advantages of adequate supply of raw materials, power, market and other advantages. In the periphery of these industries there lie the location of a few iron steel plants like Bokaro, Jamshedpur, Rourkela, Durgapur, etc. and machines and tools can be easily carried to those plants. The Damodar Valley Project provides the facility of hydel power. There are thermal power stations to supplement and ensure regular supply of power to there plants. Chandrapura-Muri-Ranchi-Rourkela Railway line and Highways link it with any part of the country which facilitate easy transportation of machines and tools.

The advantage of the capital city of Ranchi with suitable climate marketing facilities and industrial infrastructure also help engineering industries.

Cement Industry: During post independence years the increasing demand of cement for building construction, road construction, bridge construction etc. has boosted the cement industry in the country. The state of Jharkhand also holds an important place in terms of potentiality of the development of cement industry. The main raw materials required for the development of this industry are limestone, clay, coal, gypsum etc. To produce one ton of cement it needs 1.3 tons of lime stone, 1.33 tons of coal and comparatively less amount of gypsum. These heavy raw materials are available in the state. In terms of coal the Damodar Basin is worth mentioning. Even in terms of limestone this state is not far behind. Limestone is available in the Damodar Basin, the Subarnarekha Basin, Palamu, etc. There is no problem of clay. These days waste materials of blast furnaces of steel plant are being used and the steel plants of Bokaro and Jamshedpur provide this raw material. Important cement industries are located at Sindri, Banjari, Jhinkpani, Khelari, Kumardhubi, etc. Out of these Sindri cement industry uses calcium carbonate sulphate being obtained from the fertilizer industry instead of limestone. In case of Jhinkpani, the cement industry uses blast furnace waste materials of Tata Steel Plant instead of limestone. Most of these industries have per day production capacity of 1200 to 1500 tons of cement. There is an active consideration for establishing mini-cement plants to meet the increasing demand of cement.

Chemical Industry : This industry includes heavy and light chemical industries. In case of heavy chemical industry acid and fertilizer industries are associated. Light chemical industries include the manufacturing of colour, varnish oil, soap, medicines, matches etc. The heavy chemical industries also produce basic chemical materials and the higher chemical industries produce a number of materials by chemical processes.

Fertilizers industry of this region is the principal chemical industry. The Chemical industry of Sindri was established in the year 1951 with an investment of Rs. 28 crores. It has the production capacity of 3.5 lakh tons of amonia sulphate, 20 thousand tons of urea and 54 thousand tons of amonia sulphate nitrate (double salt). Besides it has product of gas, coaltar, calcium carbonate, benzene, sludge, napthalene etc.

The Sindri fertilizer plant is located in the coal belt of Damodar basin. It is situated on the bank of river Damodar. This industry has the locational advantage of coal and river Damodar, hydel power and transport. Railway wagons carrying coal for Rajasthan while in return bring gypsum for this industry. The agricultural areas lying in this state as well as in adjoining states provide good market for fertilizer produced in this plant. Waste material of this industry is used in cement industry. Its production capacity is about one lakh tons. In addition to this plant there is a chemical industry at Ghatshila where sulphur extracted from copper ore is used in chemical industries in the manufacturing of acid.

Refractory Industries : Refractory products are in good demand for constructing blast furnaces of iron and steel industry furnaces of other industries like cement, etc. For this industry clays are available in the Damodar basin and other raw materials are also available in this state. As such this industry has developed at Chirkunda, Kumardhubi, Mugma, Ranchi, etc. These centres produce 46 thousand tons of refractory materials.

Mica Industry : The mica industry is mainly confined to the mica belt. The supply of skilled labourers is primarily responsible for mica splitting industry. This industry has the advantage of huge reserve of good quality ruby mica in limited areas. In this region large industries have developed at Giridih, Kodarma, Jhumri Tilaiya etc. This industry has also developed at Chalkari, Domchanch, Masnodih, Dhab, Tisri, Maheshri, etc. The importance of this industry has considerably increased due to growing industrialization and increasing demand of electrical appliances. The micanite factory at Jhumri Tilaiya produces 305 tons of micanite, micalinen, mica silk, mica paper, etc.

Coal Washeries : Coal washeries have well developed in this coal belt to remove impurities. Some of these impurities might intimately mixed with coal and might be mechanically inseparable. But there may be some impurities which are mechanically separable like shale, fire clay, bones, pyrite, gypsum and calcite. Coal washeries undertake the separation of the above mentioned impurities. The mechanical washing of coal leads to (1) uniformity in the composition and properties of coal, (2) reduction in ash content, (3) greater economy in the production of gas and other products and better yield of coke. Important coal washeries are located at Kargali, Jamadoba, Pashchimi Bokaro, Lodna, Dugda, Bhojudih, Patherdih, Karanpura (two washery plants), Kathara (two washery plants), etc.

Coal washery of Jamadoba and Pashchimi Bokaro supply coal to Burnpur Steel plant. Kargali Coal Washery provides coal to Rourkela iron and steel factory (about 1 million metric tons) and Bhilai Steel plant (5 lakh tons). Dugda coal washery purifies coal of Mohuda-Bhaga-Katras Area. It has the capacity of 500 tons per hour. Bhojudih Coal Washery has the capacity of purifying coal of

250 tons per hour and Pathardih Coal Washery has the capacity of purifying 800 tons of coal. This coal washery purifies the coal of Jharia and Pathardih. Karanpura Coal Washery has capacity of purifying 3.5 million tons. Kathara Coal Washeries purifies the coal of Jarangdih, Sawang and Kargali. These coal washeries have the capacity of purifying 3.2 million tons of coal. Besides these coal washeries there are coke oven plants for converting coal into coke by carbonisation process. This process is also known as distillation. This process consists of subjecting raw coal to different grades of temperature in the coke oven plants. In this process a number of coal by products, hard metallurgical coal and soft coke for domestic consumption are obtained. Coal is carbonised at three different stages of temperature, low (600-700^{0}C), medium (800^{0}C) and high (1000-1100^{0}C). In the process of purifying coal and preparing coal for different uses, by-products like coaltar, gas, amonia sulphate, benzol are obtained.

Glass Industry : In the development of glass industry, the state of Jharkhand has also the advantages of raw materials, power, labour and other favourable conditions. Location of this industry is mainly influenced by natural factors. The industry consumes large quantities of sand and coal. Other raw materials are lime, refractories, soda, ash, fuel and borax. Sand provides silica, the chief constituent of glass. Sand should be free from impurities particularly iron oxide. Pure sand without iron oxide enables the production of fleecy white glasses. The average sand grains should preferably be o.4 mm in diameter otherwise, the cost on crushing and sieving may increase the cost of production. Other raw materials like soda is mixed with sand and fused to produce soft glass. In case of harder glass lime is added. If a clear optical glass is required borax is added. If the glass contains some tint decolorising materials like manganese, dioxide are used. In case of glass of particular colour, colouring materials like manganese, nickel, cobalt etc. are used. Refractories are used in constructing the furnaces. The future of glass industry in the state is bright but this industry has to depend upon other areas for soda and borax. As such the location of this industry becomes essential where heavy raw materials like sand and coal are available. This industry also requires skilled labourers. The state of Jharkhand has abundant rocks containing quartz which give rise to sand particles. Coal is also adequately available. There are four large glass factories. There is a large factory run under Japanese assistance at Bhadaninagar (Bhurkunda) near Barakakana. The second factory is at Asibona near Dhanbad. Third factory is at Kandra west of Jamshedpur and the fourth factory is near Ramgarh. Apart from these large industries there are a number of small factories specialising in electrical incadescent lamps, glass sheets, jarware, glassware and lampware.

Lac and Shellac Industry : In terms of lac and shellac industry the state of Jharkhand has leading position in the country producing 41% of the country's lac output. Lac worms are reduced on the leaves of the trees like Ber, Kusum and Palas. The adult female insects belonging to various genera of family Lacciferidae is a creature of the tropical belt. The popular name, 'lac insect' denotes the species of the genus Laccifer. Its favourite habitat is the lac hosts of the oriental region. Kerr, the principal member of the family of Laccifer lacca, secretes the true lac and contributes to all the lac of commerce.[1] Two strains of the lac insects, one flourishing on the Kusum tree, known as Kusumi and the second on all other hosts known as the Rangini because of its deeper colour. These strains undergo through two generations in a year and produce two crops. The Rangini produces Baishakhi and Katki and Kusumi produces Jethwi and Aghani crops.

The state of Jharkhand has ideal environmental conditions for the rearing of lac worms and development of lac and shellac industry. Most areas especially forest clad areas where all the three major hosts, Ber, Palas and Kusum occur, provide ideal ground. However Ber dominates in areas of Ranchi and Hazaribag Plateaus where as Palas predominates in other areas. Kusum is confined to the districts of Ranchi, Lohardaga, Gumla and parts of Pashchimi Singhbhum.

On an average the Baisakhi crop accounts for more than 70% and Katki for 17% of the state's total product. The districts of Ranchi, Lohardaga, Gumla and Pashchimi Singhbhum produce all the four lac crops but districts of Hazaribag, Chatra, Kodarma produce all the crops except Jethwi. The districts of Palamu, Garhwa, Dumka, Pakur, Godda etc. produce Baisakhi and Katki.

In the production of lac important role is played by the rural market centres or Hats where almost all lac produced are sold.

There are numerous such Hats where hundred of quintals of sticklac are collected per week during Baishakhi season. There are over 100 lac factories in lac producing areas. They process sticklac into seedlac and shellac. Only a few of them use power for crushing and washing of lac. Most of the remaining factories are small units.

Lac and shellac industry provides trade materials not only to the people of this region but to the nation also. The country gains foreign exchange with the export of the lac industrial products because of its varied uses like French polish or floor polish, manufacture of gramophone records and printing inks, electrical use, micanite leather goods, sealing waxes, gasket cement, bangles, etc. Its annual product is about 15,000 tons.

The important lac manufacturing centres are located at Ranchi, Bundu, Murhu, Khunti, Tapkara, Garhwa, Daltonganj, Latehar, Pakur, Chaibasa, Chakradharpur, Chandil, etc.

Besides aforesaid industries the state has recorded industries related to forest products, agricultural products and other resources. In case of forest products almost all small and big towns including rural service centres have recorded the growth of saw mills where from woods for building construction and furniture are supplied. Biri industry has developed in almost all small and large towns because this area has easy availability of Kendu leaves. Chakradharpur and Manoharpur are noted for the concentration of Biri industry where from large quantity of Biri are supplied to different areas. These days bee keeping has become attraction of some people in the rural areas. Presence of variety of flowers in the forest facilitates bee keeping. Rope making, malt making and other cottage industries have developed with the help of forest products like bamboo, sabai grass, etc.

No doubt this region lacks major industries related to agricultural products like rice mills, flour mills, oil mills but the region is dotted with quite a large number of small mills preparing rice, wheat flour, oil, etc. Virtually almost all rural markets and even large villages have the facility of small rice mills, wheat flour mills and oil pressing mills. Dairy and poultry industries have developed in many areas of the state for meeting local demands. Cottage industries producing bed sheets, rough cloths and other cloth materials have developed in Ranchi, Gumla, Daltonganj,

Dumka, Hazaribag and other towns etc. There are weaving communities who have kept this profession still running. Stone crushing industry has become a profitable industry because of huge demand of stones of different size for buildings, roads, bridges and other construction works. Stones are easily available on account of the presence of hilly terrain all over the state. Granite rock industry has been increasing these days because of its demand for floor and decorations. Local wine is made with the help of rice and Mahua. Toys are made with the help of wood.

The table 10.2 provides a list of registered small industrial units in the State of Jharkhand. It portrays that Ranchi district has got the largest number of such units followed by E. Singhbhum, Dhanbad, W. Singhbhum & Giridih and others. It reveals the fact that the comparatively developed districts have registered greater number of such industrial units. Of course, these areas provide necessary requirements of such industries more conveniently.

These districts having large scale industries produce a number of articles used in small scale industries. These areas have the facilities of transport, other industrial infrastructures and abundant countryside labourers. Deoghar, Hazaribagh & Dumka districts also contain a good number of small scale industries. These districts are also relatively developed than the remaining districts of the state.

Table 10.2
Small Scale Industries : Jharkhand
(2006-07)

Sl. No.	District	No. of Registered & Operational Units	Investment (In Rs. Lakh)
1.	Bokaro	964	360.39
2.	Chatra	336	105.31
3.	Deoghar	1896	1189.34
4.	Dhanbad	2400	2542.17
5.	Dumka	1000	125.86
6.	Garhwa	709	214.45
7.	Giridih	2271	2158.02
8.	Godda	761	448.21
9.	Gumla	625	406.09
10.	Hazaribagh	1548	4248.49

11.	Jamtara	78	94.51
12.	Koderma	589	384.22
13.	Latehar	108	97.91
14.	Lohardaga	309	109.81
15.	Pakur	612	293.27
16.	Palamu	1070	508.81
17.	W. Singhbhum	2381	1297.79
18.	E. Singhbhum	3042	1974.41
19.	Ranchi	5075	4154.43
20.	Sahebganj	517	211.29
21.	Saraikela-Kharsawan	278	378.77
22.	Simdega	53	48.49
	Jharkhand	**26,622**	**21352.04**

Source : Ministry of Labour & Employment, Govt. of India.

Industrial Regions : The State of Jharkhand is endowed with industrial resources in different forms almost everywhere. Certain areas have potentialities for mineral based industries based on the concentration of minerals. Some areas have immense forest resources for the development of forest based industries. Even agricultural resources have potentiality for the development of a number of agro-based industries. Besides, the Jharkhand state has cor siderable attractions for the growth of horticulture, dairy, poultry and tourist industries. Despite wide spread presence of industrial resources, some of the areas having better locational as well as other industrial infrastructures have enabled the growth and development of varied types of small and big industries which have given rise to a few industrial pockets. In addition to mineral resource based industrial pockets there has been emergence of other industrial centres of repute.

On the basis of concentration of industries, industrial labourers, raw materials, industrial infrastructure etc., the following industrial regions can be enumerated :

1. The Damodar Basin Industrial Region
2. The Upper Subarnarekha Basin Industrial Region
3. The Lohardaga Bauxite Industrial Region
4. The Jamshedpur Industrial Region

5. The Mica Industrial Region
6. The Ghatshila Industrial Region
7. The North Koel Basin Industrial Region and
8. The Deoghar Industrial Region

1. **The Damodar Basin Industrial Region :** The Damodar Valley is the largest industrial region in the state. It has also leading position in the industrial scenario of the country. Indeed, in the lower part of the valley, Dhanbad and Bokaro districts, the industrial landscape resembles to a considerable extent a mineral based industrial landscape of some Western countries. It is variable land of multi-industries. Almost every year a leaf is added to complete the story of its industrial development. No doubt this region has good concentration of mining activities which are excluded from the list of industries but based on the concentration of coal mines a lot of manufacturing, coal processing and medium and small size varied industries have emerged. The large Sindri fertilizer factory, the cement industry, coal washeries, coke ovens, refractory works, thermal plants, steel plants, etc. are worth mentioning.

The fertilizer and chemical industry at Sindri produce variety of fertilizers and chemicals. The production capacity of the Sindri fertilizer and chemical factory is 3.5 lakh tons of amonia sulphate, 20 thousand tons of urea, 54 thousand tons of amonia sulphate nitrate (double salt), etc. The Sindri fertilizer and chemical industry has brought about at Sindri, the existence of two other undertakings, a cement factory and a coke plant.

There is one lead and silver refining and smelting industry in this region, the only one of its kind in India, run by the Metal Corporation of India Ltd. at Tundoo. It produce lead and silver from the ores received from the Zawar mines in Rajasthan and other raw materials from different places.

Refractory and ceramic industries have also developed in this region. These industries include Kumardhubi fire clay and silica works, the Reliance Fire Bricks and Pottery co. at Chanch, the Bihar fire bricks and potteries Ltd. at Mugma, the Gulfabari fireclay and silica works, the Bengal-Bihar fire bricks and pottery works, the Nirsa refractory, the Jharia fire bricks and pottery works. These industries are meeting the increasing demand of the country in refractories, ceramics, fire bricks, potteries, etc.

There are thermal plants at Bokaro, Sindri, Patratu, Chandrapura, etc. These plants are supporting the growth of industries in the region with their power supply.

A number of coal washeries have developed in this region. Some of these are located at Kargali, Pathardih, Dugda, Bhojudih, Kathara, etc. These coal washeries have the capacity of purifying 3.5 million tons of coal. Besides the above coal washeries there are a number of coke oven plants for converting coal into coke by carbonisation process. Important coke oven plants are located at Sindri Coking Plant, Bararee Coke Company at Kusumda, Loyabad Coking Plant at Bansjora, Lodna Coke Plant at Lodnaek. The Coke Plant at Bhowra is also important. In the process of purifying coal and preparing coal for varied uses a number of by products like coaltar, gas, amonia sulphate, benjol, etc. are obtained.

The Engineering works at Kumardhubi is also important one. It repairs electric drills, coal cutting machines, haulages, pumps and steam plants, etc. The industry undertakes fabrication works also like, headgears, steel trusses, transmission posts and other steel structures like mixing cones, ventilation fittings, etc. Manufacture of coal tubs are also made.

There is glass factory at Chota Ambona. The main product of this industry is glass lamps, jar and bottles, etc. The most commonly used raw materials are sand, borax, soda ash, salt coke, dolomite, limestone, salt petre, sulphur and other materials.

The steel plant of Bokaro is the largest in the country on account a number of factors. There is presence of coal and iron ores in the environs of Bokaro. The plant is facilitated by water supply from Damodar river. This plant has also the advantages of hydel power, thermal power, proximity of Kolkata market and port and cheaper labour from the countrysides.

There are different types of small scale and cottage industries in the region. These industries include tar chemicals, soap industries, sodium silicate, timber and furniture works, electrical good and manufacturing industries, rerolling mills, engineering industries, food processing industries, printing press industries, stone crushing industries, Tusser industries, etc.

The aforesaid industrial development has been attributed to presence of varieties of metallic and non-metallic minerals and other industrial raw materials, several hydel and thermal plants, dense net work of transport and communication and other industrial infrastructures.

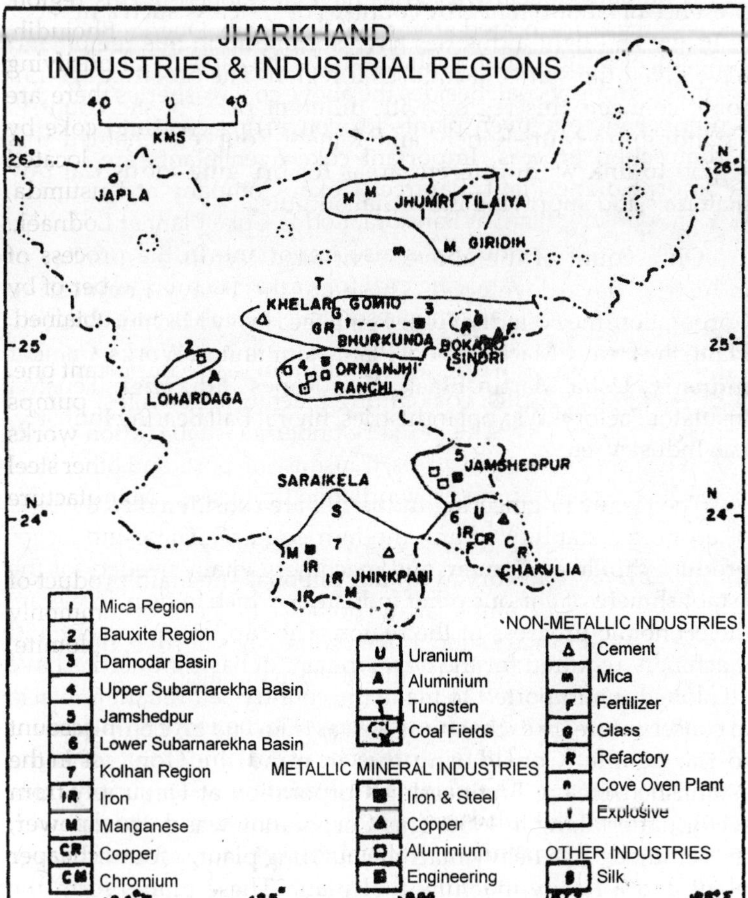

Fig. 12

2. The Upper Subarnarekha Basin Industrial Region : The Upper Subarnarekha Basin Industrial Region mainly includes the industrial areas lying in the upper portion of the river Subarnarekha. It includes industrial areas developed around Ranchi and Muri. This region is endowed with metallic and non-metallic minerals and forest products and many other resources worth using for industries. This region avails the advantages of good transport systems, the proximity of raw material and power resources of the Damodar Valley, congenial climate, abundant presence of labour in nearby countrysides, etc. As such a number of heavy and light industries have sprung up in this region and have given the status of industrial region. The broad gauge rail route connects this region with different parts of the country. Chandrapura-Muri-Ranchi and Ranchi-Rourkela enabled this region to link with different areas for bringing industrial raw materials and supplying industrial outputs.

On account of the above mentioned factors a number of industries have developed. These are the Heavy Engineering Corporation, the Foundry Forge Plant, the Heavy Machine Building Plant, the Heavy Machine Tool Plant, Aluminium Works, Cement Industry, Usha Martin Black Wire Ropes Ltd., High Tension Insulator Factory, Waxpol Industries, Bharat Ball Bearing Industry, Lac Industry, etc.

The Heavy Engineering Industries are considered as the base of economic stability of any modern country. These industries produce capital equipment and machinery vitally needed for the establishment of various other industries which in turn accelerate the economic progress of the region. The capital equipment and machinery required for the development of basic industries have all along been imported to make the country self-reliant and also to conserve foreign exchange as well as train and give employment to the people. Considering this in mind the Govt. of India established Heavy Engineering Corporation at Hatia as public sector undertaking in 1958. This Corporation was entrusted with the setting up of a heavy machine building plant, a foundry forge plant and a heavy machine tool plant. These plants form the foundation for the development of basic industries in the country and an unique industrial complex in the world.

The foundry forge plant was set up in collaboration with Czechoslovakia for the manufacture of a variety of castings and forgings required for any heavy equipment as machinery either meant for steel industry or equivalent for mining, heavy earth moving machine tools, power generation, etc. This plant produces all the castings and forgings required by the other two units of heavy engineering corporation viz. heavy machine building plant and heavy machine tool plant. It also fulfils the requirements of private sector industries developed in the region.

The Heavy machine building plant was set up in collaboration with U.S.S.R. and has capacity to produce annually 80 thousand tons of heavy machinery items for steel plants and other industries like cement, fertilizers, mining, oil, drillings, etc. Upto date equipments have been installed in the plant which enable to carry on various technological operations required for the manufacture of latest machines like boring and turning machines, cranes, coke oven equipments, blast furnace equipments, crushing and grinding equipments, rolling mills equipments, metallurgical equipments, etc.

The Heavy Machine Tool Plant was also set up in collaboration with Czechoslavakia to produce ten thousand tons of machine tools like central lathes, radial drilling machines, double column planning machines, horizontal boring machines, vertical boring and turning mills, plan milling machines, precision cylindrical and Roll grinding machines etc.

There is an industry for producing alumina at Muri based on bauxite resource of Lohardaga. The alumina of this region is provided to the smelter at Alwaye in Kerala and Hirakund. The main raw material, bauxite, comes from the company's own mines at Bagru hill near Lohardaga by narrow gauge railways and trucks. The other raw materials, caustic soda, coal, furnace oil, filter cloth, bags and starch are all indigenously procured. Water supply requirements are met by water supplies from adjacent Subarnarekha river. The process of alumina manufacturing requires large ponds where residue after extraction of alumina is transferred in slurry form.

There is a cement factory at Khelari on the Gomo-Dehri Section of Railway. The quarries of limestone, principal raw material are situated close by in Khelari and Churi hills. Coal is available from the local collieries. Gypsum is brought from Rajasthan. Electric power consumed is made available from the D.V.C.

The Usha Martin Black Wire Company, a public Ltd. concern, is situated at Tati Silbai, about 12 kms. east of Ranchi. It was started in 1961 and went on production in 1962. The bulk of raw materials, i.e. the high carbon wire rods are brought from abroad. The plant is located at this place due to the neighbourhood of market in the engineering industries and mining areas.

The High Tension Insulator Factory was established in 1961 at Namkum. The purpose of developing this factory was to manufacture high tension and low tension insulators for electrical transmission. The factory is run by electric power. The raw materials used are China clay, E.P. Ball, Kotri or plastic clay, felspar, silica sand and quartz.

The Waxpol Industries Ltd. also set up at Tatisilbai in 1962 in a public Ltd. concern. It manufactures waxpol chemicals to meet the demand of home market. The raw materials used are waxes, solvents, vegetable oil, fatty acids, tin sheets, lithographic ink and miscellaneous chemicals. The bee-wax is available from the forest of the region. The industry provides finished products as a main industrial raw material for consumers in the cosmetics and pharmaceuticals trade, manufacture of tin containers for packing and manufacture of specialized foundry chemicals.

Bharat Ball Bearing Company, a public Ltd. concern, is located at Ratu, 10 kms. west of Ranchi. It started production in 1963. The ball and roller bearing industry occupies a key position in the industrial development of the country. The automobile industry of the country is one of the main consumers of this industry.

Lac industry is important in this region since ancient times. With the beginning of the manufacture of lac dye in 19th century, it made its debut in international market. It processes properties of such values as to find ready use in a large number of industries. It is a commodity of considerable importance in the country's economy as its cultivation is largely done by poor community of the countryside. This industry earns a lot of foreign exchange. The manufacturing of lac was carried on principally at Bundu. This industry on modern scale was established at Doranda. Jharkhand state produces the largest quantity of lac in India. Seed lac and

Shellac are used for protective and decorative finishes, particularly for wood, manufacture of gramophone records and printing inks and electrical industry for the manufacture of micanite. The Indian Lac Research Institute at Namkum has been established to research on pests and lac for the development of lac cultivation.

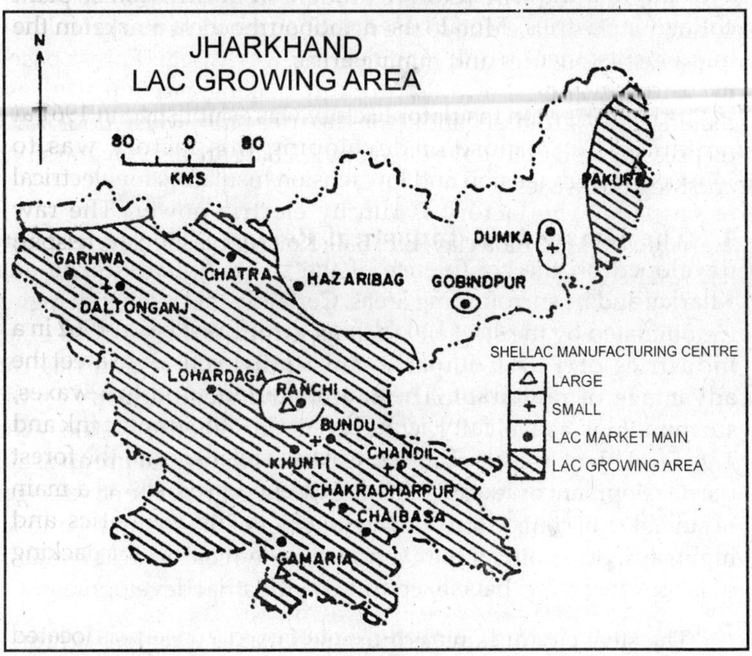

Fig. 13

Besides, the Upper Subarnarekha Basin Industrial Region accounts for the development of refractories, fire-fighting equipments, Bharat Spun Pipe Co., the Industrial Gases Ltd. and a number of forestand agro-based cottage industries like saw mills, weaving, rope-making, basket making, broom (Jharu) making, biri industry, toy making, furniture making, pottery, oil industry, rice and wheat flour milling, etc.

3. **The Lohardaga Industrial Region** : This industrial region lies in the north-western portion of the Ranchi Plateau. The region is, undoubtedly, noted for huge reserve and quarrying of bauxite but other metallic and non-metallic minerals and forest products participate in the industrial development of this region. The bauxite

quarrying is included in the list of industries. The bauxite mined in this area is sent to Muri aluminium industry. This region is linked through railways with Ranchi through which mining and industrial products are sent out the region. The countryside provides abundance of cheap labour for the development of industries. Virtually the industries developed are of either smaller size or cottage industries. Metal based industries include agricultural implements, steel trunk, engineering works, etc. Forest based industries include saw mills, furniture making, biri industry, toys making, etc. Agro-based industries like rice mills, wheat flour mills, oil pressing mills, etc. have developed. There are also several stone crushing industries.

4. **The Jamshedpur Industrial Region** : This region has developed on the confluence of the rivers Subarnarekha and Kharkai and its surrounding areas. Certainly this industrial region is dominated by the steel city of Jamshedpur and concerned other industries of Tata Company. This industrial region has the advantage of raw materials for iron and steel industry in the surroundings as well as raw materials for other industries. The Tata Steel plant itself provides a number of industrial products for the development of several industries in the region. The advantages of suitable rail connection with major cities of the country, national highways, proximity of the Damodar Valley coal area, abundance of labour force, etc. have encouraged industrial development.

The steel city of Jamshedpur itself makes a region. Within a period of more than 90 years TISCO commands veritable satellites of Agrico, Telco and a large number of factories dealing with manufactures of Chasis, Truck engine, Engineering materials, Tin plate, Indian cable, Tubes, Oxygen plants, etc.

The iron and steel industry at Jamshedpur was established by Jamshedjee Tata at Sakchi. The locational advantages of the steel plant include the presence of superior quality of iron ore from mines of Noamundi, Gurumahisani, Badampahar, Sulaipat mines, etc. Manganese is obtained from Gua and Manoharpur and good quality of coal comes from Jharia coal field. The city is duly linked with good transport system and is also getting water facility from Subarnarekha river and cheap labour from countrysides. This plant comprises coke oven, fire glass furnaces, three steel melting shops, blooming mills, rail and structural mills, steel bar and biller mills, plate mills, sleeper press, sheet mills, wheel, tyre and axle plants, etc.

The Tata Locomotive Company has three divisions i.e. locomotive, automobile and foundry. The Indian Steel and Wire Products produce wire, rods and other products. The Indian Hume Pipe Company started in 1926 on the bank of Subarnarekha river for manufacture of reinforced cement, concrete, hume pipes on hume centrifugal process. It also produces concrete spun poles used for electric transmission lines.

The Tin Plate Company situated at Golmuri, 5 km. from Tatanagar produces packed Tin plate. The principal raw materials used in the manufacture of Tin plate are steel, tin and coal. Steel is brought from Tata Iron and Steel company.

The Jamshedpur Engineering and Machine Manufacturing Company was established in 1921. It manufactures different qualities of rolls, such as plain chill roll, rolls, sand cast grain rolls ,etc. meant for locomotives.

The Indian Oxygen and Acetylene Company was established in 1955. It produces Oxygen and Acetylene for other industrial uses.

The Indian Cable Company was started in the year 1923. It produces copper wire and insulated cables for electrical purposes.

The Indian Tube Company was formed in 1953 with the collaboration of TISCO.

Besides these industries this region has given rise to a number of small scale private industries concerning machine tools, repair shops, printings, furnitures, steel trunks, etc.

5. **The Mica Industrial Region** : This belt coincides with mica mining belt existing in the districts of Giridih and Kodarma. Principal industries are mica splitting industries. Chief raw materials of this industry are abundantly available. Other advantages being rail link, road link, skilled labour, power from the Damodar Valley, etc. have encouraged development of mica industry and other industries. Important mica industrial centres are Domchanch, Jhumri-Tilaiya, Giridih, Pachamba, Bandih, etc. This region has also other cottage industries like engineering works, printing press, furniture making, saw mills, oil pressing mills, etc.

Stone crushing is being done at several places.

6. **The Ghatshila Industrial Region** : This region is located in the south-eastern portion of Purbi Singhbhum. This region has developed industries based on copper and forest resources. The Indian Copper Corporation established copper industry at Ghatshila in 1924. To produce brass sheet and yellow metal, rolling mill industry was established in 1930. The industries obtain raw materials from mines of Mosabani and Badra. Copper sheet, copper and other goods are manufactured at Ghatshila. In this factory 5 thousand tons of copper is processed per day. Brass is manufactured by mixing zinc with copper and tin. Ghatshila produced nearly one third copper of the country and is the largest producer in India.

Besides the production of copper in this region this area has also experienced the developments of industries associated with forest and agriculture. Biri industry, saw mills, furniture making, etc. have developed throughout the region. Rice mills and wheat flour mills, oil pressing mills, etc. are also found everywhere.

7. **The North Koel Basin Industrial Region** : This industrial region is an emerging industrial region. Important industries are concerned with metal works, forest products, stone products and agriculture. This zone is growing around Daltonganj, Latehar and Garhwa. This region has the facilities of raw materials like coal iron ore, limestone, graphite, dolomite, felspar, etc. The upper portion of the valley has dense forest cover which provides raw materials for forest based industries. This region is linked with rail route and roads. Cheap village labourers are available everywhere in the region.

Based on above facilities several centres of industries have emerged. In the extreme north where this basin forms part of the Son Valley, cement industries have developed at Japala. It started in 1922. This plant receives limestone from the Kaimur hills where lies huge deposit of limestone. The Huttar coal field of Koel Basin provides required amount of coal. Clay is made available both from the Damodar valley and the North-Koel Basin.

This region also has lac and shellac industries at Daltonganj, Garhwa, Latehar and Tori. There are several saw mills at

Daltonganj, Garhwa, Latehar, Chandhwa, Rehla, Barbadih, etc. These saw mills easily obtain timber from the neighbouring forests and facilitate the manufacturing of building requirements and furnitures. This region has also wide spread Biri making industry. Kendu leaves are easily available from the forests.

The Engineering Works at Japala produces cast iron materials. The weaving of cotton and wool is an important occupation to the majority of tribes. The hand oven cloth, which though course but durable, is mostly seen amongst the aboriginals and semi-aboriginal tribes. The other cottage industries are rope making, black smithy, pottery, goldsmithy, bee-keeping, carpentry, basket and bamboo wares, palm gur making, Tusser rearing, match box making, leather tanning and leather goods making. Mostly these industries are running in a disorganised form, located in the extreme interior lacking the suitable facilities for transport, technical and financial help.

The Koel Basin has vast natural resources but these have not been fully exploited. The forest of this region have abundant bamboos. There is great potentiality for this bamboo pulp industry in the neighbourhood of Chipadohar. A large number of bamboo pulps are exported from Chipadohar to Dehri-on-Son and Titagarh. This industry can be developed in this region if it is properly exploited. There is also potentiality for the development of plywood. There is tremendous resources for fire bricks factory also.

8. **The Deoghar Industrial Region** : In the north-eastern portion of the state, despite having less minerals and forest resources, several industries are growing considerably and giving rise to an industrial region, different from other industrial regions of the state. In this region industries are mainly related to products required in religious functions. This region has locational advantages for the development of industries. This region is well connected by railways and roadways. This region has well reputed Shiva temple at Deoghar since remote past. It attracts lakhs of pilgrims and tourists annually from all over the country. As such people coming from distant areas as well as neighbouring areas prefer to purchase goods of household requirements and these requirements have boosted the growth and development of concerned industries. The area is surrounded by small and large industries involved in the

production of worshipping materials, goods of household needs, engineering goods, repair shops, agricultural implements, handloom products, etc. The surrounding areas having minerals provide the facility of metallic and non-metallic industries. Forest resources provide timber for wooden items. Skilled labourers for Tusser making, weaving etc. are available throughout the region. There are a good number of small scale industries producing cooking pans, buckets, knives, axes, spears, pottery, toys, utensils, etc. There are a number of "small bakeries" in the urban areas and some in larger villages. The consumption is entirely local. Preparation of milk products like butter, ghee, chhena, curd, sweets, etc. are famous in this region. There is a heavy concentration of carpenters at Shyam Bazar Road in Deoghar. There are more than 40 units which are engaged in the manufacture of wood, furnitures, etc. There is a branch of famous Dabur industry manufacturing medicines. This region is noted for blacksmithy works. Previously blacksmiths of this region used to smelt iron of their own with traditional method and used to make iron goods for household requirements but these days techniques have improved. They manufacture iron goods with the help of iron available in the markets. Iron smithy industry is distributed in the urban and surrounding rural areas of Deoghar.

Agro-based industries have also developed in this region. There are businessmen who purchase food grains and process in industries for bulk supply. There are a few flour mills at Deoghar which purchase wheat and grind it and supply in bags. In addition to these large mills there are wide spread small rice and wheat flour mills. There are also numerous oil mills. The oil mills of the region purchase oil seeds from U.P. and other provinces and prepare oil.

This area has also given rise to a number of soap factories which produce variety of soaps for local consumptions. There are a number of industries manufacturing bricks and English tiles. Biri industry is also well developed in the region. It has spread in far flung villages also. There are a large number of persons who manufacture Biri in their spare time and supply it to the contractors. The Baidyanath Biri Company of this region is famous for making Biris.

Besides above mentioned industries tourism has taken

stupendous stride in the region because the pilgrims coming from distant places prefer to visit places of religious, historical and scenic importance. As such transport business has flourished. A number of hotels have also sprung up to facilitate boarding and lodging facilities to the pilgrims. Besides, hotels, restaurants, motels there are a number of inns developed by resourceful traders and religious minded people.

Such Inns have largely been developed on the way of Kamarias and in Deoghar also. Industries related to vermillion, alta, cosmetics, bangles, lady garments have also developed in the region for large number of females coming to perform religious activities.

11

Transport

Since the state of Jharkhand has uneven land surface and localized concentration of mineral resources and associated mining and industrial development, transport network has experienced unbalanced development. The mineralized, industrialised and densely populated even areas contain dense network of transport. On the other hand, the areas having rough terrain, dense forest cover, poor mineral resources and industrial development and sparse population are mostly inaccessible. Important means of transport are railways and roads. Almost all the rivers of the region are unfit for navigation due to rocky beds and scarcity of water during winter as well as summer season. Only the state capital is linked with air route. However, there are air strips in several district headquarters for landing of small planes. Hence, roadways and railways become the chief means of transport. The development of transport network in this region has great regional diversity based on nature of land surface ranging from plains, narrow valleys, rugged plateaus, steep scarps, ridges, etc. Though, the nature of their traffic, transportation media reveal a great deal of geography of the areas which they serve. The variations of traffic from place to place reflect areal differences, in population distribution, settlement patterns and economic activities and the flow of traffic measures the movement of people and goods between regions. Particularly the economy of this region depends on the interplay of several transportation systems such as railways and roadways.

Railways : Railways are the chief means of transport for the movement of raw materials and finished goods particularly in the mining and industrial areas of the region which become obvious when the development of rail route of the mining areas of the Damodar Basin is observed and compared with rail network of other areas. The highest density of railways is found in the lower Damodar valley where coal mining and mineral based industries are mostly concentrated. Rail network has developed in unbalanced way. The area is linked with main rail route connecting Howrah-

Delhi Grand Chord Line, Howrah-Mumbai Rail route but there are localized dense rail net based on mineral resources and industries.

The total railway-route length in Jharkhand is 1984 Km (2010-11). The rail density is 26.8 Km per 1000 sq. km. Ranchi, Dhanbad, Jamshedpur, Bokaro, Daltonganj, Barka kana are the major railway stations. The state has extensive goods handling facilities available at Ranchi, Bokaro, Dhanbad & Jamshedpur station. The proposed dedicated freight corridor (Part of Eastern Corridor extension to Kolkata) will pass through the state and is expected to benefit the industry significantly.

The most developed net of railways lie, in the Lower Damodar Basin. The electrified Grand Chord Line of the eastern railway runs through the northern segment of the valley but branch lines although small, thread the rectangle south of the Grand Chord Line and north of the Damodar. Important rail routes developed in the Damodar Valley region are Kumardhubi-Gomo (68 km),

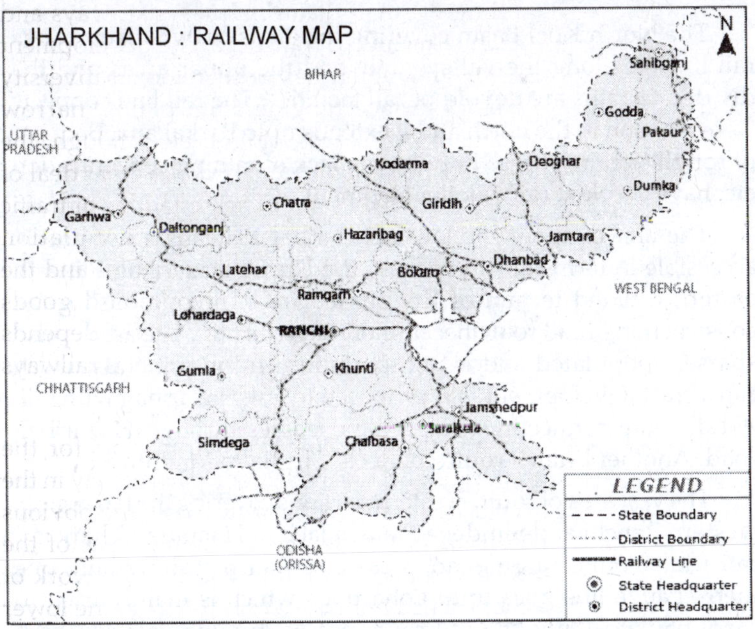

Source : Govt. of Jharkhand.

Fig. 14

Pradhankanta-Pathardih (15 km), Dhanbad-Phularitand (22 km) and Dhanbad-Pathardih (15 kms.) In addition to these there are rail routes of southern railways like Sudamdih-Gomo and a few branch lines for carrying coal from Nichitpur to Katrasgarh and Malkera, from Malkera to Mahuda, from Mahuda to Chandrapura, from Kharkhari to Malkera, from Kerkend to Jamada Bharbharia to Sudamdih, from Kerkeen to Bhowra, Bhutgoria to Jamadoba, Tetulmari to Kusunda, Jharia, etc.

The Subarnarekha Basin also provides dense pattern of railways. This region is also on the main rail route of Howrah-Mumbai. This rail route touches several mineral exploiting centres as well as Jamshedpur industrial complex. It carries a heavy traffic consisting of coal, iron ore, iron and steel and textile. Muri-Jamshedpur rail route also passes through the Subarnarekha basin. The mineralised areas of the Subarnarekha valley specially the metallic mineral mines of iron ore, copper are linked with main rail route. Almost all mines are connected with railways like Gua, Noamundi, Ghatshila.

The North Koel Basin covering the areas of two districts has rail linkage along the valleys only and the upper areas and the divided terrains are devoid of rail facilities. The rail line connects Dehri-on-son in the north and it extends upto Barkakana. Because of rough terrain, sparse population, lack of mining concentration etc. have checked railway development.

The area of upper and lower Hazaribag Plateaus is also poorly accessible in terms of rail because the Upper Hazaribag Plateau, eastern divided terrain of Koel basin are unfavourable for the construction of rail route because of rough terrain. The area is also sparsely populated and is not so resourceful in minerals which can attract development. It is proposed to link Hazaribag with rail. A rail route connecting Gaya-Gomo passes through Hazaribag road. Another branch route connects Giridih to Madhupur.

The vast area of south-western portions covering the Pat region, western Ranchi and Simdega Plateau lack rail facilities. There is a rail route connecting Gomo, Muri and Ranchi from where one metre gauge line goes upto Lohardaga which is mainly used to carry bauxite. Only one rail route passes through vast southern portions of Ranchi and Simdega Plateau which connect Ranchi with Rourkela. In this portion the terrain is less favourable for the rail route and the area is also poor in mining, industrial development and human concentration.

The north-eastern portion of the region consisting of the areas of Santhal-Parganas has rail facility mainly in the peripheral parts and the vast interior areas having hilly landscape with sparse population and a few isolated mines are devoid of rail connection. This portion is traversed on the north-east by the loop line and south west by the chord line of the eastern railways. A short branch (12 kms.) connects Rajmahal with Tinpahar on the loop line. Another branch also runs from Madhupur on the main line to Giridih, a distance of 36 kms. Again there is a branch line connecting Jasidih to Deoghar. From Sahibganj a short line runs to Sakrigali ghat from where ferry service is available for crossing the Ganga.

Roads : No doubt, the development of transport considerably depend on the nature of terrain resourcefulness of areas, industrial, mining and urban developments, etc. but the development of road remains comparatively mere convenient to link even areas having high elevation and rough terrain. It becomes obvious by the comparative analysis of the pattern of railways and roadways of the region. This state has a good network of roads. The state has 1844 km National Highways and 6880 km State Highways. These roads link about 8184 villages of the state. In addition to this fair weather roads link 3935 villages. This net is much denser than the railways net in the region so much so that the pockets of inaccessibility seen on the railway map of the region are observed contracted. A significant feature of this pattern is the complementary nature which the road net has with railway net. Most areas which are devoid of railways are made accessible by roads. Certainly not all areas have experienced uniform growth of roadways. In certain areas having bunches of urban centres developed on mines and industries have dense net of roads where easy accessibility is witnessed but in remaining areas road net is not so developed because of limited mining pursuits, industrial developments and sparse human distribution. But inaccessible and poorly accessible areas are much less in case of roads than the railways. The development of roads in this region is more convenient because of hard base and availability of construction materials. Certainly the length of roads increases much than the crow fly distance because roads are constructed following the nature of terrain and its slope. A keen observation of the road pattern of the region reveals the location of several nodal centres in this region from where metalled roads or unmetalled roads radiate in all directions like Ranchi, Hazaribag, Dumka, Dhanbad, Barhi, Chaibasa, Daltonganj, etc.

The state has 12 national Highways (NH) running through the state; NH-2, NH-6, NH-23, NH-31, NH-32, NH-33, NH-75, NH-78, NH-80, NH-98, NH-99 and NH-100 measuring 1844 km. The road infrastructure of the state is divided basically into 3 categories.

1. National Highways
2. State Highways & Major District Roads
3. Rural Roads & Minor District Roads

National Highways constitute the primary system, state highways & major district roads provide secondary system and the rural roads & minor district roads provide tertiary system. The total length of different types of roads in Jharkhand is as follows :

Category of Roads	Lengths (Km)
1. National highways including NH-2 (Transferred to NHDP)	1844
2. State Highways & PWD Roads including Major District Roads including Urban Roads	6880

The ambitious Golden Quadrilateral also passes the state and its total length is 192 km. which caters the high density traffic and a six lane project.

The Damodar Basin having immense mineral resource, comparatively less rough terrain, concentrated growth of urban and industrial centres has highly developed road transport. The area is linked with Grand Trunk Road. This road passes though Dhanbad, Giridih, Barhi and enables the movement of heavy vehicles carrying coal and industrial products of this region to eastern and western regions of the nation. Besides the G.T. Road, the National Highway connecting Ranchi, Hazaribag, Barhi and Nawada also passes through the Damodar Basin. Another National Highway connects Gobindpur, Chas and Jamshedpur. Besides there are several state highways and other roads in the coal belt of Bokaro and Dhanbad districts. Important roads are Gobindpur-Beliapur-Sindri, Dhanbad-Pathardih-Sindri, Raghunathpur-Chandankyari-Chas, Pathardih-Chandankyari-Baramasia, Giridih-

Tundi-Gobindpur, Jharia-Balia-roads, etc. The G.T. Road passes through the coal belt of the Damodar basin and also traverses the undulated areas of lower Hazaribag plateau and Hazaribag plateau and enables the heavy vehicular movements in the region. The National Highway connecting Govindpur-Jamshedpur also establishes linkage between the coal mining areas with that of Jamshedpur industrial areas. Govindpur-Baliapur-Sindri road connects the township of Sindri, 24 kms. from Dhanbad with the G.T. Road where important factories of fertilizer and cement have developed. Dhanbad-Pathardih-Sindri road connects Sindri with Dhanbad and passes through Jharia, the most important business centre in the coal belt. Raghunathpur-Chandankyari-Chas Road connects Hazaribag district and Dhanbad district with west Bengal through Chas and Chandankyari. Rajganj-Katras-Jamdina road connects G.T. Road with Katras. Pathardih-Chandankyari-Baramasia road connects the Pathardih coal washing with West Bengal and passes through Chandankyari. Giridih-Tundi-Gobindpur Road connects G.T. Road with Tundi and Giridih.

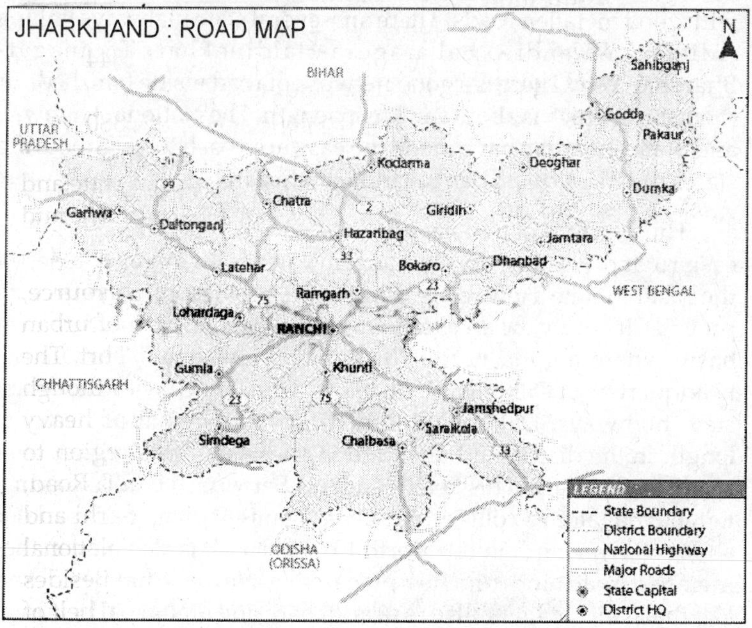

Source : Govt. of Jharkhand.

Fig. 15

Besides there are several municipal roads in urban areas and forest roads in the forest covered areas.

The mineralised and industrialised belt of the Lower Subarnarekha basin is also developed in terms of road network. The National Highway No. 33 connects Ranchi, Tamar, Chandil and Jamshedpur and passes through the valley and joins N.H.6 which connects major cities of Kolkata and Mumbai. State highway connecting Ranchi, Chaibasa, Jamshedpur also passes through the valley of Subarnarekha. The areas along the river Subarnarekha becomes more accessible because this river cuts the Dalma Range at Chandil and enables the development of transport but the areas west of this having highlands of Porhat and Dalma range provide rough terrain which unables easy development of transport facilities. The southern scarp of Ranchi Plateau creates steep scarpment and inaccessible areas where development of transport again becomes difficult. The network of transport in Dhalbhum Plain has been convenient because of relatively less undulations and occurrence of mines and related industries. Important mines and industries like Ghatshila, Moubhadar, Musabani are linked with good metalled roads. There are several state highways linking Adityapur-Kandra, Baharagora-Haldipokhar, Tatanagar-Bharbharia etc. There is a good network of road inside Jamshedpur townships which is about 240 km in length. The National Highway and State Highway have made the easy approach to neighbouring states of Orissa, West Bengal and Chattisgarh.

The North Koel Basin presents varied road transport. The eastern and western divided terrains and hilly upper courses of the basin contain a few roads where acute problem of inaccessibility prevails. It is only the areas adjoining the river valley and the lower basin where a good net of roads have developed. The district headquarters of Daltonganj and Garhwa are linked with important state highways. Daltonganj-Ranchi Road covers about 100 km length in the district of Palamu. It is one of the finest all weather roads in the state. This road connects Satwara, Manika, Sasang, Latehar, Jagaldaga colliery, etc. Daltonganj- Aurangabad road has about 68 km length in the North Koel Basin. It is also all weather metalled road and touches places like Nawa, Chatarpur and Hariharganj. It links also Aurangabad and Patwa. Chandwa-Balumath-Gonia Road connects Chandwa with Dobhi on the G.T. Road. This road has only 35 km length in this valley. Rehla-Garhwa-

Ranka-Godarmana Road starts from Rehla near Bisrampur and connecting Garhwai, Ranka Godarmana goes to Ramanujganj of Chattisgarh. This road passes through the water divide existing between the river Kanhar along the south western border and the river North Koel. As such this road bears significance because of its route in rough terrain and inaccessible areas. Daltonganj-Barwadih-Bhandaria and Godarmana road begins at Bhusar, 10 km from Daltonganj and passes through Kutmu and then takes a turn to Barwadih and Godarmana touching Hutar coal field and crossing North Koel near Husar. This Broad follows the water divide from Hutar to Godarmana. The divide lies between river Tahle and Bigka. In addition to these roads there are roads linking Garhwa, Nagar Utari and Mirjapur of Uttar Pradesh. This has a length of about 45 km Daltonganj-Lesliganj-Panki road is about 40 km in length passes through the eastern divided terrain and connects places like Lesliganj, Parsia, Panki and Balumath where it joins a road connecting Ranchi-Chatra. There is another all weather road connecting Lesliganj and Manatu in the eastern divided having undulating land surface. There is also a road linking Garhwa, Mohmadganj, Haiderganj and Hussainabad in the lower reaches of the river North Koel. There is one road from Betla reserve forest to Barwe connecting places like Garu, Baresand, Anksi, Mahuadaur. This road passes through rugged areas of the upper course of the river North-Koel and this single road passing through this area gets due importance for movement of people and transportation of goods.

The western higher areas of Pats, higher Ranchi Plateau and the rugged Simdega Plateau having high elevation, narrow valleys, steep scarpments and rough terrain have caused poor development of transport. Certainly the Pats region having scenic places of Netarhat and other important places have been linked with roads. Important roads transversing this region are Ranchi-Netarhat Road, Gumla-Raidih-Chainpur Road, Raidih-Sambalpur Road, Simdega-Kurdeg Road, etc. Ranchi-Netarhat road originate from Ghaghara and terminates at Netarhat. The entire length of 58 km is metalled. It passes through rough hilly landscape and has importance because it links Netarhat, a health resort cum tourist cum educational centre. Gumla-Raidih-Chainpur Road starts from Gumla and passing through Patratoli headquarter of Raidih block and terminates at Chainpur. A branch of this road goes to Jaspur in Chattisgarh and crosses the river Sankh. Simdega-Kurdeg Road, about 42 km in length connects Simdega with Kurdeg which leads to Sambalpur in Orissa. This road also passes through rugged terrain. The road connects the Thaitargar-Bolba road passes through the Simdega Plateau and also connects Birmitrapur in Orissa.

The plateau areas of Ranchi and Hazaribag have also a good net of roads. Ranchi forms a nodal point where from roads radiate in different directions like Jamshedpur, Chakradharpur, Simdega, Daltonganj, Ramgarh, etc. Similarly Hazaribag, Chatra and Barhi also, form nodal points where from roads radiate in different directions. Two National Highways No. 31 and 33 touch the nodal point of Ranchi. National Highway No. 31 connects Patna with Ranchi via Hazaribag. This road also meets the G.T. Road at Barhi of Hazaribag Plateau. It is one of the most important roads as it serves a number of collieries in Ramgarh areas, a large volume of vehicular traffic from Jamshedpur to the plain areas of Bihar through this road. Another National Highway is Ranchi-Bahragora Road, No. 33. It connects Namkum, Rampura, Tamar, Chandil, Jamshedpur, Ghatshila and meets the Kolkata-Mumbai National Highway at Bahragora. This highway has reduced the distance almost between Ranchi and Jamshedpur because previously road used to pass via Chaibasa and traversing dangerous ghats over a distance of 33 kms. Besides these two national highways there are several state highways like Ranchi-Purulia road of which 58 km falls in Ranchi plateau, Ranchi-Itki road which passes through Bero, Namkum-Doranda road, Ranchi-Pithoria road, Ranchi-Chaibasa Road passing through Khunti, Murmu and Chakradharpur, Ranchi-Daltonganj Road passing through Mandar and Kuru, Tangar-Khelari Road connecting Khelari Cement Factory with Ranchi town, etc.

Even the Ranchi Plateau is well served by good roads may be because the region is insignificant in terms of rail route. It was 155 miles of National Highway, 233 miles of State Highways and 921 miles of roads managed by district board. More over there are 66 miles of roads under the Hazaribag mines boards, 330 under the forest department. Some small roads maintained by the D.V.C. The Grand Trunk Road, National Highway No. 2 runs for 75 miles in Lower and Upper Hazaribag Plateau. Some important places on it being Chauparan, Barhi, Bagodar and Dumri. The Barhi-Hazaribag

Ranchi Road (22.5 miles) is a part of N.H. 33 and the Barhi-Rajauli Road (31.4 miles), a part of N.H. 31. Other important roads are Dumri-Giridih Road, Hazaribag-Bagodar-Sarai (40.5 miles), Giridih-Jamua-Sarwan (14.5 miles), Jamua-Kodarma (43 miles), Gomia-Chatra-Gosaidih (42 miles), Hazaribag-Simaria (36 miles), Chatra-Chauparan (30.5 miles), Hazaribag-Simaria-Chatra (42 miles), Hazaribag-Katkam-Sandi-Chatra (35.5 miles), Bishnungarh-Petarbar (31.5 miles), Hazaribag-Tandwa-Simaria (45miles) Ramgarh-Dakagarha (28 miles), Sarai-Dhanwar (18.5 miles), Putridih-Gobindpur (37 miles), etc.

The north-eastern portion consisting of the hilly terrains of Santhal parganas has road nodal point at Dumka where rail facility is non-existent. This area has comparatively better road net in the adjoining plains at relatively less undulated areas than the hilly areas. Important roads are Dumka-Rampur Hat, Dumka-Bhagalpur, Dumka-Pakur, Dumka-Deoghar, Godda-Hansdiha, Dumka-Madhupur, Dumka-Jamtara, Deoghar-Chakai road, etc. Dumka-Rampur Hat road is about 58 kms. and connects Sikaripara, Haripur and Rampur Hat. Dumka-Bhagalpur Road is 44 kms. from Dumka to the border of Bhagalpur. It passes through Nomhar and Hansdih. Dumka-Suri road is a portion of Bhagalpur-Suri Road. The distance from Dumka to Armjora ghat on the river Mor is 25 kms. Dumka-Deoghar road is 63 km in length. It passes via Jarmundi and Ghormara Dumka-Pakur Road is 100 km. in length and passes through the centre of Damini Koh hilly areas and connect places like Amarapara, Litipara and Hiranpur. Godda-Hansdiha road is only 30 km long. Deoghar-Jasidih Road is 6 km in length and connects the railway station of Jasidih with the religious town of Deoghar. Dumka-Madhupur road is 67 km in length and it passes through Chorekatha and Sareth. Dumka-Jamtara road is 66 km long and passes via Fatehpur. Deoghar-Chakai road, 21 km in length, leads to Giridih. Litipara-Sahibganj road is 81 km and passes via Dharampur, Burhait, Borio and Banjha.

Airports :

Jharkhand has a dometic airport, located in the state capital Ranchi. From here, direct links are available to prominent cities of India like Delhi, Mumbai, Kolkata, Patna etc. Chartered Flights are also available between the Steel city of Jamshedpur and Kolkata. Jamshedpur, Dumka, Bokaro, Giridih, Deoghar, Hazaribagh,

Daltonganj and Noamandi also have airstrips. The Department of Civil Aviation of Jharkhand plans to acquire land to develop 12 airstrips in Chatra, Simdega, Lohardaga, Godda, Pakur, Latehar, Sahebganj, Jamtara, Saraikela, Garhwa, Koderma and Khunti district.

12

Population

The state of Jharkhand characterised by undulating land surface, considerable vegetal cover, huge natural resources and its exploitation and concentration of aboriginals has portrayed varied demographic phenomena. The analysis of various demographic features like the growth trend, distributional pattern, population composition, occupational structure and other aspects reveals interesting features. These population characteristics attract the attention of scholars of different disciplines to peep into inter relationship between changing environmental conditions and increasing pressure of human being on confined natural resources.

It is true that the analysis of different aspects of population was concerned only to scholars of limited fields because there were less problems related to mankind and natural resources. These days resources have become limited in comparison to overcrowding of population. As such scholars of different disciplines have developed interest in solving the problems which have been caused by multiplying pressure of people on limited natural resources. Virtually the study of population has become inter disciplinary topic and varied aspects of population have been analysed. In the present investigation only limited aspects of population have been discussed considering the adequacy of population data and other considerations. Aspects related to geography have been given due consideration.

While analysing the early accounts of people in the state of Jharkhand it becomes difficult to portray early accounts due to lack of authentic data. Census enumeration initiated only in the year 1872 and since then limited population data are available for analysing population characteristics. Prior to the census enumeration, the long history of human habitation has no documented accounts and other authentic sources of information. Certainly, on the basis of historical records, archeological investigation, carbon dating of the remains of fossils and some description in the religious books and epics, scholars have given sketchy accounts of population. More over, the state of Jharkhand being considered as land of forest, relatively less suitable for ideal human habitation and untapped resources, remained unmolested by the incoming migrants who inhabited the vast plain and brought significant transformation in economic pursuits and cultural developments. This region remained occupied for long mainly by the aboriginals and only in later historical periods, some migrants from the neighbouring plain areas intruded in the forest clad region may be for safety or some other reasons and settled there.

Historical records, archeological evidences, cultural attributes substantiate human habitation in this region in the paleolithic stone age (200,000-20,000 B.C.), relics of human habitation have been discovered in the Jharia coal fields and other areas of the state. The discovery of quarzite axes and spearheads in the Jharia coal fields, chert flaks and knives along the banks of the river at Chaibasa and Chakradharpur and at Buradih in Singhbhum and eastern part of Tamar Pargana in Ranchi attest the Paleolithic habitation in this region[1]. The habitation of Neolithic age is evidenced by an earthing of a pot, full of Indo-Scythian copper coins at Gulka, south of Chaibasa and traces of Neolithic culture at Chakradharpur, Parasnath, Ranchi and Hazaribag. At the time of aryanisation of the Ganga plain the northern dissected fringe of Chotanagpur Plateau was only influenced when several kingdoms developed in the Ganga Plain. The pre-historic stage of human habitation in this region is also shrouded in mystery and it is not so easy to provide definite account by a few hazy legends of the aboriginals which give some clues about the Asurs and indicate the triumph of that race over the original inhabitants of the country, possibly the ancestors of the small iron smelting tribe of iron age of that name which still speaks a language akin to Mundari. Certainly clear cut evidences are lacking to fix the date at which the Mundas, Oraons, Santhals, Hos, etc. inhabited in this region. It seems probable that they were the first to come and enter the region from west and the north-west. It is opined that Mundas belong to the Austro-Asiatic race. They are found in considerable number in the south-eastern portion. The aboriginal races such as Kharwars, Oraons, Cheros, Hos, etc. practically ruled over this tract. At the time of Aryan and non-Aryans rivalries in the plains non-aryan encroached in this region especially in accessible areas. Afterwards, some Aryan habitation also developed in the adjoining plains of the Rajmahal, and the northern portion of the Koel Basin.

During Budhist Period this region did not get much impact of the habitation of Budhists followers except the areas around Parasnath hill, parts of the Koel Basin. Most areas remained to a considerable extent untouched because of very poor accessibility and difficulties in spreading philosophy of Budhism and Jainism. During the regime of Ashoka, the Atavi or forest estates acknowledged Magadh supremacy and several preaching sites were developed. This is why there are villages having nomenclature related to Budhism or Jainism in parts of Palamu and Hazaribag Districts. In some of these villages relics of Budhists statue are also found.

With the downfall of Budhism the region again underwent unrest in the region and distributional pattern got jolted. Stability in habitation was restored only after the Muslim conquest over major areas of the country. Even this region did remain aloof of Mughal administration. Towards the end of the 10th century A.D. this region got habitation of some Rajput immigrants. These Rajput immigrants scattered over the area ignoring the poor accessibility because they had gone for living in secluded places and they also got opportunity to develop their estates. Certainly some areas got settled during Muslim period also by the Muslims whose evidences are available in most areas like the existence of mosque near the Palamu Fort near Daltonganj, the existence of mosque near Akbarnagar, Kasimchak in the north-eastern portions of Rajmahal highlands, etc.

Distributional pattern and other aspects of population also got considerably influenced by the development of transport network, establishment of administrative headquarters, development of service amenities, partial control over epidemics, etc. during the British regime. The role of post-independence development particularly in terms of land tenancy reforms, increasing resource appraisal, growing industries and urban centres, etc. became more effective in the determination of demographic features of the region. The impact of these post-independence changes become obvious when Pre and Post-independence demographic attributes are compared.

Trend of Population Growth : Though rough enumerations were made in 1872, 1881 and 1891 but these figures are not available in detail. As such trend of population growth has been analysed with effect from 1901 though estimated data are available even for earlier censuses.

Table - 12.1
Growth of Population : Jharkhand

Census Years	Population	Growth in %
1881	49,54,979	–
1891	53,35,141	09.50
1901	56,15,938	06.00
1911	66,23,329	10.99
1921	62,43,349	00.16
1931	72,44,846	16.04
1941	81,56,009	12.57
1951	94,66,153	16.06
1961	1,07,36,428	13.42
1971	1,42,28,238	32.52
1981	1,75,85,661	23.59
1991	2,18,48,860	24.25
2001	2,69,09,428	23.16
2011	3,29,88,134	22.42

The analysis of the table 12.1 reveals fluctuating growth rate of population. It has been noted earlier that population growth remained almost stagnant for long period because of high fertility and high mortality and negligible immigration. Even in the period of 100 years the growth rate exhibits fluctuation. The year 1921 is regarded as demographic break in the history of Indian population because in most areas population declined because of the occurrence of epidemics and droughts. Even the state of Bihar with which it was combined had recorded a decline in population by -0.3% but the state recorded negligible increase of 0.16% because this region remained comparatively less affected by the epidemics. In between 1881 and 1891 the population of this region increased by 9.5% and again in between 1891 and 1901 the population increased by only 6% revealing the decline of 3.5% over the growth of previous decade. In the following decade of 1901-11 population again increased with higher rate with 10.99% which again got jolted in the following decade of 1911-21. It may be pointed out that prior to 1921, the growth rate remained relatively slow and unsteady but afterwards the growth has remained steady. In between 1921and31 the population of this region increased by 16.04% because exploitation of minerals was accelerated. In 1931-41, the growth of population remained only 12.57% but in the following decade of 1941-51, the growth rate slightly increased by 16.06% due to the development of industrial or mining centres after the independence. The growth rate during 1961-71 further accelerated and the region recorded tremendous by 32.52% because a good number of migrants from the neighbouring states immigrated here in connection with industries, mining activities and trade and commerce. Even the following decade of 1971-81 population increased by 23.59% and in the decade 1981-91 the population increased by 24.25%. During last three decades the population has remained uneven in the region. Certain areas register very high growth based on exploitation of mineral resources and industrial developments which become obvious by the analysis of spatial variation in population. During 2001-2011, the population increased by 22.42% in the state which shows a marginal decline from the previous decade.

Table - 12.2
Spatial Population Variation : Jharkhand

Sl. No.	Districts	1981	1991	2001	2011	Variation % 2001-2011
1.	Bokaro	-	-	17,77,662	20,62,330	16.01
2.	Chatra	-	-	7,91,434	10,42,886	29.05
3.	Deoghar	7,08,828	9,18,323	11,65,390	14,92,073	28.03
4.	Dhanbad	21,15,010	27,09,170	23,97,102	16,84,487	11.99
5.	Dumka	12,15,452	14,97,266	11,06,521	13,21,442	19.42
6.	Garhwa	-	-	10,35,464	13,22,784	27.75
7.	Giridih	17,31,462	22,24,006	19,04,430	24,45,474	21.78
8.	Godda	7,13,405	8,58,678	10,47,939	13,13,551	25.35
9.	Gumla	10,17,231	11,53,557	8,32,447	10,25,213	23.16
10.	Hazaribagh	21,98,310	28,38,836	22,77,475	17,34,495	25.79
11.	Jamtara	-	-	6,53,081	7,91,042	25.46
12.	Koderma	-	-	4,99,403	7,16,259	32.42
13.	Khunti	-	-	-	5,31,885	22.32

14. Latehar	-	-	5,60,894	7,26,978	29.61
15. Lohardaga	2,29,783	2,88,585	3,64,521	4,61,790	26.68
16. Pakur	-	-	7,01,664	9,00,422	28.33
17. Palamu	19,17,528	24,51,048	15,37,465	19,39,869	26.17
18. W. Singhbhum	14,82,850	17,89,796	12,33,945	15,02,338	21.75
19. E. Singhbhum	13,78,949	16,17,170	19,82,988	22,93,919	15.68
20. Ramgarh	-	-	-	9,49,493	13.10
21. Ranchi	18,23,415	22,05,034	27,85,064	29,14,253	24.00
22. Sahebganj	10,79,753	12,97,391	9,27,770	11,50,567	35.74
23. Saraikela-Kharsawan	-	-	8,48,850	10,65,056	25.47
24. Simdega	-	-	5,14,320	5,99,578	16.58
Jharkhand	**-**	**2,18,43,911**	**2,69,45,829**	**3,29,88,134**	**22.42**

Source : Census of 2001 & 2011.

While analysing spatial variation in population remarkable changes in growth of population at the district level becomes obvious. In the decade 2001-2011, the increase in population was the greatest in Sahebganj district (35.74%) which is followed by Koderma (32.42%), Latehar (29.51%), Chatra & Deoghar. The increase in Sahebganj district is duw to growth of religious, cultural & commercial activities forcing people to settle here in large numbers. There is also increase in mining, industrial activities in Koderma due to which it experienced 2nd maximum growth of population. District which recorded moderate growth of population above the state's average of 22.42% include Pakur, Garhwa, Lohardaga, Palamu, Hazaribagh, Saraikela-Kharsawan, Jamtara, Godda etc. Dhanbad recorded the least population growth (11.99%) in the state during 2001-2011, which is due to closure of many mining activities, surface collapse, fire in mines etc. It is followed by Ramgarh (13.10%) district in the state.

Distributional Pattern : While analysing the distributional pattern of population in this region, a few areas have emerged as focal points of human concentrations but the remaining areas portrayed sparse distribution. The agglomeration of rural population is visualised by the crowding of dots showing population and urban concentration is reflected by the increasing size of circles and concentration of urban agglomeration in limited areas. Unevenness in population distribution becomes more obvious when the relief and population distribution maps are overlapped over each other.

This also becomes obvious by the analysis of man land ratio and density of population of both rural and urban.

The analysis of the table 12.3 reveals the share of area as well as population of the district in the state. The comparison of the share of the area and population reveals the unevenness in the distribution of population. Districts consisting of more population in comparison with area included Dhanbad, Bokaro, Ranchi, E. Singhbhum, Hazaribagh, Giridih, Deoghar, Godda, Jamatara, Pakur, Palamu, Ramgarh, Sahebganj. The district of Dhanbad consisted of 2.56% area of the state but has 8.14% population share of the state. Similarly, Bokaro has 3.62%, E. Singhbhum 4.47% and Ranchi 6.39% area but the population share is 6.25%, 6.95% and 8.83% respectively.

The districts having sparse population because of more land than population share included W. Singhbhum, Gumla, Latehar, Khunti, Chatra, Dumka, Garhwa, Koderma, Lohardaga, Saraikela-Kharsawan. The W. Singhbhum district has 9.06% area but 4.55% population of the state because of considerable forest cover. Similarly, Gumla has 6.72%, Latehar 5.38%, Khunti 3.18% area but the population share is 3.11%, 2.20%, 1.61% respectively.

The table 12.3 also reveals share of rural-urban population in the state. The district of Dhanbad has got only 4.49% rural population of the state but has 19.67% urban population of the state. Similarly, the district of E. Singhbhum has 4.07% rural population and 16.07% urban population of the state. The district of Ranchi has 6.61% rural population of the state but 15.85% urban population. The districts having high concentration of rural population included Giridih (8.93%), Palamu (6.84%), Ranchi (6.61%), Hazaribagh (5.82%) etc. The districts having high concentration of urban population included. Dhanbad (19.67%), E. Singhbhum (16.07%), Ranchi (15.85%), Bokaro (12.40%). The controlling factors for larger concentration of urban population in Dhanbad, E. Singhbhum and Ranchi are industrial developments as well as the development of other economic and cultural characteristics. The percentage of rural population is large in Giridih, Palamu, Hazaribagh etc. which are comparatively large in area and less developed.

The above mentioned facts also become obvious by the comparison of district-wise rural and urban population of the state. The table 12.4 explain the higher percentage of rural population in Godda, Garhwa, Chatra, Gumla, Dumka, Latehar, Simdega, Pakur, Khunti, Giridih and Jamtara, where 90% people of these districts live in rural areas. The districts having higher percentage of urban population include Dhanbad, E. Singhbhum, Bokaro, Ramgarh, Ranchi where more than 40% people live in urban areas. In Dhanbad and E. Singhbhum districts more than 55% people live in urban areas.

Table - 12.3

Distribution of Population : Jharkhand (2011)

Sl. No.	District	Area of the State (%)	Population of the State (%)		
			Total Population (%)	Rural Population (%)	Urban Population (%)
1.	Bokaro	3.62	6.25	4.31	12.40
2.	Chatra	4.66	3.16	3.91	0.79
3.	Deoghar	3.11	4.52	4.92	3.26
4.	Dhanbad	2.56	8.14	4.49	19.67
5.	Dumka	4.72	4.01	4.91	1.14
6.	Garhwa	5.13	4.01	5.00	0.88
7.	Giridih	6.22	7.41	8.93	2.62
8.	Godda	2.84	3.98	4.99	0.81
9.	Gumla	6.72	3.11	3.83	0.82
10.	Hazaribagh	4.41	5.26	5.82	3.47
11.	Jamtara	2.27	2.40	2.85	0.95
12.	Koderma	3.19	2.17	2.29	1.78
13.	Khunti	3.18	1.61	1.94	0.57
14.	Latehar	5.38	2.20	2.69	0.65
15.	Lohardaga	1.88	1.40	1.61	0.72
16.	Pakur	2.27	2.73	3.32	0.85
17.	Palamu	5.51	5.88	6.84	2.85
18.	W. Singhbhum	9.06	4.55	5.13	2.75
19.	E. Singhbhum	4.47	6.95	4.07	16.07
20.	Ramgarh	1.68	2.88	2.12	5.28
21.	Ranchi	6.39	8.83	6.61	15.85
22.	Sahebganj	2.59	3.49	3.95	2.01
23.	Saraikela-Kharsawan	3.33	3.23	3.22	3.26
24.	Simdega	4.73	1.82	2.22	0.54

Jharkhand	100.00	100.00	100.00	100.00
	79716.00 Sq. Km.	32988134 Person	25055073 Person	7933061 Person

Source : Census of India, 2011.

Table - 12.4

Rural-Urban Population : Jharkhand (2011)

Sl. No.	District	Total Popu lation	Rural Popu-lation	Urban Popu-lation	Rural Population (%)	Urban Popu-lation (%)
1.	Bokaro	20,62,330	10,78,686	9,83,644	52.30	47.70
2.	Chatra	10,42,886	9,79,932	6,2,954	93.96	06.04
3.	Deoghar	14,92,073	12,33,712	2,58,361	82.68	17.32
4.	Dhanbad	26,84,487	11,24,093	15,60,394	41.87	58.13
5.	Dumka	13,21,442	12,31,264	90,178	93.18	06.82
6.	Garhwa	13,22,784	12,53,114	69,670	94.73	05.27
7.	Giridih	24,45,474	22,37,450	2,08,024	91.49	08.51
8.	Godda	13,13,551	12,49,132	64,419	95.10	04.90
9.	Gumla	10,25,213	9,60,132	65,081	93.65	06.35
10.	Hazaribagh	17,34,495	14,59,188	2,75,307	84.13	15.87
11.	Jamtara	7,91,042	7,15,296	75,746	90.42	09.58
12.	Koderma	7,16,259	5,75,013	1,41,246	80.28	19.72
13.	Khunti	5,31,885	4,86,903	44,982	91.54	08.46
14.	Latehar	7,26,978	6,75,120	51,858	92.87	07.13
15.	Lohardaga	4,61,790	4,04,379	57,411	87.57	12.43
16.	Pakur	9,00,422	4,32,910	67,512	92.50	07.50
17.	Palamu	19,39,869	17,13,866	2,26,003	88.35	11.65
18.	W. Singhbhum	15,02,338	12,84,304	2,18,034	85.49	14.51
19.	E. Singhbhum	22,93,919	10,19,328	12,74,591	44.44	55.56
20.	Ramgarh	9,49,443	5,30,488	4,18,955	55.87	44.13
21.	Ranchi	29,14,253	16,56,918	12,57,335	56.86	43.14
22.	Sahebganj	11,50,567	9,90,901	1,59,666	86.12	13.88
23.	Saraikela-Kharsawan	10,65,056	8,06,310	2,58,746	75.71	24.29
24.	Simdega	5,99,578	5,56,634	42,944	92.84	07.16
	Jharkhand	**3,29,88,134**	**2,50,55,073**	**79,33,061**	**75.95**	**24.05**

Source : Primary census Abstract, Total Table for Jharkhand 2011.

Density of Population : While analysing man-land relations in earlier pages it has been observed that human population is unevenly distributed. It also becomes obvious by the analysis of density of population in the region. It has been found that higher density of population is recorded near the cities, industries and mining areas of Dhanbad, Bokaro, Suburbs of Jamshedpur and Ranchi. The district of Dhanbad has registered the highest density of population i.e. 1316 (2011) persons per square km. but there are development blocks where per sq. km. density has been very high. Dhanbad-cum-Kenduadih-cum-Jagata development block had the density of population of over 3000 persons per sq. km. in 2011. It may be noted that development blocks of Dhanbad have very high density because of urban and industrial influence. In the development Block of Golmuri-cum-Jugsalai in Purbi Singhbhum district again shows density of 1441 persons per sq. km. because of the influence of Tata Nagar industrial centres. Besides the mining and industrial areas of Dhanbad and Jamshedpur Regions there are areas near Ranchi, Hazaribag, Bokaro where the density of population is also very high. In the development block of Kanke the density of population is 703 persons per sq. km. Apart from the areas having the influence of industrial and mining developments recording high density there are productive river basins of North-Koel, lower Barakar, parts of Subarnarekha and flat plateaus of Ranchi which also has high density of population. The lowest density of population is found in the hilly areas of Palamu, Upper Hazaribag Plateau, Gumla, Godda, etc. These areas exhibit lesser development of meagre economic resources. The table 12.5 provides the density of population on the basis of district level and remarkable changes in the density of population during last 10 years.

In the year 1901 the region had density of 73 persons per sq. km. which has shoot up to 414 in 2011. The districts recording higher density than the regional average in the year 1901 included Godda (314), Sahibganj (142), Deoghar (132), Dhanbad (130), W. SInghbhum (119) etc. because of agricultural suitability of the land. Districts having less density were Gumla (49), Lohardaga (64), Palamu (49), hazaribagh (56) etc. In the year 1951, the density of the population in the region increased 123 based on natural growth and in in-migration from other areas due to industrial and rising employents. In the year 2011, the density was recorded highest in Dhanbad (1316), followed by Bokaro (715), Ramgarh (708), E. SInghbhum. These districts have large urban population. The highest density was recorded in Dhanbad because of accelerated industrial & mining activities which attracted a large no. of people from other areas. Districts having the density of population less than the regional average included Simdega (159), Latehar (169), Gumla (191), W. Singhbhum (208), Khunti (210), Chatra (281), Koderma (282) etc.

On the basis of the density figure of 2011, the density of the state can be categorised into 3 groups:

1. High Density Areas
2. Moderate Density Areas
3. Low Density Areas

1. **High Density Areas :** Areas having high density lie mainly in Dhanbad district where per sq. km. density has been even more than 3000 persons in some development block which has been discusssed earlier. While considering the density of the district the figure stands only 1316 persons per sq. km. Certainly there are a few development block in othe districts like in E. Singhbhum, Ranchi etc. where the density exceeds more than 1000 persons per sq. km. These areas contain a large number of mining and industrial centres. Apart from the district of Dhanbad, the district of Bokaro, E. Singhbhum and Ramgarh have sizeable density having density more than 600 persons per sq. km.

2. **Moderate Density Areas :** Districts having moderate density ranging from 300 to 600 persons per sq. km. include Ranchi, Sahebganj, Godda, Pakur, Palamu, Hazaribagh, Saraikela-Kharsawan, Jamtara, Dumka, garhwa and Lohardaga. Even in these districts development blocks having mining and industrial centres have high density of population. but the remaining development blocks have moderate density based on suitability of the land for agricultural purposes like flat plateaus and other forest based resources.

3. **Low Density Areas :** Low density areas having less than 300 persons per sq. km. include those districts of the state where considerable areas have rough terrain, negligible mineral resources, vast forest stretch and few economic & industrial activities. These districts include Simdega, Latehar, Gumla, W. Singhbhum, Khunti, Chatra, Koderma. As such neither

agriculture resource based population have concentrated nor non-agricultural base people are found in considerable proportion.

In addition to the general density or crude density other types of density like rural, urban, agricultural, physiological density, etc. can also be analysed and spatial variations can be worked out.

The density of rural or urban population can be calculated by dividing rural and urban population by rural and urban areas respectively. Physiological density is calculated by dividing the population by cultivated area and agricultural density is calculated by dividing agricultural population with total cultivable area. Similarly nutritional density is calculated by dividing total population with total food cropped area.

Table - 12.5
Population Density : Jharkhand

Sl. No.	District	Area (In Sq. Km.)	Density (2001) Person/ Sq.Km.	Density (2011) Person/ Sq.Km.	Net Change
1.	Bokaro	2883.00	621	715	94
2.	Chatra	3718.00	213	281	68
3.	Deoghar	2477.00	468	602	134
4.	Dhanbad	2040.00	1167	1316	149
5.	Dumka	3761.00	290	351	61
6.	Garhwa	4093.00	256	323	67
8.	Godda	2266.00	496	580	84
9.	Gumla	536000	148	191	43
10.	Hazaribagh	3555.00	370	488	118
11.	Jamtara	1811.00	364	437	73
12.	Koderma	2540.00	380	282	98
13.	Khunti	2535.00	-	210	-
14.	Latehar	4291.00	154	169	15
15.	Lohardaga	1502.00	244	307	63
16.	Pakur	1811.00	388	497	109
17.	Palamu	4393.00	296	442	146
18.	E. Singhbhum	3562.00	560	644	84
19.	W. Singhbhum	7244.00	240	208	32
20.	Ramgarh	1341.00	-	708	-
21.	Ranchi	5097.00	362	572	210

22. Sahebganj	2063.00	580	558	22
23. Saraikela-Kharsawan	2657.00	308	401	93
24. Simdega	3774.00	138	159	21
Jharkhand	**79716.00**	**338**	**414**	**7.6**

Source : Census of India, 2011.

Sex Ratio : So far as the sex ratio is concerned the state of Jharkhand presents interesting features because of considerable concentration of aboriginals who hardly distinguish the importance of male and female. But in plain areas males are given more preference. At the same time increasing urbanisation in the region has also influenced sex ratio in different areas distinctly. Impact of out-migration of male workers from this region also influences sex ratio. This is why the areas having concentration of more urban population have comparatively poor sex ratio than the areas having considerable concentration of tribals having faith in Christianity. It may be noted that areas having Christian religion generally possess better sex ratio in comparison to areas having people of other religions.

In the year 2011 india recorded a sex ratio of 943 but this state had sex ratio of 948 (No. of Females per thousand males) slightly more than the average of the country. Certainly, the Jharkhand state provided remarkable spatial variation based on the percentage of tribal population. The best sex ratio has been recorded in the district of W. Singhbhum (1005), followed by Khunti (997), Simdega (997), Gumla (993), Pakur (989) and Lohardaga (985) etc. Districts having sex ratio less than that of the state's average include Dhanbad (909), Ramgarh (921), Bokaro (922), Deoghar (925), Palamu (928), Garhwa (935), Godda (938), Giridih (944), Hazaribagh (947) etc. (Table 12.6).

Table - 12.6
Sex Ratio : Jharkhand (2011)

Sl. No.	District	Male Population	Female Population	Sex Ratio (2011)
1.	Bokaro	T - 10,72,807 R - 5,54,954	T - 9,89,523 R - 5,23,732	T - 922 R - 944

		U - 5,17,853	U - 4,65,791	U - 899
2.	Chatra	T - 5,33,935	T - 5,08,951	T - 953
		R - 5,00,211	R - 4,79,721	R - 959
		U - 33,724	U - 29,230	U - 867
3.	Deoghar	T - 7,75,022	T - 7,17,051	T - 925
		R - 6,38,136	R - 5,95,576	R - 933
		U - 1,36,886	U - 1,21,475	U - 887
4.	Dhanbad	T - 14,05,956	T - 12,78,531	T - 909
		R - 5,81,956	R - 5,42,137	R - 932
		U - 8,24,000	U - 7,36,394	U - 894
5.	Dumka	T - 6,68,514	T - 6,52,928	T - 977
		R - 6,20,928	R - 6,10,336	R - 983
		U - 47,586	U - 42,592	U - 895
6.	Garhwa	T - 6,83,575	T - 6,39,209	T - 935
		R - 6,46,947	R - 6,06,167	R - 937
		U - 36,828	U - 33,042	U - 902
7.	Giridih	T - 12,58,098	T - 11,87,376	T - 944
		R - 11,49,608	R - 10,87,842	R - 946
		U - 1,08,490	U - 99,534	U - 917
8.	Godda	T - 6,77,927	T - 6,35,624	T - 938
		R - 6,43,907	R - 6,05,225	R - 940
		U - 34,020	U - 30,399	U - 894
9.	Gumla	T - 5,14,390	T - 5,10,823	T - 993
		R - 4,81,135	R - 4,78,997	R - 996
		U - 33,255	U - 31,826	U - 957
10.	Hazaribagh	T - 8,90,881	T - 8,43,614	T - 947
		R - 7,46,934	R - 7,12,254	R - 954
		U - 1,43,947	U - 1,31,360	U - 913
11.	Jamtara	T - 4,04,830	T - 3,86,212	t - 954
		R - 3,65,043	R - 3,50,253	R - 959
		U - 39,787	U - 35,959	U - 904
12.	Koderma	T - 367,222	T - 349,037	T - 950
		R - 2,93,595	R - 2,81,418	R - 959
		U - 73,627	U - 67,619	U - 918
13.	Khunti	T - 2,66,335	T - 2,65,550	T - 997
		R - 2,43,494	R - 2,43,409	R - 1000
		U - 22,841	U - 22,141	U - 969
14.	Latehar	T - 3,69,666	T - 3,59,312	T - 967
		R - 3,42,566	R - 3,32,554	R - 971
		U - 27,100	U - 24,758	U - 914
15.	Lohardaga	T - 2,32,629	T - 2,29,161	T - 985
		R - 2,03,255	R - 2,01,124	R - 990
		U - 29,374	U - 28,037	U 954
16.	Pakur	T - 4,52,661	T - 4,47,761	T - 989
		R - 4,18,251	R - 4,14,659	R - 991

	U - 34,410	U - 33,102	U - 962
17. Palamu	T - 10,06,302	T - 9,33,567	T - 928
	R - 8,87,551	R - 8,26,315	R - 931
	U - 1,18,751	U - 1,07,252	U - 903
18. E. Singhbhum	T - 11,76,902	T - 11,17,017	T - 949
	R - 5,14,498	R - 5,04,830	R - 981
	U - 6,62,404	U - 6,12,187	U - 924
19. W. Singhbhum	T - 7,49,385	T - 7,52,953	T - 1,005
	R - 6,37,657	R - 6,46,647	R - 1,014
	U - 1,11,728	U - 1,06,306	U - 951
20. Ramgarh	T - 4,94,230	T - 4,55,213	T - 921
	R - 2,72,167	R - 2,58,321	R - 949
	U - 2,22,063	U - 1,96,892	U - 887
21. Ranchi	T - 14,94,937	T - 14,19,316	T - 949
	R - 8,40,528	R - 8,16,390	R - 971
	U - 6,54,409	U - 6,02,926	U - 921
22. Sahebganj	T - 5,89,391	T - 5,61,176	T - 952
	R - 5,05,892	R - 4,85,009	R - 959
	U - 83,499	U - 76,167	U - 912
23. Saraikela-Kharsawan	T - 5,44,411	T - 5,20,645	T - 956
	R - 4,08,848	R - 3,97,462	R - 972
	U - 1,35,563	U - 1,23,183	U - 909
24. Simdega	T - 3,00,309	T - 2,99,269	T - 997
	R - 2,78,425	R - 2,78,209	R - 999
	U - 21,884	U - 21,060	U - 962
Jharkhand	**T - 1,69,30,315**	**T - 1,60,57,819**	**T - 949**
	R - 1,27,76,486	**R - 1,22,78,587**	**R - 961**
	U - 41,53,829	**U - 37,79,232**	**U - 910**

In case of rural population the highest sex ratio has been recorded in W. Singhbhum (1014), being followed by Khunti (1000), Simdega (999), Gumla (996), Pakur (991) and Lohardaga (990) etc. The state's average is 961. Comparatively, poor sex ratio in rural areas have been recorded in Palamu (931), Dhanbad (932) and Deoghar (933).

In case of urban population, the state average is 910. The highest sex ratio has been recorded in Khunti (969), being followed by Simdega (962), Pakur (962) and Lohardaga (954). Other districts having poor sex ratio in urban areas included Chatra (867), Ramgarh (887), Godda (894), Dhanbad (894), Dumka (895), Bokaro (899) etc.

Literacy : While analysing peopling of the area in earlier pages it has been emphasized that limited areas remained under the influence of Aryans and the people who firstly came here and developed habitation, and most areas remained sparsely populated because of considerable forest cover and very poor accessibility. As such the region also remained very backward in terms of the development of literacy. Even today the region is marked by level land of literacy particularly in rural areas. Certainly the development of industries and mining activities in mineralised and industrialised pockets of the region, the arrival of educated migrants considerably accelerated the literacy rate. The increase in the rate of literacy makes the role of development in the region obvious. In the year 1901 this region had only 1 to 3% literate people. The same has gone upto 66.41% in 2011. The highest literacy has been recorded in Ranchi (76.06%) district where one of the largest urban centre and state capital Ranchi is located. Other districts having higher percentage of literacy i.e. more than the state's average include E. Singhbhum (75.49%), Dhanbad (74.52%), Ramgarh (73.12%), Bokaro (72.01%), Hazaribagh (69.75%), Simdega (67.99%), Saraikela-Kharsawan (67.70%). These districts have a considerable urban growth where industrialization and urbanisation process has a significant bearing on the literacy. Districts having poor literacy rate in comparison to that of state's average include Pakur (48.82%), Godda (56.40%), W. Singhbhum (58.63%), Latehar (59.51%), Chatra (60.18%), Garhwa (60.33%), Dumka (61.02%) etc. (Table 12.7).

Table - 12.7
Literacy : Jharkhand (2011)

Sl. No.	District	Number of Literates			Literacy Rates (%)		
		Total	Male	Female	Total	Male	Female
1.	Bokaro	12,73,520	7,59,088	5,14,432	72.01	82.51	60.63
2.	Chatra	5,10,061	3,03,902	2,06,159	60.18	61.92	49.92
3.	Deoghar	7,93,538	4,89,837	3,03,701	64.85	76.85	51.80
4.	Dhanbad	17,22,204	10,14,950	7,07,254	74.52	83.81	64.29
5.	Dumka	6,72,409	4,06,275	2,66,134	61.02	72.96	48.22
6.	Garhwa	6,53,476	4,05,161	2,48,315	60.33	72.19	47.58
7.	Giridih	12,53,475	7,83,736	4,69,739	63.14	76.76	48.72
8.	Godda	6,04,519	3,76,212	2,28,307	56.40	67.84	44.14
9.	Gumla	5,59,720	3,21,795	2,37,925	65.73	75.55	55.90
10.	Hazaribagh	1,013,249	5,96,113	4,17,136	69.75	80.01	58.95

11. Jamtara	4,26,312	2,58,260	1,68,052	64.59	76.46	52.15
12. Koderma	3,90,249	2,38,780	1,51,469	66.84	79.78	53.23
13. Khunti	2,84,575	1,64,741	1,19,834	63.86	74.08	53.69
14. Latehar	3,50,682	2,09,706	1,40,976	59.51	69.97	48.68
15. Lohardaga	2,59,707	1,49,568	1,10,139	67.61	77.41	57.69
16. Pakur	3,52,881	2,06,989	1,45,892	48.82	57.06	40.52
17. Palamu	10,24,563	6,21,706	4,02,857	63.63	74.30	52.09
18. W. Singhbhum	7,27,561	4,39,273	2,88,288	58.63	71.13	46.25
19. E. Singhbhum	15,07,699	8,56,526	6,51,173	75.49	83.75	66.81
20. Ramgarh	5,96,497	3,50,031	2,46,466	73.12	82.44	63.09
21. Ranchi	19,11,433	10,85,244	8,26,159	76.06	84.26	67.44
22. Sahebganj	4,83,263	2,87,303	1,95,960	52.04	60.34	43.31
23. Saraikela-Kharsawan	6,12,993	3,65,332	2,47,661	67.70	79.03	55.88
24. Simdega	3,43,483	1,91,991	1,51,492	67.99	76.08	59.92
Jharkhand	**1,83,28,069**	**1,08,82,519**	**74,45,550**	**66.41**	**76.84**	**55.42**

Source : Primary Census Abstract, Total Table for Jharkhand, 2011.

In case of male litercy the highest percentage has been recorded in Ranchi (84.26%) being followed by Dhanbad (83.81%), E. Singhbhum (83.75%), Bokaro (82.51%), Ramgarh (82.44%), Hazaribagh (80.01%) and Koderma (79.86%) etc. Districts having less than that of state average include Pakur (57.06%), Sahebganj (60.34%), Chatra (61.92%), Godda (67.84%), Latehar (69.97%), W. Singhbhum (71.13%), Dumka (72.96%) etc. It may be noted that male literacy of Jharkhand excluding the children upto 7 years consisted of 76.84%.

It is true that female literacy is low in most regions of the country and Jharkhand is not an exception. Here, the female literacy is much lower because of more people having poverty, social backwardness and tribal heritage. Some people even did not give due importance to female literacy. It becomes obvious by the comparison of female literacy of the state. In this part female literacy is very much unevenly distributed. Industrialised, urbanised and cosmopolitan urban centres have higher female literacy than the remaining areas. The highest female literacy has been noted in Ranchi (67.44%) being followed by E. Singhbhum (66.81%), Dhanbad (64.29%), Ramgarh (63.09%), Bokaro (60.63%), Simdega (59.92%), Hazaribagh (58.95%) etc.

Occupational Structure : The state of Jharkhand exhibits variation in the percentage of workers in different occupations based on some diversities in mining, industrial, trade and commerce and other

Table - 12.8
Sex-wise Occupational Structure : Jharkhand (2011)

Sl. No.	District	Total Population	Workers (%)			Non-Workers	Category of Workers (Main & Marginal) (%)			
			Total Workers	Main Workers	Marginal Workers		Cultivators	Agri. Labourers	Household Workers	Other Workers
1.	Bokaro	T-20,62,330	T-33.23	18.44	14.79	66.77	20.07	18.44	3.16	58.33
		M-10,72,807	M-47.39	30.56	16.76	52.68	16.41	13.42	2.14	67.75
		F-9,89,593	F-17.96	5.30	12.66	82.04	30.51	32.80	5.30	31.40
2.	Chatra	T-10,42,886	38.13	18.75	19.39	61.87	30.00	47.40	3.57	19.03
		M-5,33,935	47.05	27.26	19.79	52.95	31.40	42.50	2.81	23.29
		F-5,08,951	28.78	9.81	18.97	71.22	27.60	55.80	4.88	11.72
3.	Deoghar	T-14,92,073	36.96	19.91	17.04	63.04	23.33	38.02	7.14	31.51
		M-7,75,022	49.51	30.74	18.77	50.49	23.52	33.41	4.58	38.49
		F-7,17,051	23.40	8.22	15.18	76.60	22.89	48.58	13.00	15.53
4.	Dhanbad	T-26,84,487	31.46	20.37	11.09	68.54	11.50	12.25	3.07	73.18
		M-14,05,956	46.86	33.92	12.95	53.14	8.09	8.45	2.48	80.97
		F-12,78,531	14.52	5.46	9.06	85.48	23.59	25.75	5.13	45.53
5.	Dumka	T-13,21,442	47.20	17.97	29.31	52.72	30.96	48.51	3.70	16.84
		M-6,68,514	54.91	25.70	29.21	45.09	31.74	43.24	3.29	21.74
		F-6,52,928	39.46	10.06	29.41	60.54	29.84	56.02	4.29	9.85
6.	Garhwa	T-13,22,784	43.20	13.24	29.96	56.80	19.87	57.52	2.54	20.07
		M-6,46,947	49.07	19.79	29.28	50.93	19.93	50.49	2.20	27.38

		F-6,06,167	36.92	6.23	30.68	63.08	19.80	67.50	3.02	9.69
7.	Giridih	T-24,45,474	42.38	16.88	25.49	57.62	35.70	34.99	3.14	26.17
		M-12,58,098	49.59	26.04	23.55	50.41	30.83	30.37	2.63	36.17
		F-11,87,376	34.73	7.18	27.55	65.27	43.06	41.99	3.91	11.04
8.	Godda	T-13,13,551	42.04	17.95	24.09	57.96	26.19	55.74	3.39	14.69
		M-6,77,927	50.75	26.91	23.83	49.25	28.67	49.37	2.90	19.06
		F-6,35,624	32.75	8.38	24.37	67.25	22.08	66.25	4.20	7.47
9.	Gumla	T-10,25,213	47.55	27.21	20.34	52.45	65.90	20.09	2.62	11.39
		M-5,14,390	51.47	34.23	17.25	48.53	65.93	16.85	2.61	14.62
		F-5,10,823	43.60	20.14	23.47	56.40	65.87	23.93	2.64	7.56
10.	Hazaribagh	T-17,34,995	37.74	18.87	18.87	62.26	37.04	26.33	2.47	34.16
		M-8,90,881	47.53	29.41	18.13	52.47	32.39	20.54	2.27	44.81
		F-8,43,614	27.40	7.74	19.66	72.60	45.56	36.95	2.83	14.65
11.	Jamtara	T-7,91,042	41.29	15.48	25.81	58.71	32.45	41.54	3.54	22.47
		M-4,04,830	53.04	23.74	29.30	46.96	31.87	37.27	3.16	27.70
		F-3,86,212	28.97	6.81	22.16	71.03	33.58	49.73	4.25	12.44
12.	Koderma	T-7,16,259	35.94	20.01	15.93	64.06	32.29	24.97	3.08	39.66
		M-3,67,222	47.38	32.10	15.28	52.62	27.01	20.51	2.50	49.97
		F-3,49,037	23.90	7.29	16.61	76.10	43.30	34.27	4.29	18.14
13.	Khunti	T-5,31,885	48.88	28.78	20.10	51.12	63.96	19.02	3.56	13.46
		M-2,66,335	52.99	35.22	17.76	47.01	64.45	15.28	3.27	17.01
		F-2,65,550	44.76	22.32	22.44	55.24	63.38	23.47	3.90	9.25
14.	Latehar	T-7,26,978	43.11	16.14	26.97	56.89	33.09	45.55	2.94	18.43
		M-3,69,666	48.72	22.42	26.30	51.28	33.78	40.70	2.37	23.14
		F-3,57,312	37.30	9.63	27.66	62.70	32.15	52.09	3.70	12.06
15.	Lohardaga	T-4,61,790	47.93	22.30	25.62	52.07	45.08	36.52	2.01	16.38
		M-2,32,629	52.27	30.52	21.75	47.73	46.38	29.35	1.98	22.29

		F-2,29,161	43.52	13.96	29.55	56.48	43.50	45.27	2.04	9.19
16.	Pakur	T-9,00,422	44.93	27.99	16.94	55.07	26.04	37.08	4.81	32.07
		M-4,52,661	51.70	37.69	14.61	48.30	30.41	32.74	1.99	3486
		F-4,47,761	38.09	18.79	19.30	61.91	20.03	43.03	8.69	28.25
17.	Palamu	T-19,39,869	36.76	14.62	22.14	63.24	18.73	53.42	2.79	25.06
		M-10,06,302	47.01	22.39	24.61	52.99	19.25	48.28	2.30	30.17
		F-9,33,567	25.72	6.25	19.47	74.28	17.72	63.54	3.76	14.98
18.	W. Singhbhum	T-15,02,338	46.25	22.83	23.42	53.75	35.14	38.49	3.57	22.79
		M-7,49,385	52.40	32.01	20.40	47.60	37.16	29.26	3.31	30.26
		F-7,52,953	40.13	13.70	26.43	59.87	32.51	50.49	3.91	13.09
19.	E. Singhbhum	T-22,93,919	36.50	23.09	13.41	63.50	13.39	25.06	2.56	58.99
		M-11,76,902	52.36	37.85	14.52	47.64	13.05	17.49	2.13	67.33
		F-11,17,017	19.78	7.54	12.24	80.22	14.33	46.18	3.75	35.74
20.	Ramgarh	T-9,49,443	32.87	23.29	9.59	67.13	29.08	15.00	2.67	53.25
		M-4,94,230	47.39	37.06	10.33	52.61	24.25	12.02	2.40	61.33
		F-4,55,213	17.12	8.34	8.78	82.88	43.59	23.96	3.49	28.96
21.	Ranchi	T-29,14,253	39.22	25.95	13.27	60.78	27.78	23.39	3.36	45.46
		M-14,94,937	49.77	37.06	12.71	50.23	24.88	17.88	2.98	54.26
		F-14,19,316	28.11	14.25	13.86	71.89	33.20	33.67	4.08	29.05
22.	Sahebganj	T-11,50,567	42.61	28.04	14.57	57.39	22.43	42.21	8.81	26.55
		M-5,89,391	50.17	38.47	11.70	49.83	25.80	40.41	2.65	31.14
		F-5,61,176	34.66	17.08	17.58	65.34	17.33	44.94	18.17	19.57
23.	Saraikela-Kharsawan	T-10,65,056	40.38	21.46	18.91	59.62	27.14	33.24	2.38	37.23
		M-5,44,411	52.03	33.26	18.78	47.97	26.78	24.64	2.01	46.57
		F-5,20,645	28.19	9.13	19.06	71.81	27.84	49.85	3.10	19.20

24. Simdega	T-5,99,578	48.14	25.94	22.20	51.86	54.75	31.26	2.78	11.22
	M-3,00,309	53.23	35.20	18.04	46.77	58.28	24.70	2.79	14.23
	F-2,99,269	43.03	16.64	26.39	56.97	50.35	39.40	2.76	7.48
Jharkhand	**T-3,29,88,134**	**39.71**	**20.67**	**19.04**	**60.29**	**29.12**	**33.87**	**3.47**	**33.53**
	M-1,69,30,315	**49.76**	**30.92**	**18.84**	**50.24**	**27.20**	**27.80**	**2.68**	**52.33**
	F-1,60,57,819	**29.10**	**9.87**	**19.24**	**70.90**	**32.60**	**44.81**	**4.91**	**17.67**

Source : Primary Census Abstracts, Census of India, Jharkhand-2011.

non-agricultural activities. Virtually agriculture related involvement of workers is dominant in all districts especially in rural areas but the mineralised and industrialised areas have involved workers engaged in mines and industries respectively. In the census of 2011 people have been grouped into workers, marginal workers and non-workers and workers have further been categorised into 1. Cultivators, 2. Agricultural labourers, 3. Household Workers and 4. Other Workers.

In the census records workers are considered those persons who are participating in any economically productive activity. Such participation may be physical or mental in nature. Work involves not only actual work but also effective supervision or direction of work. Main workers are those who have worked for the major part of the year preceding the date of enumeration. Major part of the year means six months i.e. 180 days or more during the reference year. Marginal workers are those workers who have not worked for the major part of the year i.e. those who have worked for less than 6 months. Non workers are those who have not worked any day during the concerned year. The above mentioned categories of workers are also broadly grouped into Primary, Secondary and Tertiary workers. In case of primary workers human activities concern with natural products in which no transformation of natural gift is done. But in case of Secondary activities construction and manufacturing activities are included in which natural production are transformed into different articles and tertiary activities include all other services and works other than primary and secondary (Table 12.8 and 12.9).

While considering the workers involvement in different categories of occupation different areas have portrayed significant variations. It becomes obvious by the analysis of the table.

In case of the percentage of workers not much difference has been recorded among the districts of the region. The highest percentage of workers has been recorded in the districts of Khunti (48.88%) and districts having the percentage of workers above 40% include Simdega (48.14%), Lohardaga (47.93%), Gumla (47.55%), Dumka (47.28%), W. Singhbhum (46.25%), Pakur (44.93%), Garhwa (43.20%), Latehar (43.11%), Sahebganj (42.61%), Giridih (42.38%), Godda (42.04%), Jamtara (41.29%) and Saraikela-Kharsawan (40.38%). Districts having higher percentage of workers are more

Table - 12.9
Occupational Structure : Jharkhand (2011)

Sl. No.	District	Total Population	Workers (%)			Non-Workers	Category of Workers (Main & Marginal) (%)			
			Total Workers	Main Workers	Marginal Workers		Cultivators	Agri. Labourers	Household Workers	Other Workers
1.	Bokaro	T-20,62,330	33.23	18.44	14.79	66.77	20.07	18.44	3.16	58.33
		R-10,78,686	38.26	15.40	22.87	61.74	30.56	28.51	3.29	37.24
		U-9,83,644	27.71	21.78	5.93	72.29	3.57	3.20	2.96	90.27
2.	Chatra	T-10,42,886	38.13	18.75	19.39	61.87	30.00	47.40	3.57	19.03
		R-9,79,932	38.78	18.48	20.30	61.22	31.34	49.39	3.48	15.92
		U-62,954	28.04	22.93	5.11	71.96	3.85	4.65	5.62	85.87
3.	Deoghar	T-14,92,073	36.96	19.91	17.04	63.04	23.33	38.02	7.14	31.51
		R-12,33,712	38.75	18.79	19.96	61.25	26.46	43.62	7.46	23.69
		U-2,58,361	28.39	25.27	3.12	71.61	2.93	4.49	5.05	87.53
4.	Dhanbad	T-26,84,487	31.46	20.37	11.09	68.54	11.50	12.25	3.07	73.18
		R-11,24,093	36.58	17.36	19.22	63.42	21.87	22.80	3.30	27.71
		U-15,60,394	27.77	22.53	5.24	72.23	5.39	7.20	9.25	78.06
5.	Dumka	T-13,21,442	47.28	17.97	29.31	52.72	30.96	48.51	3.70	16.84
		R-12,31,264	48.52	17.56	30.96	51.48	32.17	50.35	3.64	13.84
		U-90,178	30.39	23.59	6.80	69.61	4.47	8.48	4.87	82.17
6.	Garhwa	T-13,22,784	43.20	13.24	29.96	56.80	19.87	57.52	2.54	20.07
		R-12,53,114	49.93	12.95	30.98	50.07	20.42	58.70	2.51	10.84

	U-69,670	30.07	18.39	11.68	69.93	5.50	26.41	3.12	64.97
7. Giridih	T-24,45,474	42.38	16.88	25.49	57.62	35.70	34.99	3.14	26.17
	R-22,37,450	43.65	16.31	27.33	56.35	37.58	36.79	3.07	22.55
	U-2,08,024	28.72	23.02	5.70	71.28	4.83	5.60	4.28	85.30
8. Godda	T-13,13,551	42.04	17.95	24.09	57.96	26.19	55.74	3.39	14.69
	R-12,49,132	42.79	17.68	25.10	57.21	26.77	57.23	3.37	12.63
	U-64,419	27.07	23.07	4.47	72.93	8.70	10.83	4.08	78.11
9. Gumla	T-10,25,213	47.55	27.21	20.34	52.45	65.90	20.09	2.62	11.39
	R-9,60,132	48.92	27.52	21.39	51.08	68.14	20.62	2.45	8.79
	U-65,081	27.41	22.53	4.88	72.59	6.90	5.95	7.12	80.03
10. Hazaribagh	T-17,34,495	37.74	18.87	18.87	62.26	37.04	26.33	2.47	34.16
	R-14,59,188	39.65	18.04	21.61	60.35	41.45	29.38	2.45	26.72
	U-2,75,307	27.61	23.26	4.35	72.39	3.43	3.15	2.60	90.82
11. Jamtara	T-7,91,042	41.29	15.48	25.81	58.71	32.45	41.54	3.54	22.47
	R-7,15,296	42.42	14.69	27.73	57.58	34.69	44.27	3.58	17.46
	U-75,746	30.66	22.93	7.73	69.34	3.30	5.84	2.94	87.95
12. Koderma	T-7,16,259	35.94	20.01	15.93	64.06	32.29	24.97	3.08	39.66
	R-5,75,013	37.97	19.57	18.40	62.03	37.29	28.37	2.87	31.46
	U-1,41,246	27.68	21.80	26.97	72.32	4.37	5.96	4.25	85.41
13. Khunti	T-5,31,885	48.88	28.78	20.10	51.12	63.96	19.02	3.56	13.46
	R-4,86,903	50.52	29.28	21.24	49.48	26.12	19.80	3.48	9
	U-44,982	31.13	23.34	7.79	68.87	14.49	5.38	4.82	75.31
14. Latehar	T-7,26,978	43.11	16.14	26.97	56.89	33.09	45.55	2.94	18.43
	R-6,75,120	14.47	15.76	64.27	85.53	34.57	47.24	2.89	15.30
	U-51,858	30.21	21.08	9.13	69.79	4.93	13.30	3.86	77.90
15. Lohardaga	T-4,61,790	47.93	22.30	25.62	52.07	45.08	36.52	2.01	16.38

	R-4,04,379	50.52	22.12	28.41	49.48	48.40	39.06	1.89	10.65
	U-57,411	29.67	23.64	6.03	70.33	5.25	6.09	3.44	85.22
16. Pakur	T-9,00,422	44.93	27.99	16.94	55.07	26.04	37.08	4.81	32.07
	R-8,32,910	45.71	27.67	18.04	54.29	27.56	39.12	4.52	44.56
	U-67,512	35.32	31.91	3.41	64.68	1.73	4.53	9.48	84.26
17. Palamu	T-19,39,869	36.76	14.62	22.14	63.24	18.73	53.42	2.79	25.06
	R-17,13,866	37.85	14.01	23.84	62.15	20.27	57.26	2.66	19.80
	U-2,26,003	28.51	19.28	9.23	71.49	3.19	14.78	4.10	77.94
18. W. Singhbhum	T-15,02,338	46.25	22.83	23.42	53.75	35.14	38.49	3.57	22.79
	R-12,84,304	4878	22.22	26.55	51.22	38.77	42.14	3.60	5.49
	U-2,18,034	31.38	26.42	4.96	68.62	1.90	5.11	3.28	89:71
19. E. Singhbhum	T-22,93,919	36.50	23.09	13.41	63.50	13.39	25.06	2.56	58.99
	R-10,19,328	43.84	16.92	26.92	56.16	43.39	21.63	2.83	32.15
	U-12,74,591	30.62	28.02	2.61	69.38	0.54	1.66	2.09	95.49
20. Ramgarh	T-9,49,443	32.87	23.29	9.59	67.13	29.08	15.00	2.67	53.25
	R-5,30,488	36.99	23.27	13.72	63.01	43.39	21.63	2.83	32.15
	U-4,18,955	27.67	23.31	4.35	72.33	4.86	3.77	2.40	88.97
21. Ranchi	T-29,14,253	39.22	25.95	13.27	60.78	27.78	23.39	3.36	45.46
	R-16,56,918	45.34	25.78	19.55	54.66	41.07	33.53	3.22	22.19
	U-12,57,335	31.15	26.16	4.99	68.85	2.29	3.95	5.67	90.11
22. Sahebganj	T-11,50,567	42.61	28.04	14.57	57.39	22.43	42.21	8.81	26.55
	R-9,90,901	44.53	28.53	16.00	55.47	24.35	45.57	8.97	21.11
	U-1,59 666	30.65	24.96	5.69	69.35	5.13	11.92	7.39	75.56
23. Saraikela-Kharsawan	T-10,65,056	40.38	21.46	18.91	59.62	27.14	33.24	2.38	37.23
	R-8,06,310	43.58	19.56	24.02	56.42	32.95	40.19	2.45	24.41

	U-2,58,746	30.40	27.39	3.01	69.60	1.19	2.24	2.10	94.47
24. Simdega	T-5,99,578	48.14	25.94	22.20	51.86	54.75	31.26	2.78	11.22
	R-5,56,634	49.45	26.05	23.40	50.55	56.73	32.02	2.79	8.45
	U-42,944	31.18	24.41	6.78	68.82	13.89	15.59	2.57	67.95
Jharkhand	**T-3,99,88,134**	**39.71**	**20.67**	**19.04**	**60.29**	**29.12**	**33.87**	**3.47**	**33.53**
	R-2,50,55,073	**43.01**	**19.50**	**23.51**	**56.99**	**34.82**	**40.26**	**3.52**	**21.40**
	U-79,33,061	**29.26**	**24.35**	**4.91**	**70.74**	**2.68**	**4.17**	**3.27**	**89.89**

in agriculture dominated districts. In case of agricultural occupation even elderly people and teenagers remain involved especially in main agricultural season but in case of industrial and mining dominated areas only certain groups of people of working age group remain involved. In case of rural workers, the percentage of workers in different districts has not recorded much variation. The highest percentage of rural workers has been recorded in Khunti (50.52%) and Lohardaga (50.32%) being followed by Garhwa (49.93%), Simdega (49.15%), Gumla (48.92%), Dumka (48.26%) etc. The least percentage of rural workers has been recorded in Latehar (14.47%). In case of urban population, the highest percentage of workers has been recorded in Pakur (35.32%). All the districts have urban workers varying between 27% to 35%. Here, the percentage of workers is comparatively smaller because urban centres also have good number of children and teenagers associated with studying in educational institutions and elderly people rarely participate in active works.

Marginal workers are not very significant in this region. The highest percentage has been recorded in Garhwa (29.96%) being followed by Dumka (29.31%), Latehar (26.97%), Jamtara (25.81%), Lohardaga (25.62%), Giridih (25.49%) etc. In case of rural marginal workers the highest percentage has been in Latehar (64.27%), being followed by Garhwa (30.98%), Dumka (30.96%) etc. In urban areas, the percentage of marginal workers is highest in Koderma district (26.97%). Although, all other districts have a lesser percentage (less than 12%), of urban marginal workers. This is because of the fact that in urban areas only a few persons prefer to work irregularly but in rural areas a good number of persons work in the main seasons and remain idle during non-agricultural reasons.

In rural areas cultivators and agricultural labourers share above 70% of the total workers (main & marginal) and even less than 30% are engaged in other activities in agriculture dominated districts. The district of Simdega has (86.01%) as cultivators and agricultural labourers being followed by Gumla (85.99%), Khunti (82.98%), Godda (81.93%) etc. It is interesting to note that districts having considerable urban, mining, industrial growth have poor percentage of cultivators. The district of Dhanbad has only 11.50% cultivators. In rural areas this district has 21.87% cultivators. But in most districts rural areas have high percentage of cultivators and urban areas have low percentage of cultivators. Only those districts have significant percentage of cultivators in urban areas where urban centres are of smaller size which are overgrown villages and rural character still persists. Similar to cultivators, agricultural labourers have higher percentage in those districts where non-agricultural activities are comparatively less developed. The highest percentage oof agricultural labourers has been recorded in Garhwa (57.52%) where rural areas have 58.70% agricultural labourers and urban areas have 26.41%. The lowest percentage of agricultural labourers has been in Dhanbad (12.25%). In this district rural areas have only 22.80% agricultural labourers and urban areas have only 7.30% agricultural labourers. Other districts having lower percentage of agricultural labourers are Ramgarh (15.00%), Bokaro (18.44%) Khunti (19.02%) etc. It may be noted that districts having more small cultivators have lower percentage of labourers because they manage their agricultural works themselves but in districts where resourceful cultivators remain in considerable numbers the need of labourers are more realised.

No doubt, some areas of the state have registered phenomenal increase in household industries but the percentage of workers engaged in these activities is not very high. The highest percentage of workers has been in Sahebganj (8.81%) followed by Deoghar (7.14%). These two districts have presence of good number of industrial estates and several private handlooms manufacturing, processing and related industries have engaged a good number of workers in household works. Lohardaga (2.01%) have least percentage of people engaged as household workers.

Other workers engaged in services activities are found mostly in urban centres and rural market. As such districts having considerable urban population record high percentage. The highest percentage has been recorded in the district of Dhanbad (73.18%) where services like education, helath, recreation, entertainment and administration etc. have developed in good proportion which is followed by E. Singhbhum (58.99%), Bokaro (58.33%), Ramgarh (53.25%), Ranchi (45.46%). These districts have urban centres having urban amenities in substantial proportion. Even rural areas of all these districts have other workers more than 27% because even in rural areas facilities like educational institution health care and other civil amenities have developed.

While comparing occupational structure of males and females, some diversities have emerged. The percentage of workers among female in much lower. Only 5 districts consisting of Khunti (44.74%), Gumla (43.6%), Lohardaga (43.5%), Simdega (43.03%) and W. Singhbhum (40.13%) have female workers slightly more than 40%. Only 7 districts have female workers ranging between 30% to 40% and the remaining districts have less than 30%. But in case of male workers all districts have male workers more than 46%. First of all, female worker is smaller because even today females are supposed to look after domestic affairs but female of poor families are participating in economic activities. Districts having less development have higher percentage of female workers such as Khunti, Gumla, Lohardaga etc. However in case of marginal workers females of all districts have comparatively much higher percentage. Garhwa (30.68%) has the maximum percentage of female marginal workers in the state. It may be noted that female have higher percentage of marginal workers because they do not participate in economic activities throughout the year. They help the farmers only in main season and remain idle in the remaining period but in case of males most of them remain engaged throughout the year.

In case of percentage of cultivators not much contrasting picture has emerged between male and female cultivators. The highest percentage of male cultivators is 65.93% in Gumla district and the lowest percentage has been reported in Dhanbad district (8.09%). But in case of female cultivators 65.87% has been the highest again in Gumla district and 14.33% is the lowest in the E. Singhbhum district. The highest percentage of male agricultural labourers is in Garhwa (50.49%) and lowest in Dhanbad (8.45%). But female agricultural labourers has been recorded again in the Garhwa district (67.50%). It may be noted that females are more involved in agricultural related activities than non-agricultural activities. But males are also involved in non-agricultural activities like mining, industries, transport etc. and have insignificant differences.

The household workers are found both in males and females. They do not reveal much contrasting percentage. The highest male household workers percentage is found in Deoghar (4.58%) whereas lowest percentage has been recorded in Lohardaga (1.98%)

district. Similarly, the highest female household workers percentage has been recorded in Sahebganj (18.17%) and lowest in Lohardaga (2.04%) district. Since, household industries are mostly run in houses male and female both participate without much hesitation. In household industries, females are generally preferred for their fine works, embroidery and hence their participation is generally higher in comparison to the males.

Schedule Caste Population :

The total number of scheduled caste population in the state of Jharkhand is 39,85,644 which constitutes 12.08% of the total population of the state. Out of the total scheduled caste population of 39,85,644 of Jharkhand 31,52,863 (79.11%) reside in the rural areas and 8,32,781 (20.89%) in urban areas of the state. The distribution of these people in the state is also uneven. The concentration of scheduled caste people is more marked in the districts located in northern half of the state like Palamu, Dhanbad, Giridih, Chatra, Hazaribagh, Simdega etc. They have comparatively lower percentage in the tribal dominated areas in the south like Gumla, Lohardaga, E. & W. Singhbhum etc. The highest percentage of schedule caste population of the total population of the district has been found in Saraikela-Kharsawan (34.05%) being followed by Chatra (32.65%), Palamu (27.65%), Garhwa (24.91%), Latehar (21.31%) etc. Other districts having less than 20% and more than 12% schedule caste population include Hazaribagh (17.50%), Dhanbad (16.29%), Koderma (15.22%), Bokaro (14.51%), Giridih (13.31%) and Deoghar (12.74%). The remaining districts have schedule caste population less than 12%. The lowest percentage has been recorded in Pakur (3.16%) being followed by Gumla (3.17%), Lohardaga (3.32%), W. Singhbhum (3.79%), Khunti (4.52%) and E. Singhbhum (4.86%). All these districts have good concentration of aborginals and less number of scheduled caste migration has taken place from the neighbouring states. In rural areas schedule caste population is higher than the urban areas in some districts and in the remaining districts urban areas have higher percentage of schedule caste than the urban areas. Districts having higher percentage of schedule caste in rural areas than in urban areas include Chatra, Palamu, Garhwa, Latehar, Koderma etc. Districts having more schedule castes in urban areas than the rural areas in terms of percentage consisted of Dhanbad, Ramgarh,

Jamtara, Sahebganj, Dumka, Khunti etc. In urban areas schedule caste people are employed in municipality or corporation for cleaniness and labour work. The analysis of the uneven distribution also becomes obvoius by their uneven distribution in the state. The highest concentration of schedule caste population in the state is found in Palamu where 13.46% scheduled caste of the state is found. This is followed by Dhanbad (10.97%), Chatra (8.54%) Giridih (8.17%), Hazaribagh (7.62%) and Bokaro (7.50%). These six districts have more than 56% schedule caste population of the state. (Table 12.10).

Table - 12.10
Schedule Caste Population : Jharkhand (2011)

Sl. No.	District		Total Population	SC Population	%SC Population to Total population of the District	%SC Population to total population of State
1.	Bokaro	T	20,62,330	2,99,227	14.51	7.50
		R	10,78,686	1,75,690	16.29	5.57
		U	9,83,644	1,23,537	12.56	14.83
2.	Chatra	T	10,42,886	3,40,553	32.65	8.54
		R	9,79,932	3,32,431	33.92	10.54
		U	62,954	8,122	12.90	0.98
3.	Deoghar	T	14,92,073	1,90,036	12.74	4.77
		R	12,33,712	1,59,077	12.89	5.05
		U	2,58,361	30,959	11.98	3.72
4.	Dhanbad	T	26,84,487	4,37,309	16.29	10.97
		R	11,24,093	1,63,090	14.51	5.17
		U	15,60,394	2,74,219	17.57	32.93
5.	Dumka	T	13,21,442	79,614	6.02	2.00
		R	12,31,264	71,695	5.82	2.27
		U	90,178	7,919	8.78	0.95
6.	Garhwa	T	13,22,784	3,19,946	24.19	8.03
		R	12,53,114	3,08,843	23.35	9.80
		U	69,670	11,103	15.94	1.33
7.	Giridih	T	24,45,474	3,25,493	13.31	8.17
		R	22,37,450	3,01,327	13.47	9.56
		U	2,08,024	24,166	11.62	2.90
8.	Godda	T	13,13,551	1,15,567	8.80	2.90
		R	12,49,132	1,11,055	8.89	3.52
		U	64,419	4,512	7.00	0.54

9. Gumla	T	10,25,213	32,459	3.17	0.81
	R	9,60,132	29,506	3.07	0.94
	U	65,081	2,953	4.54	0.35
10. Hazaribagh	T	17,34,495	3,03,515	17.50	7.62
	R	14,59,188	2,72,868	18.70	8.65
	U	2,75,307	30,647	11.13	3.68
11. Jamtara	T	7,91,042	72,885	9.21	1.83
	R	7,15,296	63,026	8.81	2.00
	U	75,746	9,859	13.02	1.18
12. Koderma	T	7,16,259	1,09,003	15.22	2.73
	R	5,75,013	93,885	16.33	2.98
	U	1,41,246	15,118	10.73	1.82
13. Khunti	T	5,31,885	24,037	4.52	0.60
	R	4,86,903	21,223	4.36	0.67
	U	44,982	2,814	6.26	0.34
14. Latehar	T	7,26,978	1,54,910	21.31	3.89
	R	6,75,120	1,46,213	21.66	4.64
	U	51,858	8,697	16.77	1.04
15. Lohardaga	T	4,61,790	15,330	3.32	0.38
	R	4,04,379	13,035	3.22	0.41
	U	57,411	2,295	4.00	0.27
16. Pakur	T	9,00,422	28,469	3.16	0.71
	R	8,32,910	25,177	3.02	0.80
	U	67,512	3,292	4.88	0.40
17. Palamu	T	19,39,869	5,36,382	27.65	13.46
	R	17,13,866	5,03,315	29.37	15.96
	U	2,26,003	33,067	14.63	3.97
18. W. Singhbhum	T	15,02,338	56,986	3.79	1.43
	R	12,84,304	38,215	2.98	1.21
	U	2,18,034	18,771	8.61	2.25
19. E. Singhbhum	T	22,93,919	1,11,414	4.86	2.80
	R	10,19,328	44,771	4.39	1.42
	U	12,74,591	66,643	5.23	8.00
20. Ramgarh	T	9,49,443	1,06,356	11.20	2.67
	R	5,30,488	50,227	9.47	1.59
	U	4,18,955	56,129	13.40	6.74
21. Ranchi	T	29,14,253	1,52,943	5.25	3.84
	R	16,56,918	84,270	5.09	2.67
	U	12,57,335	6,873	0.55	0.83
22. Sahebganj	T	11,50,567	72,341	6.29	1.82
	R	9,90,901	57,490	5.80	1.82
	U	1,59,666	14,851	9.30	1.78
23. Saraikela-Kharsawan	T	10,65,056	56,195	34.05	5.28
	R	8,06,310	43,683	5.42	1.39

	U	2,58,746	12,512	4.84	1.50
24. Simdega	T	5,99,578	44,674	7.45	7.45
	R	5,56,634	42,751	7.68	1.36
	U	12,944	1,923	4.48	0.23
Jharkhand	**T**	**3,29,88,134**	**39,85,644**	**12.08**	**100.00**
	R	**2,50,55,073**	**31,52,863**	**12.58**	**100.00**
	U	**79,33,061**	**8,32,781**	**10.50**	**100.00**

*Primary Census Abstract, 2011.

Migration :

Accounts of human migration and its socio-economic implications are immense. Certainly during earlier periods migration phenomena did not play important role in influencing socio-economic conditions of the region but during last 100 years impact of human migration has become more complex and have influenced development of the region in terms of economy, social ecology and even way of life of the people. In early days this region was out of focus of the invaders, outside rulers because of intense forest cover and least development of transport network. During those days only aboriginals used to move from one place to the another based on forest products and hunting materials. Encroachment of people from outside region began when the vast Ganga Plain started getting Aryanised. The struggle between the Aryans and the Non-Aryans (Dravidians) caused movement of Dravidians from the plain areas to the forest clad areas. The evidence of movement of people from the plain to the northern portion of Palamu, Hazaribag and peripheral areas are available during Budhist period. Even during Medieval period this region did not attract Mughals in the interior areas of the state. Only the peripheral parts were occupied by them. It is during British regime when the entire region got traversed by transport network for administration and revenue collection and tapping resources of the region. In this process a good number of people involved in administration mining, education, trade and commerce and other activities came across different portions of the region and some of them got attracted by the resource potentials, cheaper land, chances of accumulating wealth and developing trade and commerce. All centres of administration like district, subdivision, revenue thana, nodes of transport and other places attracted people from different states. More remarkable immigration in this region took place during independence days when mining, industrial, urban growth got accelerated. During this period, the mineralised, industrialised and urbanised areas have got so much assemblage of people of different areas having varied cultural background that traditional socio-economic conditions have considerably been transformed. Some places have lost the original character and given rise to cosmopolitan nature. It is true that these areas have not only got the migrants of only other states but have also got the aboriginals having proficiency in industrial, mining, educational and other activities. Uneducated people have also moved to these areas for involving themselves in unskilled work.

No doubt, this region has been attracting migrants from other states as well as neighbouring countries from the very beginning but during recent years the rate of immigration has increased enormously. This region also witnessed out migration of people from this region. Previously poor people especially labourers used to go to free growing areas of the country and some persons used to go to the industries and urban centres of Hooghly region for getting labour work. Educationists have out-migrated to different parts of the country because they could get job easily on the basis of reservation. These days directional changes in migration have taken place. Workers are hardly moving towards north-east and east either in tree growing areas or to the industries of Hooghly region but they are now moving in more number towards north and north-west. Good number of workers of this region regularly go to Punjab and Haryana for labour work and also in the major cities of Delhi, Kanpur for driving rickshaw and getting construction work. Labourers from this region are getting labour work in brick industries developed in the Ganga Plain especially along the river Ganga.

Similar to other geographical regions this state has witnessed different types of human migration. Normally migration is categorised on the basis of space, time, volume, direction, cause, etc. Important types are inter-continental, international, inter-state, local consisting of short distance, inter-village, inter district, rural to rural, rural to urban, urban to urban, urban to rural, etc. Among these, rural urban migration has been more important. Besides, labour out-migration has become an important phenomenon.

Local migration in the form of village to village migration or district to district migration has been important from the beginning itself. Inter tribal rivalries used to motivate people to move from one village to another. Cultivators used to shift from one place to another in search of virgin agricultural land and abandoning the infertile land. In this process they used to move from one village to another. These days job opportunities or business or some other work also have caused movement of people from their native places to other destinations. Sometimes marriage also motivates even male persons to desert their native places for their wives places. Most of the females migrate from their native places to other destinations after marriage. In this process less educated and poor people manage marriage locally and female migration caused by marriage remains confined locally. But in case of educated families and socially conscious families they arrange marriage even crossing the districts and long distance migration takes place.

Inter-district migration has also been not less remarkable but this type of migration has been in the field of industrial labour work or labour in construction works or concerning business, professional work, etc. This state has recorded lop-sided development in the field of mining, industrial and urban growth and the mineralised, industrialised and urbanised districts have attracted workers or migrants from the adjoining districts of the state. A few urban centres having better development in terms of educational, health, recreational and other fields have graviated resourceful villagers to these urban centres.

Rural-urban migration has also become an important phenomenon in this region because of fastly growing several cities and towns. These urban centres have job opportunities for persons related to construction work, business, industrial work, professional work, etc. and Ranchi, Bokaro, Dhanbad, Hazaribag, Jamshedpur etc. have become focal point of rural out- migrants. Rural-urban migration has intensified during recent years because of several push factors in the villages, and several attractions in the urban areas. It may be noted that rural areas of this region have also not remained very peaceful and self dependent. Increasing pressure of man on land is reducing per capita agricultural land. Not all families manage to grow food crops for their requirements. Virtually in all villages a good number of persons remain with little land with which they can not manage proper living. In such case they prefer to move to urban centres where they manage to get work and earn more money. In rural areas villagers sometimes loose their land in the process of taking loan from the local money lenders and fail to pay back the loan due to heavy interest. Ultimately they become landless. Sometimes some poor farmers manage to move out for other destination for earning more money so that they can pay the loan. In few cases out migration has been also due to social crime like extra-marital relations or some other social rules. In such cases these people are socially outcasted from the society and social inter-action of such persons are restricted. In such case these socially out caste people prefer to move out for other destination so that they can manage their living like others. Main cause of out migration from the village to the towns or industrial centres has been the intention of earning more money, getting non-agricultural work, and getting freedom from the problems of rural areas and transforming their socio-economic conditions. It may be noted that rural out-migrants have experienced comparatively better development than non-migrant families in the rural areas. Migrants manage to earn more money elsewhere and solve the economic problems of the family, enable the family members to purchase land, construct better house, provide better education to their children and to live in a better way.

It is not only the push factors which are compelling people to move out from their native places to urban or industrial or some other centres. Virtually there are several attractive forces prevailing in the urban centres which attract rural people. These attractions are the availability of white collar job, attractive business, lucrative industries, urban amenities, better chances of leading comfortable life and proper opportunities for the prosperity of the children, etc. This is why rural people are rushing towards urban centres. The resourceful persons are purchasing land and constructing house in the urban areas for keeping their family member for availing urban opportunities. Some of them are even disposing or industries or in some other economic gainful works. In this process the urban centres are gaining the capital of rural areas and development of rural areas get adversely affected. These days people have become more conscious towards education, health and modernised living which are only possible in the urban centres.

These attractions are becoming more effective and motivating rural people to move to urban centres.

Table - 12.11

Reasons for Migration : Jharkhand (2001)

Sl. No.		Males	Females	Total Persons	% of Total Migration
1.	Work / Employment	6,18,162	55,137	6,73,299	9.06
2.	Business	28,007	5,785	33,792	0.45
3.	Education	52,381	18,612	70,993	0.96
4.	Marriage	49,346	46,85,650	47,34,996	63.74
5.	Moved after birth	39,518	27,423	66,941	0.90
6.	Moved with household	2,15,768	4,83,742	7,79,510	10.49
7.	Others	5,17,155	5,51,473	10,68,628	14.39
	Jharkhand	16,00,337	58,27,822	74,28,159	100.00

Source : Census of India, 2001.

The analysis of the table 12.11 reveals the fact that there were 74,28,159 migrants in the districts of the state, enumerated in the census of 2001. About 78.45% migrants were females which mainly occured as girls move to groom's house after marriage in our society. Marriage is still dominant reason for migration in our society as this along constitute 63.74% of total migrants in the state. people migrate to other places in search of job, business, education also, but these reasons are not very significant in the case of Jharkhand. Sometimes due to bad law & orders problems, people move with household properties in safer places for better livelihood & safety considerations.

Table - 12.12
Population classified by Place of Birth : Jharkhand
(States in India beyond the state of enumeration) (2001)

Sl. No.	State	Males	Females	Total Migrants
1.	Bihar	5,19,623	6,03,760	11,23,383
2.	W. Bengal	87,788	2,62,318	3,50,106
3.	U.P.	54,112	54,986	1,09,098
4.	Orissa	22,906	59,709	82,615
5.	Chattisgarh	8,394	20,815	29,209
6.	Others	39,977	46,867	86,844
	Jharkhand	7,32,800	10,48,455	17,81,255

Source : Census of India, 2001.

The analysis of the table 12.12 reveals the fact that people from Bihar are maximum in numbers in the state enumerated in the census of 2001. This is quite obvious due to the fact that Jharkhand and Bihar were once united before 2000 and people from present day Bihar worked here in large number in jobs, business, mines and industries and in course of time they settled here in large numbers. W. Bengal and U.P. comes next from where people moved in large number to work in jobs & mines.

A large number of people especially from eastern U.P. worked in coal mines of Jharkhand. The state of Orissa & Chattisgarh owing to their proximity & measures provided a large number of people to work in mines & industries.

West Bengal serves as a house to the highest number of migrants. It has a developed industrial sector and a sprawling metropolitan capital. Bihar and Uttar Pradesh have been identified as states with the second and third highest inflow of migrants from Jharkhand respectively. A high net migration, greater than 15% was observed in the districts of Dhanbad, Lohardaga and Gumla. More than 90,000 people travel from Dhanbad to Howrah each year, which has been identified as the top route for migration in India.

POPULATION CHARACTERISTICS, 2011 :

Variation in Population (2001-2011) :

Despite efforts of the government and the private agencies for curbing the explosive growth of population the areas has registered 22.42% variation in population during 2001-2011 decade. The highest increase has been recorded in Sahebganj district (35.74%) which may be due to the immigration of population from other areas. Other districts having variation in population more than the average of the state included Latehar (29.61%), Chatra (29.05%), Pakur (28.33%), Deoghar (28.03%), Garhwa (27.75%) and Lohardaga (26.68%); Dhanbad district (11.99%) experienced the least population variation in the state. Other districts experiencing comparatively lower growth of population consisted of Ramgarh, E. Singhbhum, Bokaro, Simdega, Dumka etc. Out of these districts some districts exhibit lower growth due to lack of further development of mines and industries which usually attract migrants. Awareness towards family planning has been more effective in some districts in restricting the population growth. The table no. 12.13 reveals detail information about the variation of population during 2001-2011.

Table 12.13

Variation in Population : Jharkhand (2001-2011)

Sl. No.	District	Population 2001	Population 2011	Difference (2011-2001)	% Variation
1.	Bokaro	17,77,662	20,62,330	2,84,668	16.01
2.	Chatra	7,91,434	10,42,886	2,51,452	29.05
3.	Deoghar	11,65,390	14,92,073	3,26,683	28.03
4.	Dhanbad	23,97,102	26,84,487	2,87,385	11.99
5.	Dumka	11,06,521	13,21,442	2,14,921	19.42
6.	Garhwa	10,35,464	13,22,784	2,87,320	27.75
7.	Giridih	19,04,430	24,45,474	5,41,044	21.78
8.	Godda	10,47,939	13,13,551	2,65,612	25.35
9.	Gumla	8,32,447	10,25,213	1,92,766	23.16
10.	Hazaribagh	22,77,475	17,34,495	5,42,980	25.79
11.	Jamtara	6,53,081	7,91,042	1,37,961	25.46
12.	Koderma	4,99,403	7,16,259	2,16,856	32.42
13.	Khunti	-	5,31,885	-	22.32
14.	Latehar	5,60,894	7,26,978	1,66,084	29.61
15.	Lohardaga	3,64,521	4,61,790	97,269	26.68

16.	Pakur	7,01,664	9,00,422	1,98,758	28.33
17.	Palamu	15,37,465	19,39,869	4,02,404	26.17
18.	W. Singhbhum	12,33,945	15,02,338	2,68,393	21.75
19.	E. Singhbhum	19,82,988	22,93,919	3,10,931	15.68
20.	Ramgarh	-	9,49,493	-	13.10
21.	Ranchi	27,85,064	29,14,253	1,29,189	24.00
22.	Sahebganj	9,27,770	11,50,567	2,22,797	35.74
23.	Saraikela-Kharsawan	8,48,850	10,65,056	2,16,206	25.47
24.	Simdega	5,14,320	5,99,578	85,258	16.58
	Jharkhand	**2,69,45,829**	**3,29,88,134**	**60,42,305**	**22.42**

Source : Census of India, 2001 & 2011.

Distributional Pattern :

The analysis of the table no. 12.14 reveals the features concerning distributional pattern of population. It further reveals the fact that no significant changes have emerged in the last census of 2011 except increase in population. The largest share of population in 2011 has been recorded in Ranchi (8.83%), followed

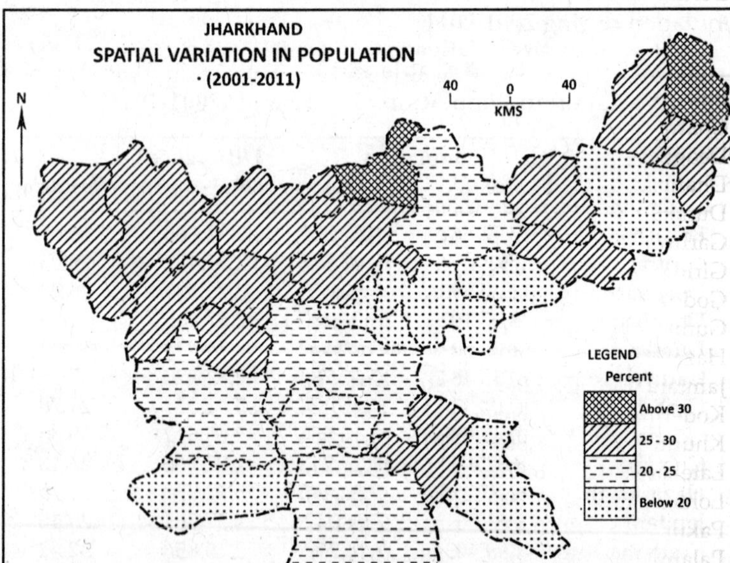

Fig. 16

by Dhanbad (8.14%), Giridih (7.41%), E. Singhbhum (6.95%), Bokaro (6.25%), Palamu (5.88%) etc. Districts having moderate share of population consisted of Hazaribagh, Deoghar, Dumka, Garhwa, Sahebganj etc. Lohardaga is the smaller district in terms of total population which accounts just 1.40% population of the state in the census year 2011. It is closely followed by Khunti (1.61%). Districts having higher population share are comparativly more developed in mining, industrial and other developments. In case of the share of male population the highest position is again held by Ranchi (8.93%) being followed by Dhanbad (8.30%), Giridih (7.43%), E. Singhbhum (6.95%) etc. Some of these districts have better concentration of urban population where males remain more in number than females. The share of females has registered highest in the district of Ranchi (8.93%). Other districts consisting of comparatively better share in terms of female population consist of Dhanbad, Giridih, E. Singhbhum, Bokaro etc.

Table - 12.14
Distribution Pattern : Jharkhand (2011)

Sl. No.	District	Total Population	% State Population	Male Population	% State Population	Female Population	% State Population
1.	Bokaro	20,62,330	6.25	10,72,807	6.34	9,89,523	6.16
2.	Chatra	10,42,886	3.16	5,33,935	3.15	5,08,951	3.17
3.	Deoghar	14,92,073	4.52	7,75,022	4.58	7,17,051	4.47
4.	Dhanbad	26,84,487	8.14	14,05,956	8.30	12,78,531	7.96
5.	Dumka	13,21,442	4.01	6,68,514	3.95	6,52,928	4.07
6.	Garhwa	13,22,784	4.01	6,83,575	4.04	6,39,209	3.98
7.	Giridih	24,45,474	7.41	12,58,098	7.43	11,87,376	7.39
8.	Godda	13,13,551	3.98	6,77,927	4.00	6,35,624	3.96
9.	Gumla	10,25,213	3.11	5,14,390	3.04	5,10,823	3.18
10.	Hazaribagh	17,34,495	5.26	8,90,881	5.26	8,43,614	5.25
11.	Jamtara	7,91,042	2.40	4,04,830	2.39	3,86,212	2.42
12.	Koderma	7,16,259	2.17	3,67,222	2.17	3,49,037	2.17
13.	Khunti	5,31,885	1.61	26,335	1.57	2,65,550	1.65
14.	Latehar	7,26,978	2.20	3,69,666	2.18	3,57,312	2.23
15.	Lohardaga	4,61,790	1.40	2,32,629	1.37	2,29,161	1.43
16.	Pakur	9,00,422	2.73	4,52,661	2.67	4,47,761	2.79
17.	Palamu	19,39,869	5.88	10,06,302	5.94	9,33,567	5.81
18.	W. Singhbhum	15,02,338	4.55	7,49,385	4.43	7,52,953	4.69
19.	E. Singhbhum	22,93,919	6.95	11,76,902	6.95	11,17,017	6.96
20.	Ramgarh	9,49,443	2.88	4,94,230	2.92	4,55,213	2.83

21.	Ranchi	29,14,253	8.83	14,94,937	8.83	1,49,316	8.93
22.	Sahebganj	11,50,567	3.49	5,89,391	3.48	5,61,176	3.49
23.	Saraikela-Kharsawan	10,65,056	3.23	5,44,411	3.22	5,20,645	3.24
24.	Simdega	5,99,578	1.82	3,00,309	1.77	2,99,269	1.86
	Jharkhand	**3,29,88,134**	**100.00**	**1,69,30,315**	**100.00**	**1,60,57,819**	**100.00**

Density of Population :

The uneven distribution of population also becomes obvious by the analysis of density if density of population vide table no. 12.15. As per 2011 census, the no. of persons living in one sq. km. of area in the state of Jharkhand is 414. It was 338 as per 2001 census. The highest density of population has been noted in the district of Dhanbad 1316 (person per sq. km.) because of concentration of a good number of urban centres which have developed on account of the development of coal resources and related industries. Other districts having higher density of population consist of Bokaro (715), Ramgarh (708), E. SInghbhum (644), Deoghar (602), Godda (580), Ranchi (572), Sahebganj (558) etc. Some of these districts have

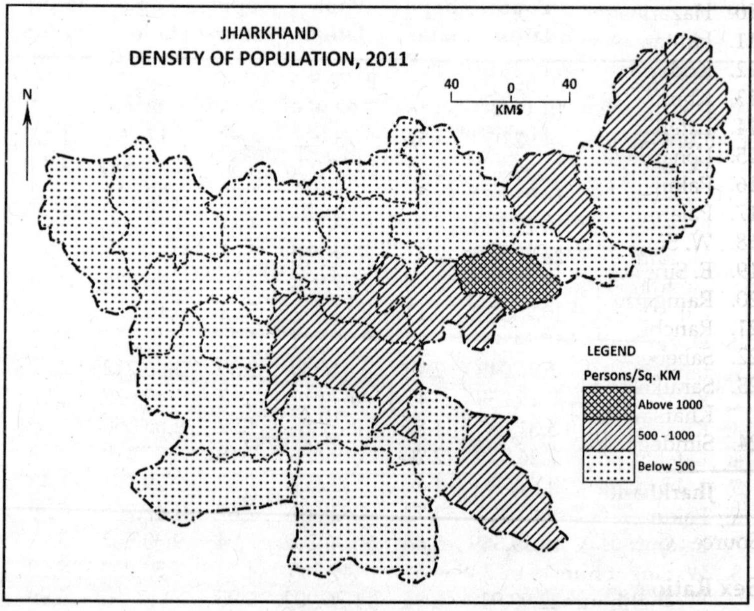

Fig. 17

comparatively better industrial development. Districts having density of population less than the average of the state (414) consist of Simdega (159), latehar (169), Gumla (191), W. Singhbhum (208), Khunti (210), Chatra (281), Koderma (282), Lohardaga (307), Garhwa 9323), Dumka (351) and Saraikela-Kharsawan (401). These districts have comparatively poor development in terms of industries and other economic activities.

Table 12.15
Population Density : Jharkhand (2011)

Sl. No.	District	Area in Sq. Km.	Total Population	Density Person/Sq. Km.
1.	Bokaro	2833.00	20,62,330	715
2.	Chatra	3718.00	10,42,886	281
3.	Deoghar	2477.00	14,92,073	602
4.	Dhanbad	2040.00	26,84,487	1316
5.	Dumka	3761.00	13,21,442	351
6.	Garhwa	4093.00	13,22,784	323
7.	Giridih	4962.00	24,45,474	493
8.	Godda	2266.00	13,13,551	580
9.	Gumla	5360.00	10,25,213	191
10.	Hazaribagh	3555.00	17,34,495	488
11.	Jamtara	1811.00	7,91,042	437
12.	Koderma	2540.00	7,16,259	282
13.	Khunti	2535.00	5,31,885	210
14.	Latehar	4291.00	7,26,978	169
15.	Lohardaga	1502.00	4,61,790	307
16.	Pakur	1811.00	9,00,422	497
17.	Palamu	4393.00	19,39,869	442
18.	W. Singhbhum	7224.00	15,02,338	208
19.	E. Singhbhum	3562.00	22,93,919	644
20.	Ramgarh	1341.00	9,49,493	708
21.	Ranchi	5097.00	29,14,253	572
22.	Sahebganj	2063.00	11,50,567	558
23.	Saraikela-Kharsawan	2657.00	10,65,056	401
24.	Simdega	3774.00	5,99,578	159
	Jharkhand	**79716.00**	**3,29,88,134**	**414**

Source : Census of India, 2011.

Sex Ratio :

The table 12.16 contains sex-ratio of the state of Jharkhand.

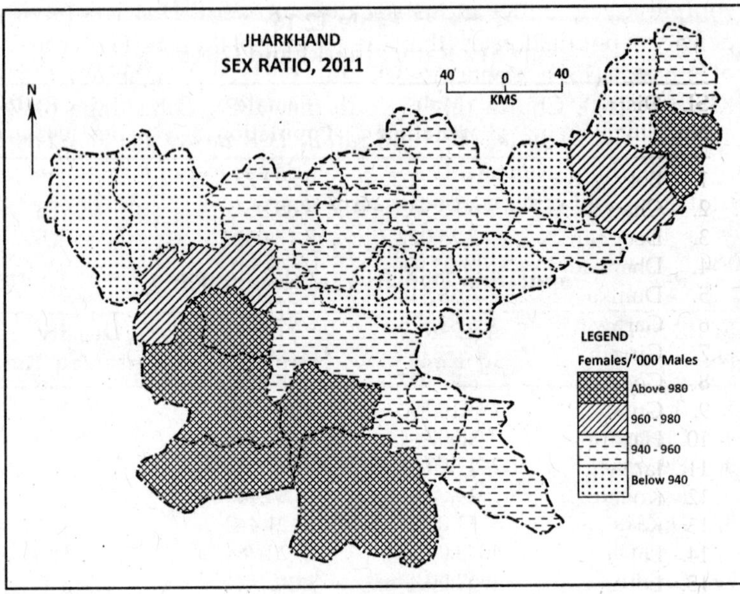

Fig. 18

The average sex-ratio of the state has been recorded 948 females per thousand males as per 2011 census an increase of 07 females per thousand males in 2001 census (941). The best sex-ratio has been registered in W. Singhbhum district as 1005 females per thousand males. Other districts having comparatively better sex ratio, more than the regional average consist of Khunti (997), Simdega (997), Gumla (993), Pakur (989), Lohardaga (985), Dumka (977), Latehar (967), Saraikela-Kharsawan (956), Jamtara (954), Chatra (953), Sahebganj (952), Koderma (950) and Ranchi (949). The lowest sex ratio has been recorded in Dhanbad (909). Other districts having poor sex ratio has been recorded in the districts of Ramgarh (921), Bokaro (922), Deoghar (925), Palamu (928) etc. Poor sex ratio is more marked in the districts having urban and industrial growth where sex selective migration has taken place in favour of males.

Table 12.16
Sex Ratio : Jharkhand (2011)

Sl. No.	District	Male Population	Female Population	Sex Ratio Females/1000 males
1.	Bokaro	10,72,807	9,89,523	922
2.	Chatra	5,33,935	5,08,951	953
3.	Deoghar	7,75,022	7,17,051	925
4.	Dhanbad	14,05,956	12,78,531	909
5.	Dumka	6,68,514	6,52,928	977
6.	Garhwa	6,83,575	6,39,209	935
7.	Giridih	12,58,098	11,87,376	944
8.	Godda	6,77,927	6,35,624	938
9.	Gumla	5,14,390	5,10,823	993
10.	Hazaribagh	8,90,881	8,43,614	947
11.	Jamtara	4,04,830	3,86,212	954
12.	Koderma	3,67,222	3,49,037	950
13.	Khunti	2,66,335	2,65,550	997
14.	Latehar	3,69,666	3,57,312	967
15.	Lohardaga	2,32,629	2,29,161	985
16.	Pakur	4,52,661	4,47,761	989
17.	Palamu	10,06,302	9,33,567	928
18.	W. Singhbhum	7,49,385	7,52,953	1005
19.	E. Singhbhum	11,76,902	11,17,017	949
20.	Ramgarh	4,94,230	4,55,213	921
21.	Ranchi	14,94,937	14,19,316	949
22.	Sahebganj	5,89,391	5,61,176	952
23.	Saraikela-Kharsawan	5,44,411	5,20,645	956
24.	Simdega	3,00,309	2,99,269	997
	Jharkhand	**1,69,30,315**	**1,60,57,819**	**948**

Literacy :

The state of Jharkhand has a literacy rate of 66.41% as per the 2011 census, whereas it was 54.13% in 2001 census. No significant change has been noticed between 2001-2011 period. The most literate district is Ranchi (76.06%) in terms of total literacy as per the 2011 census. It is being followed by E. Singhbhum (75.49%), Dhanbad (74.52%), Ramgarh (73.12%), Bokaro (72.01%), Hazaribagh (69.75%) etc. All these districts have a large urban population base which have a better education amenities. Pakur district (48.82%) occupies the lowest position in terms of total literacy in the state of Jharkhand. In case of male literacy, the state has a literacy level of 76.84% in 2011, which was earlier 67.94% in 2001. In case of male literacy, the highest percentage has been recorded in Ranchi (84.26%), which is closely followed by Dhanbad (83.81%), E. Singhbhum (83.75%), Bokaro (82.51%), Ramgarh (82.44%) and Hazaribagh (80.01%) etc. Districts having comparatively lower male literacy include Pakur, Sahebganj, Chatra, Godda etc. So far as the female literacy is concerned. The state has a female literacy of 55.42%. Female literacy has been the highest in the Ranchi district (67.44%) in the whole of the state, which is being followed by E. Singhbum (66.81%), Dhanbad (64.29%), Ramgarh (63.09%), Bokaro (60.63%) etc. Districts having female literacy less than the regional average include Pakur, Sahebganj, Godda, W. Singhbhum, Garhwa, Dumka, Latehar, Giridih, Chatra, Deoghar, Palamu, Jamtara, Koderma, Khunti. (Table 12.17).

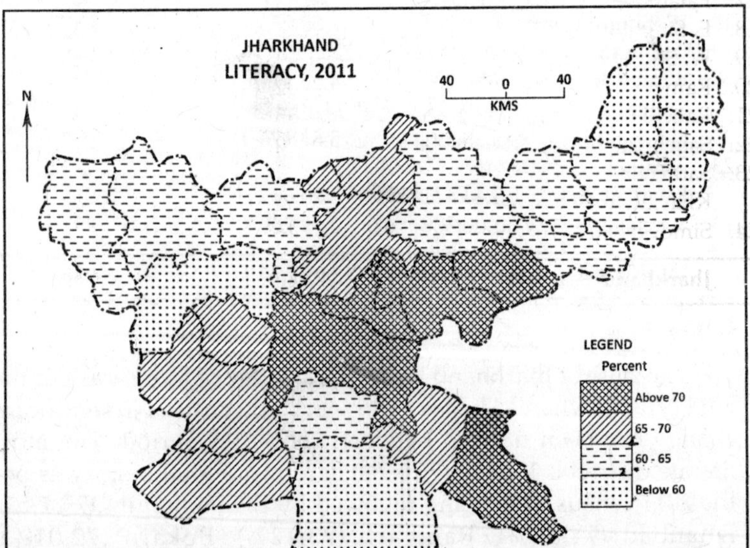

Fig. 19

Table 12.17
Literacy : Jharkhand (2011)

Sl. No.	District	Total (%)	Rural (%)	Urban (%)	Male (%)	Female (%)
1.	Bokaro	72.01	63.50	81.01	82.51	60.63
2.	Chatra	60.18	58.78	80.94	61.92	49.92
3.	Deoghar	64.85	60.49	84.36	76.85	51.80
4.	Dhanbad	74.52	68.17	78.97	83.81	64.29
5.	Dumka	61.02	59.28	83.72	72.96	48.22
6.	Garhwa	60.33	59.33	77.69	72.19	47.58
7.	Giridih	63.14	61.55	79.45	76.76	48.72
8.	Godda	56.40	55.01	82.16	67.84	44.14
9.	Gumla	65.73	64.16	87.79	75.55	55.90
10.	Hazaribagh	69.75	66.64	85.49	80.01	58.95
11.	Jamtara	64.59	62.58	82.62	76.46	52.15
12.	Koderma	66.84	63.69	79.15	79.78	53.23
13.	Khunti	63.86	61.95	83.71	74.08	53.69
14.	Latehar	59.51	57.99	78.33	69.97	48.68
15.	Lohardaga	67.61	64.98	85.37	77.41	57.69
16.	Pakur	48.82	47.33	66.50	57.06	40.52
17.	Palamu	63.63	61.31	80.54	74.30	52.09
18.	E. Singhbhum	75.49	74.37	85.27	83.75	66.81
19.	W. Singhbhum	58.63	54.31	82.44	71.13	46.25
20.	Ramgarh	73.12	66.83	81.02	82.44	63.09
21.	Ranchi	76.06	67.81	86.55	84.26	67.44
22.	Sahebganj	52.04	48.45	73.20	60.34	43.31
23.	Saraikela-Kharsawan	67.70	63.53	80.54	79.03	55.88
24.	Simdega	67.99	66.59	85.46	76.08	59.92
	Jharkhand	**66.41**	**61.11**	**72.60**	**76.84**	**55.42**

13

Tribal Population

The earth surface is endowed with a wide variety of tribal people varying in concentration on environmental affinity, physiological construction, economic pursuits and cultural traits. Significant areas having considerable tribal people include the Eskimo in the far North, Red Indians in the Americas, Pygmies, Hottentots, Bushmen and Negroes in Africa, Australian aborigines and Papuans in Australia. Indian sub-continent is also important in this respect. India comprises more than 450 tribes portraying diverse tribal attributes contributing to the anthropologists paradise. Each of these is a proud community participating in the national life stream. They still retain their identity and reveal the basic values of Indian civilization.

The tribal people in India is dispersed in almost parts of the country but their main concentration lies in difficult areas in terms of accessibility. They are known as Vanvasi, Adivasi, Janjati and Anusuchit Janjati. Broadly the tribes are concentrated in the Himalayan Region, North Eastern India, Central Plateau Region, Western India, Hilly areas of Southern India and Islands. These tribes belong to different Racial Groups. The tribes found in north-west India are supposed to be primitive Australoids who came here from the west. At the time of Aryan invasion indigenous Dravidians were pushed to inaccessible areas of the South. It is therefore, the plain areas of Bihar is densely populated with people of Aryan stock. In process of Aryan expansion in the plains some Dravidians adopted the way of life of Aryans. At the same time some secondary Mangolian tribes were also entering into north-eastern part of the country. They mostly occupied the Himalayan region and later on also moved to the Tarais and other areas.

The state of Jharkhand having congenial environment abounds in such communities. There are thirty tribal groups of which big tribal communities are the Santhals, the Oraons, the Mundas, etc. Most of these are racially believed to be Pre-

Dravidian of Proto-Australoids. They are dark complexioned, squat and stocky with wavy hair, long head (Dolichocephalic) and broad nose. They are well built with strong limbs.

The state of Jharkhand has 26.21% tribal population consisting of 31.40% in rural areas and 9.79% in urban areas. This population has recorded comparatively lower growth than the non-tribal population. In the decade 2001-2011, the population of the state increased by 22.42% but the tribal people registered an increase of 21.98%. It is notable that tribal people do not take much care of health and more depend upon local quacks and traditional methods of treatment. As such, their longevity is comparatively lower. These people are comparatively more alcoholic which is injurious to their health.

As per the census record of 2011, the state of Jharkhand has 86,45,042 tribal population which accounts for 26.21% of the state. Over, the years the no. of tribal population has been increasing as in 1991 census, it was 60,44,010 to 70,87,068 in 2001 registering a higher percentage of growth. The percentage growth of tribal population in 1981-1991, 1991-2001 and 2001-2011 decade was 13.62%, 17.26% and 21.98% respectively. (Table 13.1).

Table - 13.1
Scheduled Tribe Population : Jharkhand

Sl.	Year	Population	Growth (%)
1.	1961	45,32,342	-
2.	1971	53,69,880	0.83%
3.	1981	53,19,370	16.41%
4.	1991	60,44,010	13.62%
5.	2001	70,87,068	17.26%
6.	2011	86,45,042	21.98%

Source : Different Census Year.

As regards the distribution of tribal population of the state, the table 13.2 reveals uneven distribution of population. The highest concentration of tribal population has been recorded in the district of Ranchi (12.05% of the state). Here, there is a considerable hilly & forested areas having dominance of tribal people. Other districts having more than 6% of tribal population included W. Singhbhum (11.70%), Gumla (8.17%), E. Singhbhum (7.56%) and Dumka (6.61%). Other prominent districts having tribal population are Khunti (4.51%) and Pakur (4.38%). These seven districts of the state have about 55% tribal population of the Jharkhand and the remaining districts have just 45% tribal population. In case of rural tribal population, the highest percentage has been recorded in the district of W. Singhbhum (12.11%), being followed by Ranchi (10.04%), Gumla (8.68%), Dumka (7.19%), E. Singhbhum (6.55%) etc. The highest percentage of urban tribal population of the state has been recorded in Ranchi (32.45%), where almost 1/3rd of the urban tribal population is concentrated. It may be due to early growth of township & tribal organisations having education, health & cultural facilities attracted tribal people. Other districts having comparatively higher share in urban-tribal population included E. Singhbhum (17.85%), Bokaro (9.45%), W. Singhbhum (7.49%), Ramgarh (7.30%) etc.

Table - 13.2

Schedule Tribe Population : Jharkhand (2011)

Sl. No.	District		Total Population	ST Population	%ST Population to Total population of the District	%ST Population to Total ST population of State
1.	Bokaro	T	20,62,330	2,55,626	12.40	2.96
		R	10,78,686	1,82,223	16.89	2.32
		U	9,83,644	73,403	7.46	9.45
2.	Chatra	T	10,42,886	45,563	4.37	0.53
		R	9,79,932	44,298	4.52	0.56
		U	62,954	1,265	2.01	0.16
3.	Deoghar	T	14,92,073	1,80,962	12.13	2.09
		R	12,33,712	1,76,140	14.22	2.24
		U	2,58,361	7,822	1.87	0.62
4.	Dhanbad	T	26,84,487	2,33,119	8.68	2.70
		R	11,24,093	1,98,079	17.62	2.52
		U	15,60,394	35,040	2.25	4.51
5.	Dumka	T	13,21,442	5,71,077	43.22	6.61
		R	12,31,264	5,65,629	45.94	7.19
		U	90,178	5,448	6.04	0.70
6.	Garhwa	T	13,22,784	2,05,874	15.56	2.38
		R	12,53,114	2,03,918	16.27	2.59

	U	69,670	1,956	2.81	0.25
7. Giridih	T	24,45,474	2,38,188	9.74	2.76
	R	22,37,450	2,35,170	10.51	2.99
	U	2,08,024	3,018	1.45	0.39
8. Godda	T	13,13,551	2,79,208	21.26	3.23
	R	12,49,132	2,76,785	22.16	3.52
	U	64,419	2,423	3.76	0.31
9. Gumla	T	10,25,213	7,06,754	68.94	8.18
	R	9,60,132	6,82,789	71.11	8.68
	U	65,001	23,965	36.82	3.08
10. Hazaribagh	T	17,34,495	1,21,768	7.03	1.41
	R	14,59,188	1,07,553	7.37	1.37
	U	2,75,307	14,215	5.16	1.83
11. Jamtara	T	7,91,042	2,40,489	30.40	2.78
	R	7,15,296	2,36,962	33.13	3.01
	U	75,746	3,527	4.66	0.45
12. Koderma	T	7,16,259	6,903	0.96	0.08
	R	5,75,013	6,152	1.07	0.08
	U	1,41,246	751	0.53	0.10
13. Khunti	T	5,31,885	3,89,626	73.25	4.51
	R	4,86,903	3,71,469	76.29	4.72
	U	44,982	18,157	40.37	2.34
14. Latehar	T	7,26,978	3,31,096	45.54	3.83
	R	6,75,120	3,24,023	47.99	4.12
	U	51,858	7,073	13.64	0.91
15. Lohardaga	T	4,61,790	2,62,734	56.89	3.04
	R	4,04,379	2,48,472	61.44	3.16
	U	57,411	14,262	24.84	1.84
16. Pakur	T	9,00,422	3,79,054	42.10	4.38
	R	8,32,910	3,76,390	45.19	4.78
	U	67,512	2,664	3.95	0.34
17. Palamu	T	19,39,869	1,81,208	9.34	2.10
	R	17,13,866	1,77,852	10.38	2.26
	U	2,26,003	3,356	1.48	0.43
18. W. Singhbhum	T	15,02,338	10,11,296	67.31	11.70
	R	12,84,304	9,53,106	74.21	12.11
	U	2,18,034	58,190	26.69	7.49
19. E. Singhbhum	T	22,93,919	6,53,923	28.51	7.56
	R	10,19,328	5,15,214	50.54	6.55
	U	12,74,591	1,38,709	10.88	17.85
20. Ramgarh	T	9,49,443	2,01,166	21.19	2.33
	R	5,30,488	1,44,484	27.24	1.84
	U	4,18,955	56,682	13.53	7.30
21. Ranchi	T	29,14,253	10,42,016	35.76	12.05
	R	16,56,918	7,89,838	47.67	10.04

	U	12,57,335	2,52,178	20.06	32.45
22. Sahebganj	T	11,50,567	3,08,343	26.80	3.57
	R	9,90,901	3,01,930	3.05	3.84
	U	1,59,666	6,413	4.02	0.83
23. Saraikela-Kharsawan	T	10,65,056	3,74,642	35.18	4.33
	R	8,06,310	3,45,187	42.81	4.39
	U	2,58,746	29,455	11.38	3.79
24. Simdega	T	5,99,578	4,24,407	70.78	4.91
	R	5,56,634	4,04,487	72.67	5.14
	U	42,944	19,920	46.39	2.56
Jharkhand	**T**	**3,29,88,134**	**86,45,042**	**26.21**	**100.00**
	R	**2,50,55,073**	**78,68,150**	**31.40**	**100.00**
	U	**79,33,061**	**7,76,892**	**9.79**	**100.00**

Source : Primary Census Abstract, Census of India, Jharkhand, 2011.

The analysis of the table 13.2 portrays districtwise share of tribal population to the total population of the district. The highest percentage tribal population to the total population of the district is shared by Khunti district (73.25%) which is a predominently a tribal district. It is closely followed by Simdega (70.78%), Gumla (68.94%), W. Singhbhum (67.31%) etc. The districts having more than 33% tribal population included Lohardaga (56.89%), Latehar (45.54%), Dumka (43.22%), Pakur (42.10%), Ranchi (35.76%) and Saraikela-Kharsawan (35.18%). Koderma district (0.96%) has the least percentage of tribla population to the total population of the district being followed by Chatra (4.37%), Hazaribagh (7.03%) and Dhanbad (4.37%). It may be noted here that the districts having considrable urban industrial and mining growth have comparatively lower percentage of tribals. So far as the percentage of rural tribal population is concerned the highest percentage is shared by Khunti (76.29%) followed by W. Singhbhum (74.21%), Simdega (72.67%), Gumla (71.11%), Lohardaga (61.14%) etc. But in case of urban tribal population, the highest percentage has been recorded in Simdega (46.39%), being followed by Khunti (40.37%), and Gumla (36.82%) etc.

Table 13.3
Schedule Tribe Population : Jharkhand (2011)

Sl. No.	District	Tribal Population (In Persons)	Total Tribal Population of the state (%)
1.	Ranchi	10,42,016	12.05
2.	W. Singhbhum	10,11,296	11.70
3.	Gumla	7,06,754	8.17
4.	E. Singhbhum	6,53,923	7.56
5.	Dumka	5,71,077	6.61
6.	Khunti	3,89,626	4.51
7.	Pakur	3,79,054	4.38
8.	Other Districts	38,91,296	45.02
	Jharkhand	**86,45,042**	**100.00**

Source : Primary Census Abstract, Govt. of India, Jharkhand, 2011.

The uneven distribution of tribal population in the state also becomes more obvios by the analysis of density of tribal population (Table 13.4). The average density of tribal population in the state is 108 persons per sq. km. as per 2011 census. Out of the 24 districts as per 2011 census, 14 districts consisted of higher density than the state average. The highest density has been recorded in the district of W. Singhbhum (284), being followed by Pakur (209), Ranchi (204), Lohardaga (175), Khunti (154), Dumka (152), Ramgarh (150), Sahebganj (149), Saraikela-Kharsawan (141), Jamtara (133), Gumla (132), Godda (123), Dhanbad (114) and Simdega (112). All the remaining districts have tribal density less than that of the state average.

Table - 13.4
Tribal Density : Jharkhand (2011)

Sl. No.	District	Area (Sq. Km.)	Tribal Population	Tribal Density (Sq. Km.)
1.	Bokaro	2883.00	2,55,626	89
2.	Chatra	3718.00	45,563	12
3.	Deoghar	2477.00	1,80,962	73
4.	Dhanbad	2040.00	2,33,119	114
5.	Dumka	3761.00	5,71,019	152
6.	Garhwa	4093.00	2,05,874	50

7.	Giridih	4962.00	2,38,188	48
8.	Godda	2266.00	2,79,208	123
9.	Gumla	5360.00	7,06,754	132
10.	Hazaribagh	3555.00	1,21,768	34
11.	Jamtara	1811.00	2,40,489	133
12.	Koderma	2540.00	6,903	3
13.	Khunti	2535.00	3,89,626	154
14.	Latehar	4291.00	3,31,096	77
15.	Lohardaga	1502.00	2,62,734	175
16.	Pakur	1811.00	3,79,054	209
17.	Palamu	4393.00	1,80,208	41
18.	E. Singhbhum	7244.00	6,53,923	90
19.	W. Singhbhum	3562.00	10,11,296	284
20.	Ramgarh	1341.00	2,01,166	150
21.	Ranchi	5097.00	10,42,016	204
22.	Sahebganj	2063.00	3,08,343	149
23.	Saraikela-Kharsawan	2657.00	3,74,642	141
24.	Simdega	3774.00	4,24,407	112
	Jharkhand	**79,716.00**	**86,45,042**	**108**

Source : Primary Census Abstract, Govt. of India, Jharkhand, 2011.

Table - 13.5

Tribal Sex Ratio : Jharkhand (2011)

Sl. No.	District	Tribal Male Population	Tribal Female Population	Tribal Sex Ratio
1.	Bokaro	1,29,233	1,26,393	978
2.	Chatra	23,141	22,422	969
3.	Deoghar	91,012	89,950	988
4.	Dhanbad	1,17,256	1,15,863	988
5.	Dumka	2,82,125	2,88,952	7024
6.	Garhwa	1,04,499	1,01,375	970
7.	Giridih	1,20,646	1,17,542	974
8.	Godda	1,38,510	1,40,698	1016
9.	Gumla	3,52,514	3,54,240	1004
10.	Hazaribagh	60,796	60,972	1002
11.	Jamtara	1,20,035	1,20,454	1003
12.	Koderma	3,594	3,309	921

13. Khunti	1,93,710	1,95,916	1011
14. Latehar	1,66,427	1,64,669	989
15. Lohardaga	1,30,814	1,31,920	1008
16. Pakur	1,86,967	1,92,087	1027
17. Palamu	92,577	88,631	957
18. W. Singhbhum	5,00,949	5,10,347	1019
19. E. Singhbhum	3,25,989	3,27,934	1006
20. Ramgarh	1,01,901	99,265	974
21. Ranchi	5,20,582	5,21,434	1002
22. Sahebganj	1,53,435	1,54,908	1010
23. Saraikela-Kharsawan	1,87,149	1,87,493	1002
24. Simdega	2,11,546	2,12,861	1006
Jharkhand	**43,15,407**	**43,29,635**	**1003**

Source : Primary Census Abstract, Govt. of India, Jharkhand, 2011.

Table - 13.6
Tribal Population Distribution : Jharkhand (2011)

Sl. No.	Tribe	Total Population	%
1.	Santhal	27,54,723	31.86
2.	Oraon	17,16,618	19.86
3.	Munda	12,29,221	14.22
4.	Ho	9,28,289	10.74
5.	Kharwar	2,48,974	2.88
6.	Lohra	2,16,226	2.50
7.	Bhumij	2,09,448	2.42
8.	Kharia	1,96,135	2.27
9.	Mahli	1,52,663	1.77
10.	Mal Paharia	1,35,797	1.57
11.	Bedia	1,00,161	1.16
12.	Chero	95,575	1.11
13.	Karmali	64,154	0.74
14.	Chick Baraik	54,163	0.63
15.	Gond	53,676	0.62
16.	Kol	53,584	0.62
17.	Sauria-Paharia	46,222	0.53
18.	Kisan	37,265	0.43
19.	Korwa	35,606	0.41
20.	Kora	32,786	0.38

21.	Parhaiya	25,585	0.30
22.	Asur	22,459	0.26
23.	Binjhia	14,404	0.17
24.	Birhor	10,726	0.12
25.	Savar	9,688	0.11
26.	Kawar	8,145	0.09
27.	Birjia	6,276	0.07
28.	Gorait	4,973	0.06
29.	Baiga	3,582	0.04
30.	Bathudi	3,464	0.04
31.	Banjara	487	0.01
32.	Khond	221	-
	Jharkhand	**86,45,042**	**100%**

Source : Primary Census Abstract, Govt. of India, Jharkhand, 2011.

As per table no. 13.5, the tribal sex ratio is also uneven. The state has 1003 tribal females per thousand tribal males as per the 2011 census. The best tribal sex ratio has been registered in the district of Pakur (1027) and being followed by Dumka (1024), W. Singhbhum (1019), Godda (1016), Khunti (1011), Sahebganj (1010), Lohardaga (1008), Gumla (1004), E. Singhbhum (1006), Hazaribagh (1003), Jamtara (1003) etc. All the remaining districts have lower sex ratio than the state average. It may be noted that sex ratio is normally better in tribal societies because least distinction is made between boys and girls. As such tribal areas have comparatively better sex ratio. Certainly, areas having considerable concentration of urban population or industrial population or mining workers etc. more people lead bachelor life and such areas consists of comparatively poor sex ratio due to above mentioned reasons and comparatively less industrialised, mineralised and urbanised districts have much better sex ratio.

The table 13.6 reveals that there is an unbalanced population size of the tribal people. The 'Santhals' are the largest tribal group in the state whose total no. is 27,547,23 as per 2011 census which accounts for 31.86% of the total tribal population of the state. The 'Oraon' (19.86%) 'Munda' (14.22%) and 'Ho' (10.74%) are the next largest groups in the state. These four tribal groups constitute more than three fourth of tribal population of the state. On the other hand, 'khond' tribe is the smallest group whose total no. is just

221 persons in the state. There are 20 tribes which have less than 1% tribal population each. The remaining eight tribes comprise 1% to 3% tribal population of the state.

The Santhals :

The Santhals are the leading tribe of the State. Their total population in the state is 27,54,723 persons, (31.86% of the tribal population of the state) as per 2011 census. They are mostly concentrated in the districts of Dumka, Godda, Pakur, Sahibganj and Deoghar (old district of Santhal Parganas). They got concentrated in Damin-i-koh. In the year 1872, Santhal Parganas had 4,55,513 persons which increased to 6,70,535 in 1901 and further increased to 7,54,804 persons in 1931. In the year 1981 this main area of Santhal consisted of 1,35,8063 persons. Besides they are also found in the districts of Hazaribag, Giridih, Bokaro, Dhanbad, Chatra, Pashchimi and Purbi Singhbhum, etc. in good number. In addition to the distribution of Santhals in the Jharkhand, they are also found in the districts of Bhagalpur, Banka, Purnea, Katihar of Bihar, Midnapur, Manbhum, Birbhum, Burdwan, Bankura districts of West Bengal, Balasore district of Orissa in some areas of Chattisgarh and Assam and in a few number in north-east states.

As regards the origin of the Santhals scholars hold different views. The Revd. L.O. Skrefsrud had conjectured that they lived successively in Persia, Afghanistan and Chinese territory and entered India from the north-west and settled in Punjab and thereafter came to the region. Colonel Dalton believed that the Santhals came from North-east India and found their way to this region. In support of this theory he cited certain remarkable coincidence of custom and language between the Santhals and some of the aboriginal tribes on the north-eastern frontier of this country. This view has also been accepted by Sir William Hunter in the Annals of Rural Bengal.

The name Santal, spelt in one way or another (e.g. Sonthal) is an English form which corresponds with the form Saontar used by the Bengali speaking people. Both Santal and Saontar have the same origin. According to Mr. Bodding the word Santhals have developed from the word Sant and Saont derived from the word Samanta. As a matter of fact all the forms (Sant, Saont, Saot and

Sat) are easily derivable from this according to common Aryan Phonetic laws. There is no doubt that the word itself is of Aryan origin. Thus, this tribe exhibits a long history of continuous migration from one area to another to keep pace with their increasing number.

The Santhals are people of medium stature but muscular sturdily built wiry and capable of great physical endurance. Their complexion is dark, varying from brown to an almost jet black colour. The latter is very rare while a light brown complexion is more common. They are mostly beardless and have coarse and sometimes curly hair. The cheek bones remain some what prominent. The nose is generally broad and depressed. The mouth is large and the lips are full and projecting. Sir H. Risley considered them as pure Dravidian stock on the basis of their physical outlook. The proportions of the skull, approaching the dolichochephalic type, conclusively refute the hypothesis of their Mongoloid descent but faces of a Mangoloid type are, however, sometimes seen.

These people speak Santhali language which has relationship with Austro-Asiatic language. In spite of that their dialect somewhat tally with the language of other tribe.

Initially there were hunters and food gatherers resorted to shifting cultivation. They preferred to live in jungles or on hill tops. But now they have also settled in the plains clearing the forest clads and preparing the same for paddy cultivation. But they have not forgotten altogether their old habit of procuring food from forest catching fish and hunting. Santhals are gradually becoming better cultivators and they are now cultivating all kinds of vegetables, seasonal fruits, etc. But they are somewhat happy-go-lucky in their habits.

Santhals villages are of slightly different nature. They have most dispersed dwellings. They prefer linear dispersed settlements on the ridge top or along the footpath in the hilly areas. One revenue village consists of several hamlets. Most of the houses face the lane which are quite wide. There is separate provision for cattle, pigs and chickens. Most of the houses contain one large room and one attached small room for keeping valuables. Houses lack windows because of the fear of theft. Walls have space for keeping domestic goods. Most of the houses are mud walled and thatch roofed. Houses are surrounded by raised platforms which are used for drying crops and using for other purpose. Santhal women are careful to keep their houses neat and clean.

Economically they are not well off. Most of them lead very simple life. Virtually they take poor diet. They prefer rice, maize, pulses and flesh. For flesh they depend on goat, sheep, pig, fowl etc. They are fond of keeping cattle and fowl. The cattle is preferred for agricultural purposes. Mostly these people depend on cultivation and daily wages. They are hardy and do not get afraid of scorching heat and biting cold. Their agricultural plots are of tiny size and fields are scattered. Certainly they keep some land in the back of their houses for growing vegetables and fruit trees. Previously they were more hesitant in using chemical fertilizer and pesticides but in the association of non-tribals they have adopted modernized agricultural practices in using proper agricultural implements, improved seeds and growing crops more than domestic requirements. Still they are exploited by the money lenders in the rural areas. In addition to agriculture they are expert in weaving arts. Some of them have developed handloom industries but the development of cotton textile industries have adversely affected their this profession. They are expert in hunting and fish catching. They hunt with their arrows and bows. At the time of hunting they beat drums in the forest for terrorising wild animals who start running hither and thither for shelter. In this process they take advantage of hunting. On such occasions they also please their Forest God with sacrifices. They also use enough alcoholic drink at the time of hunting.

Their social life is highly organised. The headman of the village is known as 'Manghi'. He is assisted by 'Jag Manghi' who keeps control over the morale of the people. The village priest is called 'Naike' and all functions are performed according to his directions. The Mukhia (Manghi) of the village is not by election but by inheritance. There is also provision of Godet who provides information about birth and death to the Mukhia. At the same time he collects contributions on social occasions. Several villages constitute Pargana having separate headman.

Adult marriages are the rule among the Santhals. Child marriage is very rare. The regular santhal name for all kinds of marriage is Bapla, a word which very probably meant originally mutual strengthening of the two families. There are two essential features of the marriage ceremony. The first is the Sindurdan and the second is a meal in which the husband and wife eat together. The marriage is negotiated through a marriage broker who finds out where an eligible girl is and arranges the day for young man's friends to come and see girl and girl's house. The dowry system prevails but the dowry price is given to bride family. More than one wife is not allowed except the woman fails in giving birth to a child. There is provision of divorce with several restrictions.

The Santhals believe in several bongas or evil spirits to whom the ills of human life are due, and that they most be appeased by blood offerings. Thakur is considered the supreme being. They also believe in witch craft. They consult ojha at the time of illness or problems. Their sacred places are Jatarthan, Manghthar, Bhittar, etc. In case of Jatarthan there remains trees of Sal and Mahua where Marang Burhu and his wife Jahar Burhu are supposed to live. Manghithar is known as burning place. Bhittarthar remains in one corner of the house which is not allowed for all except the members of house.

The Santhals celebrate many festivals. Most of the festivals are connected with agricultural operations. The chief of these is the Sohrae or Harvest festival being celebrated in Push after the rice crop has been harvested. On this occasion they prepare Handi (bear) and sacrifice fowl and enjoy festival with all relatives. Other festivals are Hariar Sim, the feast of the sprouting of the rice is held in July-August, Iri Gundhi-Nawai-the offering of the first fruits of the millets and Gundhi, MaghSim is held in the month of Magh when the jungle grass is cut. They also celebrate Sakral, Chata, Pata, Kuru etc. On all such occasions they dance, sing and celebrate with local witness.

They also celebrate birth ceremony. There is curious practice giving a child two names, his real name and second name. When a Santhal is dying the door of his house is kept open in order that his spirit may leave it and not haunt the family residence. After death the body is taken to the place where two roads meet at the end of the street and is lamented over by the women folk. From this place the dead body is taken to the burning place near the bank of river, pond etc. When the cremation is over the relatives go and pick up the bones, wash them pouring turmeric water and milk over them, and put them in a new pot. This is covered with a pot sherd with a hole in it (a breathing hole for the dead). Later on it is buried outside the village. At the end of necessary rituals they organise a great feast with a sacrifice to the dead.

Although Santhals are not rich but they lead happy life. They are lover of music and dance. They play the same upon flutes prepared by them. They do so in order to exhibit their pleasure and partly to forget their sorrow. They manage different types of dances on different occasions like marriage, religious functions, harvest, etc.

The Oraons :

The Oraons are the second leading tribe in the state. They are the people known in many parts of India as 'Dhangais' meaning a hillman. Their total population in the state is 17,16,618 persons (19.86% of the tribal population of the state) as per 2011 census. The concentration of this tribe is more remarkable in districts of Lohardaga, Gumla, Ranchi and adjoining Latehar subdivisions of Palamu district. Besides this tribe is also found in the districts of Hazaribag, Chatra, Giridih, Bokaro, Dumka, Godda, Pakur of the state. They are also living in smaller number in Purnea, Katihar, Rohtas and Pashchimi Champaran districts of Bihar. Other states having some Oraons include West Bengal, Madhaya Pradesh and Orissa.

As regards their origin scholars are of different opinions. Mr. Colonel Dalton is of the opinion that the tribe had actually migrated earlier from the western Konkan coast of India to Rohtas upland and from that highland they moved into the state of Jharkhand. It is also believed that Oraons originally came from Southern India as philologists have traced in the languages of Kanarese and other Tamilian languages a close resemblance to the Kurukh tongue. Another theory that has been put forward is that the Oraons accompanied the Mundas in their migrations from the north-west of India before the arrival of Aryans. But this opinion does not appear to be supported by Oraon tradition. Though the Oraon tradition agrees with the Munda tradition in naming Azimgarh, Pipragarh, Hardinagar, Rohtasgarh as homes of their race, it is possible that the tradition was merely borrowed from their neighbours. This appears to be more convincing that they came from Rohtasgarh in this region. While moving out from the Rohtas Plateau this tribe split up into two branches, one branch proceeded down the Ganga and settled in the Rajmahal hills and are considered the ancestors of the present male tribe. Another branch proceeded up the Son and then along the valley of North Koel into Palamu till they reached the highlands of the state. According to Oraon tradition, they were more civilised race and introduced the use of the plough and the construction of the terraced Don land. Ptolemy's geographical accounts state that this tribe was quite significant in the state even as early as 2nd century.

Oraon call themselves Kurukh a name which is sometimes traced to one of their mythical kings called 'Karakh'. It is probable that this tribe derived from some obsolete Dravidian route, meaning man or it may possibly be connected with the Dravidian word 'Karugu' and 'Eagle' and be totemistic in origin. Ancestors of this tribe consider monkey as totem. Their ancestor used to live in caves and used to depend upon hunting of wild animals.

The Oraons are of comparatively good physique capable of carrying heavy loads and travelling long distances. An Oraon can carry a load of about 80 kg. to a distance of 45 kms. in a day and continue to do so for several days. The colour of most Oraons is dark brown to black. The hair is jet black, coarse and inclined to frizzi. They have projecting jaws and teeth, thick lips, low narrow foreheads, broad flat nose, etc. The eyes are open bright and full and obliquity is traceable in the opening of the eyelids. They remain cheerful and light hearted when young delight in physical exercise and dancing.

The Oraons have retained their tribal traits and language. They speak Mundari. They are good agriculturists and have established economic relations with others also. They preferably grow paddy and vegetables because rice is their staple food. They domesticate cattle and fowls. They mainly occupied the flat plateaus for developing terraced rice fields and settlements on the Tanr lands. As such persons who developed land for cultivation are known as Bhumihar in their language and such land is known as Bhumihari land. They categorise land into three categories such as Bhumihari land, Raiyati Bhumi or Rajahaland and Zamindari land. Almost all resourceful Oraon families have Bariland attached to their dwellings where they grow different types of vegetables and fruits. Since they are good cultivators they keep oxen for ploughing the land and other domestic animals. They have also developed interest in irrigating crops, using chemical fertilizers, improved seeds and modern agricultural implements. Virtually they are growing such crops which enable them to earn more money like vegetables, seasonal fruits and other cash crops. Because Oraons are comparatively better educated than other tribes they have also engaged in non-agricultural works like services, business, industries etc. Most of them move out for labour work after the completion of their own agricultural works.

The Oraons have mostly mud walled and tile roofed houses. In areas stone walled and tile roofed are also found where comparatively softer rocks are available. Resourceful Oraons have bricks walled and tiled roofed houses and pucca roofed houses. Their house remain jumbled together in an haphazard manner. The Oraons have their village communities into groups known as Parhas under the supervision of headman of one of the villages who is known as the Parha Raja. Every village has religious headman for priest called Pahar or Paga and he is assisted in all religious matters by a Pujar. The democratic village headman is called Mahto. Seven to twenty one village in group form a democratic organisation called Panch headed by a leader known as Parla. He is assisted by Mahto and Pahans of concerned villages. The Oraons keep their houses neat and clean.

Majority of the Oraons are Hindus. They believe in the existence of supreme being and worship Vishnu, Shiva and Shakti. They have tribal religion like Sarna. Some of them have even accepted Jainism, Sikhism, Budhism also. A good number of Oraons have also accepted Christianity because of the facilities provided by the Christian missions. Marriage ceremony of the oraons are generally similar to those of the Mundas. They ignore the same Gotra for marriage. Oraons are divided into numerous exogamous septs which take their names from some birds, fishes, animals, plants or minerals. These totems are regarded as sacred by the sept named after them and are rejected by them as ancestors. They call all members of one generation by one name. This is why one Oraon calls his father and his uncles as 'Ba' and his mother and the females who may be married with his father are called 'Ayau'. Marriage is arranged by the parents, though the consent of the boy is obtained. First of all a messenger is sent to the guardian of the girl and when the answer is favourable, the messenger and the boy's guardian and friends and sometimes the boy himself proceed for the girl's village for final negotiation. When the term is agreed upon the ceremony is completed by the Pahan. A bridal feast is organised. Marriage with members of other tribes is highly prohibited. Yatra is the most important festival for boys and girls in this community. In yatra they recite songs and develop love and select their companion but marriage is arranged by mother and father of the couple.

A peculiar feature of Oraon villages is the Dhumkuria (Jankha-erpain kurukh) or bachelor's hall, an interesting survival is still found in older villages. The young boys are taught here about religion and safely measures from evil spirits. At the same time social and moral preachings are also given. Dhumkuria house lacks windows and has only one door. Normally it is one big room having one bed made of palm leaves. These leaves are offered by the married girls of the village in a year. Close to this there is provision of site for wrestling and dancing. The members of Dhumkuria remains of the age between ten to twenty five years only. To look after this institution one boy is made an incharge known as Mahto.

The dress of the Oraons is very simple and scanty. Males ordinarily wear two ends called 'Botoi' and use a piece of cloth as a wrapper for the upper portion of the body known as Barkhi or Pichowri. The women wear a long piece of cloth called 'Paria' round the waist allowing a portion of it to pass diagonally over the upper part of the body. Shoes are seldom used and head is usually uncovered. Certainly well to do Oraons bear Pagris. In the rains they often carry bamboo umbrellas. Both sexes are extremely fond of ornaments. The young men adorn themselves with a girdle (Kardhani) of twisted black cotton thread or silk which also serves of carrying keys, purse, limebox, etc. The most noticeable ornaments of the women are the heavy brass bracelets on the arms, and the ornaments of various kinds which are thrush through the lobe of the ear. At their dances their head-dresses are often adorned with plumes of heron feathers.

This tribe organise drink and dances on all festive occasions. The traditional drink is Handia which is prepared from rice and

Mahua. They assemble together on Akhara almost every evening to dance to the tune of local music and enjoy on festive occasions. These dances go on for days together. Important festivals are Sarhul, Karma, Sohrai, etc. In addition they also celebrate Durga Puja, Dipawali, Shivratri, Holy, etc. Those who have adopted other religions they observe their festivals accordingly.

The hunting is a traditional pastime of the tribal. Cock fight has also been of absorbing interest to them since time immemorial. Dance and music with Mandar and Flute are also an important source of their recreation. Hockey has been a popular sport in the countryside for about a century. Football is also a popular game among them. The tribals regale themselves over community drink. The cycle of festivals seldom leave them in want of mirth. They trek long distances to Melas and Hats to enjoy social life as well as fun and gaiety. Itinerant cinema and circus also visit rural areas occasionally. Community radio also draws good audience. In urban areas cinema is a common source of entertainment. Young persons have imbibed a craze for pictures. People working in factories often go to hills and jungles on picnic. Travel to picturesque spots is also becoming a regular feature of their life. Youth hostelling and mountaineering are providing new avenues for their enjoyment.

The Oraons belong to paternal society. Females depend upon males. Boys inherit the property after the death of his father. They also celebrate birth ceremony like Chhathiar, Mundan and other functions like the Hindus being followed by feast, drink, dance, etc. The oraons having faith in Hinduism usually cremate the dead body on the bank of river but they bury the children upto the age of six years. In case of snake bite, smallpox, leapor, they also bury the dead body. The main Saradh is usually performed on the twelfth day of the death. Christian Oraons generally bury the dead bodies.

The Mundas :

The Mundas are the third leading tribe of the state. They are sometimes called 'Kols'. They themselves called Horoko (Men) and their race Horo (Man) and consider as the earliest aboriginals of the state. Their total population in the state is 12,29,221 persons (14.22% of the tribal population of the state) as per 2011 census. The main concentration of this tribe is in the districts of Ranchi,

Lohardaga and Gumla. They are also found in Hazaribag, Chatra, Giridih, Bokaro, Purbi and Pashchimi Singhbhum. They mostly concentrate in the higher plateau areas.

As regards their origin scholars hold the view that they initially occupied the forest areas of the Vindhayan mountains. Later on they were pushed to the Rohtas plateau when the Aryan invasion took place. These people were further driven by Oraons to the deepest jungles of the North Koel valley and the higher plateau areas of Ranchi, Hazaribag and Singhbhum. Ethnically they are proto-Austroloids and speak Mundari language. The word Munda means a man of prestige and wealth in the Mundari language. The Mundas are divided into two branches. The elder branch consists of the Mahali Mundas also called Tamarias because of their habitation in Tamar Region who formed a separate tribe with whom the other Mundas including the Ho Mundas generally refuge to inter-marry. Another branch is subdivided into Kompah Mundas (generally called Mundas) and the Ho Mundas (simply called Hos). These two branches comprise the main body of the Mundas and the Hos of Singhbhum. They are divided into a number of exogamous as regards the killis, the Mundas are endogamous as regards other tribes and marriage with members of other kolarian tribes such as the kharias, Asurs, Santhals, etc. is not permitted. The only except is the Munda of Tamar. The killis are Totemistic and derive their name from their natural objects such as animals, birds, fishes, plants, reptiles, etc. Killis are descendant from a common ancestor. Important killis are Tuki, Mundu, Su, Horo, etc.

The Mundas are in build, stout and muscular and in stature somewhat below medium height. They are of the darkest colour, the colour of their skin being black brown. The head is long, the nose is thick and broad and sometimes depressed at the root, the lips are thick and the cheekbones remain prominent.

The Mundas are basically agriculturists. All of their agricultural implements required are owned and prepared by themselves. Similar to Oraons they have also developed terraced fields for rice cultivation because their concentration lies in relatively higher areas. The tilling of land is done by wooden plough and method of cultivation is traditional. It is, however, during recent years educational development has enabled these people also to copy the agricultural techniques of non-tribals in the form of developing irrigational means, growing crops, using fertilizers, pesticides, improved seeds and other improvements. They grow rice in the terraced lower fields and maize, pulses, spices, etc. on the relatively higher areas. The rice area having clayey soil is called Panku but the fertile soil area is known as Nagara where they grow several crops and soils having dominance of sand is used for growing since pulses and maize and Marua. Such land is known as Khursi. They construct houses on the Tanr lands where soils remain less fertile. The tribe is comparatively more educated and politically conscious people have adopted modernity in agricultural practices. They have started growing crops which enable them to fetch more money. The production of cash crops, vegetables, seasonal fruits has increased considerably. They have also developed interest in dairy activities and keeping fowls. Educated persons have started involving in mines, industries and other non-agricultural activities.

The Mundas prefer living areas close to their agricultural land. They construct dwellings on the Tanr lands and grow crops in terraced fields. As much their dwellings do not form concentration, they remain scattered based on the scattered agricultural patches. Moreover they do not care for safety from theft and settle anywhere. Their dwellings consist of open space in the backyard which they use for growing vegetables and fruit trees. In the front they keep fenced open space in the form of courtyard and a platform for keeping agricultural implements, agricultural products, enjoying summer heights etc. They have also keen interest in keeping cattle and fowls for which they provide space in the houses. Their houses are mostly mud walled and tile roofed. Certainly resourceful Mundas have brick walled and tile roofed houses and pucca roofed houses. Stone walled and tile roofed houses are also found in rural areas. Their houses also lack windows and have only one opening. Certainly the walls of the houses have provision of space for keeping small items. They keep their house neat and clean with regular use of mixture of soil and cow dung. They also use lime for whitening the walls. More people draw sketches of animals on the walls. Usually they live alongwith other tribes like Ghasi, Lohra, Chik Baraik, etc. so that their villages consist of several such tribes. Like an oraon village, the village of Munda tribe also contains a centrally located 'Akhara' where they sing and dance and do meetings. There is usually a secret grove where the deity is kept. Their houses contain separate space for sleeping and cooking.

Similar to Oraons there is a separate house near the main settlement for socialising unmarried boys and girls which is known as Gitiora. But this tribe does not have institution like Dhumkuria. They use the Gitiora for singing, dancing, etc. In addition to Gitiora they have also Sarna, Akhara, Sasan, etc. They keep certain forest areas reserved considering its religious importance where village God is supposed to reside. Near the settlement they have separate space in the form of burial ground where stones are kept above the bones of their ancestors.

The social system of the Mundas being of the Patriarchal types, the head of the family is at the same time the priest of the family who in case of illness or distress has to office sacrifices to propitiase the domestic and other Bongas. They consider Singa Bonga as supreme religious body who takes care in all difficulties. They also worship spirits of their ancestors. There are other Bongas also like Desauli Bonga, Jahar Bonga and Chandi Bonga, etc. Their religion to a considerable extent, is similar to Hindu religion. But the village community being composed of families descended from common ancestors looks on itself has a family and hence requires a special representatives to exercise over the whole village functions similar to those exercised by the head of the family in the household. This representative is called the Pahan and his functions concern offerings, sacrifice to please the local spirit and to preserve the village from the ravages of wild animals, to obtain good harvest and to ensure successful hunts. Social organisation like Parha is also in exercise in Munda tribe like that of Oraon tribe. Mundas decide their disputes and act accordingly as they are directed by the permanent standing Parha Panchayat consisting of Parha Raja, the Diwan, the Thakur, the Pandey, the Karta and the Lal.

Marriage is generally arranged by the parents or guardians. Certainly consent of the boy is obtained. A messenger is sent to the guardian of the girl and after getting favourable answer the messenger, the boy's guardian and friends go to the girls' village and settle the onwards formalities concerned with the marriage. The term being agreed upon the ceremony is completed by the

Pahan of the bride village. At the end they celebrate the function with feast.

Important festivals are Mage Parab, Fago Festival, Ba Festival, Soharai Dasai, Karam, etc. The Mage Parab is for pleasing the spirits of their ancestors when they celebrate with local beer. The Fago Festival is like a Holi. Ba Festival is also known as Sarhul when the Sal trees remain in blossom. On this occasion Chandi Bango is worshipped. They also celebrate the Batuli, Karam, Dasai, etc.

Fishing and hunting form a part of their hobby. Hunting is done usually on occasion of festivals and they do hunting from their self manufactured bows and arrows. They also organise dances and play by beating drums, flute, singing songs etc.

The dress of the Mundas is very simple and scanty. The males wear a loin cloth with a coloured border at the two ends called Botoi. They also use a piece of cloth as wrapper for the upper portion of the body locally known as Barkhi. The females wear a long piece of cloth called Paria round the waist allowing a portion of it for the upper portion of the body. Shoes are hardly used but during recent years males even use paint and shirt and females use saris and blouse etc. They are fond of ornaments. The poorer Munda women wear a peculiar ornament in the year called Tar, Sakom which consist of a roll of Palm leaf. The Munda girls appreciate jewellery and deck themselves out with earrings, necklace, bracelets and anklets of brass or some more precious metals.

In case of Mundas family all members have right in property but if the father is not happy with any son he can force his son to live outside the family and even after death of his father he fails to inherit property. Virtually the father has full right over property. As such sons hardly demand separation in his life time. The father has right to distribute property according to his wish. At the time of separation the eldest son and the unmarried daughters are given substantial consideration in property. In case of Munda boy marrying with girls of other caste he looses the right of property in the family.

The Mundas erect usually big memorial stones in the burial

ground called 'Sasan'. After a memorial stone is erected a sheep is slaughtered by it and a feast held in which kinsmen partake if the meat of the sheep as well as rice beer. The funeral feast is held several weeks, even a year after the death depending upon the convenience of the relatives of the deceased.

The Birhors :

The Birhors are an important nomadic tribe of the state. Despite the efforts of the Govt. nomadic character still concern with most of them. They change the place of abode with the change of season and convenience usually four times a year. As such majority of them have not developed affinity with the land and domestic animals they are far behind in terms of socio-economic development. This tribe consists of 10726 population (0.12% of the tribal population of the state) as per 2011 census. They are mostly found in the districts of Hazaribag, Chatra, Giridih, Ranchi, Lohardaga, Palamu, Purbi and Paschimi Singhbhum, etc.

The word Birhor is composed of the two words 'Bir' and 'Hor'. These two words mean forest and man respectively meaning thereby the man of the forest. Dr. L.P. Vidyarthi and anthrophologist of world repute has categorised this tribe as food gatherers. Their standard of living is confined to that level where they can carry on their shoulders and arms. They belong to the racial stock as the Hos and the Mundas and speak Austro Asiatic dialect. The Birhors are of the view that Kharwars are integral part of them. They consider themselves as descendent of the sun on the earth. They hold the opinion that they had seven ancestors who had come to this region from the hilly areas of Kaimur (Khairagarh). Out of them four moved eastward and three of them entered in the old Ramgarh district of this state. While these three brothers were proceeding for battle with the ruler of Ramgarh, the head dress of one of them got entangled in a tree. He deemed it a bad omen and did not proceed on with his brothers and he started living in jungle. He did not like to go with two brothers who became ruler of Ramgarh. Later on, they formed two groups: 'Jaghis' who adopted permanent way of life and 'Uthlu' who are still in wandering stage.

As regards the physique of this tribe they resemble with the physique of other tribes like Santhals, Mundas and Hos. They are also of short stature having dark colour, curly hair, broad nose, etc. Physically they are quite strong and remain capable of carrying their belongings from one place to another.

Since most of the Birhors have not developed closeness with permanent habitation and agricultural pursuits, negligible changes have taken place in their profession. They still lead wandering life and depend on forest products and animals trapped with the help of net. The net is prepared by ropes made of jungle creepers. Certainly a good number of them have been allotted houses, land, cattle and other facilities. They have developed involvement in agriculture. More particularly Uthulu Birhors have been settled in the residential colonies of Jehangutua, Katia and Beti of Bishunpur Block of Gumla district, Residential Colonies of Bigha, Dembu, Peta, Sultan Adara and Amnam of Hazaribag district and other areas with the allocation of 5 acres of agricultural land, 10 decimals of plantation land, one pair of bullock, goats and fowls. Cultivation is done mainly by bullocks. In Jehan Gutua and Beti of Bishunpur block and in some other areas even tractors have been given by the government in group. As such Birhors have started using tractors for cultivation. Time to time seeds and chemical fertilizers are also provided to the Birhors by the government. They have also developed interest in using organic manures like gobarkhad. But due to lack of irrigation facilities, the Birhors in mass depend on natural precipitation. However, they have developed smaller irrigation facilities like ponds, wells for irrigating this lands. Due to inadequate land they also depend on forest products. They also work as labourers. Most of them also make rope and manufacture various kinds of goods from wood. They also earn some money by selling wild animals like deer, rabits, etc. They trap animals and birds with the help of the net in the most interesting manner. Monkey trapping is rather a favourite pastime for them. They also collect herbs and honey from the forest. Some of them have adopted modern way of rope making and bee keeping.

The Birhors are divided into exogamous clans named mostly after animals and plants. Baiga clan among them is the most powerful. They have two type of organisations - one group manages economic affairs and another group looks after social affairs. They are divided into several smaller groups. Each group consist of 3 to 10 families. They remain in search of food items from one place to other. Sometimes, some of the groups stay for longer period which is known as Tanda organisation. Each Tanda organisation consists of one chief and known as Naya who cares for religious and social affairs of the Birhors. In case of death Naya, his son becomes successor and after necessary tribal formalities like offerings the spirits they go out for hunting. One member of the organisations looks after the evil spirits and remedies. Such persons is known as Mati. The married Birhors do not like to hunt such animals and birds which have traditional relationship. Before marriage they do not follow this rule because unmarried Birhors are not members of Tanda. They belong to Patriarchal society while getting separated the eldest son is given slightly more share than others. In case of two wives children of first wive get relatively more share.

The Birhors live in groups. Their settlement comprises of a dozen or sometimes more than two or three dozen of families in some temporary settlements. They occupy lands for houses on the gentle sides of hills and close to a spring or well. They construct hut with the branches of trees and leaves. Birhors settlements of huts is called 'Tanda'. They do not have beast of burden or even a cow etc. and carry their belongings themselves. Their camping huts are usually circular in nature about two metres high and three metres in diameter. The entrance door faces the east and is about two feet above the ground. A man and his wife and young children sleep together in this small hut but grown up children are provided with separate hut. They sleep on palm leaf mat on the ground.

The Birhors worship female deities and devils. They have assigned to Devi, the chief place among the former and others are supposed to be her daughters and grand daughters. She is worshipped as the creater and destroyer. The devil Biruhut is worshipped in the form of a raised semi circular land. A small round piece of wood nearly a foot in length, the top painted red is called Banhi, the Goddess of the jungle. Some of them have adopted Christianity. They celebrate the Baradin of Christian festival.

In case of marriage of Birhors parents take the lead. The father of the bridegroom pays token dowry to the father of the bride. They have no priest and the only ceremony is drawing blood from the little finger of the bridegroom and the bride and with this the Tilak is given to each by marks made above the clavicle. They do not marry in their Gotra. The women dress decently. They have marks of tattooing on their chest, arms, ankles, etc. Previously males used to bear Bhagwa and put Gamacha on their shoulder. But at present they wear Pant and Shirt, Kurta, Dhoti, Lungi, etc. Females wear Sari, Blouse, Peticot. Girls wear Skirts, Blouse, Salwar. Previously females used to wear only Sari and used to cover their body with small piece of cloth. They are also fond of ornaments. Females wear Kekana as ear ring which is made of palm tree leaves. Modern ornaments are lesser in use.

In the past Birhors were using earthern utensils but these days they are using utensils made of aluminium, brass, bronze or even steel. In some household articles are made of glass. They have also recorded change in their deity habits. They have been using rice, wheat, maize, ragi, mahua, gondali, etc.

They have recorded considerable changes in their festivals and beliefs by coming in contact with Hindus. Birhors have begun celebrating Rakshabandhan, Gomhapuja, Makar-Sankaranti, Saraswati Puja, Holi, Deepawali, Laxmipuja. Besides they are also celebrating festivals like Karma, Dusshera, Sarhul, etc. Christians Birhors follow the christian festivals.

At the time of birth Birhors keep the pregnant lady in corner of the house surrounded by other women. After the birth of the child the concerned women is not allowed to use main door of the house rather a separate door is made for her. That day Birhors go on hunting in the forest and the successful life of the infant is speculated on the availability of the hunt. In case of the death of Birhors they burn the dead body and follow the Hindu rituals.

The Hos :

The Hos are agriculturists-cum-hunting tribe. Their total population is 9,28,289 (10.74% of the total tribal population of the state) as per 2011 census. They are mainly concentrated in the Kolhan upland region, in the southern portion of the state. The Hos dominated areas lie in the Chaibasa sadar sub-division of western Singhbhum district. This area has partly very fertile land and some areas have dense forest having wild life.

As regards their origin no separate account is available because they earlier formed the part of Mundas. Virtually Hos and Mundas were in one tribal group a few centuries earlier and even today they are also found in some areas near Mundas. They also comprise similar tribal and cultural traits. It is supposed that prior to their arrival to this area Kolhan inhabited areas were occupied by the Bhuiyas and Sarawaks who were driven away from the locality and the whole of the Kolhan was in course of time occupied by the Ho tribe. These tribal people dislike to be called by others as Kol. They state emphatically that they are resident of Singhbhum, the land of lions.

Similar to the Mundas the Hos are stout and muscular and in stature somewhat below medium height. They are of dark, brown colour. The head is long. The nose is thick and broad and lips are thick. Bone cheek remains prominent. Since Hos are concentrated in the Kolhan upland area having the hilly terrain and homogenous fertile plain they portray the land utilizations varying in both areas dominate with patches of agricultural lands, isolated homestead area and limited areas engaged transport network and other land uses. The homogenous fertile areas have comparatively more habitation, limited areas under forest, more culturable land, pastures and other non-agricultural uses of land. They are basically agriculturists but also depend for their livelihood on hunting and food gathering from forest products. A good number of them are also working in mines and industrial enterprises. They do any work collectively like preparing the agricultural land, performing other agricultural works, etc. They spare a portion of agricultural products to their deity 'Bonga'. These people have also adopted modern techniques of agriculture like non-tribals. Now they do not hesitate in using improved agricultural implements, high yielding seeds, chemical fertilizer, pesticides, etc. They have also developed minor irrigation facilities for irrigating vegetables and other cash crops. Some of them are also carrying dairy activities and poultry.

Previously they used to construct their dwellings on the flat hill tops but these days they prefer settling on the river banks and lie close to springs for water facility. Their houses are well built and grouped to form Tolas. Akhara or dancing place, dormitory, Jahera or village grove for village deity and burial ground called 'Sasan' characterise landscape of the settlements of the Hos. Usually every village contains a few families of weavers, potters, cowherds and ironman. Houses of the Hos are categorised into three types (i) Large houses containing two or three rooms, (ii) Medium houses having smaller two rooms and (iii) Small huts. The houses are mud walled and tile roofed. Houses contain a Varendah.

The Ho settlement area is divided into groups of villages called 'Piro' to be controlled by a divisional headman known as 'Manki' while the headman of each village is called 'Munda'. They are assisted by constables called 'Dakuas' appointed by the Manki. At present much of the village affairs are governed by the Government.

This tribe considers the marriage as an essential feature. They do not believe in polygamy. They do marriage in their own clan. After the marriage the couple live separately. If the wife fails to give birth to a child, the male deserves right to marry with his sister-in-law. The Hos observe the following marriage rules like Andi, Diku Andi, Opartipi, Rajikushi, Anader, etc. Andi and Diku Andi marriage emphasize on the rituals.

The Hos consider their God as Bongas. The supreme body is known as Singh Bonga Goddess of water who is considered wife of Singh Bonga. Besides, there are other Gods and Goddesses like Desuli, Matang Banga, Hatu, etc.

Similar to Mundas the Hos have festivals based on agricultural practices and have also adopted festivals of the Hindus and the Christians. Some of the Hos celebrate Chhath, Dipawali, Dusshera, Shivaratri, Kalipuja, etc. They have their own festivals like Maghe, Baha, Damurai, Hiaro, Jamnama, Kolan, Boatouli. All these festivals are related to agriculture. On the occasion of festivals they drink wine and perform dances.

The Hos do not have the practice of dormitory for adults. Still children sleep in one room in group. They assemble at one place in the evening and spend times in listening stories, singing and dancing. They also perform several plays like Chhor, Kanja, Kaso, Hota-danda, Kulachal, Pete. They also celebrate birth of their children and the death of elderly people.

Kharwars :

Kharwars are a Dravidian tribe and are primarily based on cultivation. They are 2,48,974 in number and comprise 2.88% tribal of the state as per 2011 census. They are largely found in the north-western portion of the state. The main concentration lies in the districts of Lohardaga, Palamu, etc. There is a legend that reveals that they have come to this part of the state from Khari-Jhal whence the name Kharwar. Kharwars also hold the view that their original place has been Rohtasgarh. They also claim descent from Rohitasawa, the son of Harishchandra and wear Janeu or sacred thread. Colonel Dalton expressed his view that there has been traditional connection between the Cheros and Kharwars, who are supposed to have invaded Palamu from Rohtas and driven Rajput chief of the country. Some, however, allege that they came from Ramghar but non are able to give exact date. Kharwars are divided into six class or sub-castes viz. 1. Surujbansi, 2. Dualbhandi, 3. Patbandhi, 4. Kheri, 5. Bhogti or Gounju and 6. Manjihia.

They are physically strong and have medium to long stature. Rich Kharwars are very refined and have high broad appearance. They are also of brown colour. They are mostly concentrated in the hilly areas where only patches of agricultural lands are found. In these areas land under forest predominates being followed by some culturable land, homestead land, pasture, fallows, graveyard, religious places, etc.

Property descends to the sons. If one has two wives their sons get the property in equal shares. If there be only daughters the widow retains the property so long as she continues in the house. When the widow gets remarried the property goes to the nearest male relative, but a portion is given to the daughter.

Kharwars follow the local Brahaminghi and also the local tribal religious ceremonies and festivals. They have both Brahman and Baiga priests. Their main festivals are Dussehera, Chhatha, Dipawali, Holi, Mahabiri Jhanda, Sarhul and Nawakhani. They offer milk. fruits and sweets at Hindu festivals and Handia, fowl and goat at the tribal ones. The deities of Kharwars are Muchukarani or Durjagin or Pachari. She is believed to be the wife of the rascal that reigned in the region. The rascal is said to have resided on a hill in Barwadih where the ruins of his palace still exists. A sheep or goat is killed as an offering to this deity. Duarpar is supposed to govern the village. He is appeased with the killing of a pig. Dharti to whom a pig has also to be offered. No cultivation is done until he is appeased. Debi, Chandi, Darha and Dahkin all of whom have to be propitiated with offerings of either fowls, goats or pigs.

Customs are curious among Kharwars, friends and relatives meeting after long absence, salute each other in three separate ways like Ankwar (joining hands with each other), Johar (embracing each other) and Kusal Mangal (Asking good wishes or blessing of each other). When a Kharwar meets an acquaintance he simply bows and touches his right leg.

The Kharwars have their community Panchayat in which all male elders of the village participate. A number of village are grouped in a Chata and its head is known as Parohan. The head of the community Panchayat at village level is known as Mukhia. The Mukhia must inaugurate for each village the work of showing or reaping or of transplanting paddy seedlings. The Baigas are very important functionary in the community Panchayat. Almost all social matters are referred to Panchayat and its decisions are honoured. They speak Sadri and have no tribal dialect of their own. Their surnames are Singh, Sai and Ram.

Polygamy has been in practice. Infact marriages are not practised. The bridegroom must be of the age of twelve or fourteen years and able to plough and the girl should of the same age. A widow may remarry in a Sagai form. If a women is childless her younger sister may marry her husband. This is called Rijbia. A feast is organised at the end of marriage rituals. Birth and death ceremonies are similar to the Hindus.

The Kharias :

The Kharias are one of the main aboriginal tribes. They are 1,96,135 in number and consists of 2.27% tribal population of the state as per 2011 census. Originally they were living in Rohtasgarh and M.P. where from they have come to this region. They are mainly found in Purbi and Pashchimi Singhbhum, Gumla and in some number in Palamu, Lohardaga districts, etc. Besides they are also found in Mayurbhanj district of Orissa and in eastern hilly areas of Chattisgarh in Assam and West Bengal. Kharias, the Dhoh and the Dheekhi Kharias. The hill Kharias are mainly concentrated in Singhbhum district and in Mayurbhanj district of Orissa. They live in villages situated on the hills or valleys surrounded by wooded hill ranges. They are a most primitive community and subsist mainly on the forest produce. They do not like to socialise with others. Dhoh Kharias are mainly found in the Purbi and Pashchimi Singhbhum at Gumla and the Dhelki Kharias have taken to cultivation and are somewhat better than the hill Kharias. They are mainly found in Chattisgarh and do not like marital relationship with hill Kharias.

Physique of these people are of medium stature. They are of brown to black colour. Females are comparatively of fair complexion. They also have good physique. Their faces are flat and noses are depressed. The language of Kharias is a branch of Mundari language. It has developed from Austro-Asiatic language. Hill Kharias have almost forgotten their own dialect and have adopted local dialects.

The Kharias have developed closeness with agricultural pursuits and landuses distinctly. Hill Kharias still are more primitive and their cultivation is comparatively in bad condition. They still practice Jhum cultivation on the sloping land and most of them depend upon wild animals and fruits. On the other hand Duoha and Dhaki Kharias have developed landuses and agricultural practices in a better way because most of them have adopted Christianity and have associations concerning economic and social reforms. They have Kharia improvement society. They have developed interest in growing food crops, vegetables and seasonal fruits. They have little irrigation facilities. As a result, their agriculture depends largely on natural rainfall. Some of them have started using modern agricultural techniques in the application of high yielding seeds, chemical fertilizer, growing several crops, etc. Educated Kharias are in services and have good attachment with Christian missions.

Among the Kharias clan system is strongly followed. The totem objects are not injured or eaten. The clan names of Duoh and Dheki Kharias include like the following - the Dherka Kharias have clans like Muru (Tortoise), Soren (Stone), Samad (deer),

Barliha (a fruit), Charha (a bird), Hansa (eel), Topno (a bird), Mail (dish) and Duoh Kharias have clans like Dundung (eel), Kulu (Tortoise), Karketa (quail), Bilung (Salt), Soren (stone), Ba (Padd), Tote (a bird), Topo (a bird), Kiro (Tiger). Exogamy and Totemism are the main features of the Duoh and the Dhelki Kharia clans.

The Kharias have settlements of varied types. Hill Kharias avoid living in main settlement. Normally a few families of hill Kharias clear a patch of land and construct isolated huts and lead forest life. But Duoh and Dheki Kharias live in main settlement and they have good houses. Almost all villages have Akhara where they conduct singing and dancing. Houses of Pahari Kharia has only one room which is used for sleeping, eating and keeping household items. They cook food in open space. Their houses are mud walled, tiled and thatch roofed. Houses hardly contain windows. Kharias mainly eat rice. But hill Kharias supplement rice with hunts and fruits.

They hold patriarchal family system. Their nomenclature follows patrilineal descent. They do not believe in polygamy. They take enough care of wife and children. The non-christian Kharias, men and women take keen interest in personal decoration and ornaments but the Kharia bhagats discourage their women to wear ornaments except the sacred threads round their neck. They live on pure vegetarian diet while other Kharias are non-vegetarian. The Kharias Bhagats also abstain from Handia which is an important drink of the Kharias.

The head of village is Pahan and the Pahan is associated by Mahto. If any trouble arises the decision of the Pahan and Mahto is binding on the community. In coming to a decision the Pahan is assisted by the village elders. The statutory Grampanchayat has not much affected their traditional Panchayat.

The Kharias have also several religious activities related to agriculture, tradition and spirits. They obtain strength by religious activities as such they worship their ancestors, religious bodies like Gods and Goddesses, Clan spirits, etc. They consider Sun as supreme religious body. They call God in their own language as Giring, Bero, Dharam Raja. After God they consider spirits of their ancestor. Their main festivals are phago, Polo, Gaogangari, Ronol or Bangeri, etc. They also believe in magic.

Loharas :

This tribe is much different from other tribes of the state because they have less relation with hunting, food gathering, etc. They have confined themselves to black smithy works and they are, therefore, widespread. Loharas are 2,16,226 in number and comprises 2.50% of tribal population of the state as per 2011 census. The Loharas consist partly of immigrants from Bihar who are known as Kanaujia Loharas and partly of indigenous blacksmiths who are known as Kol Loharas or Nagpuria Loharas or Loharas.

The Nagpuria Loharas are divided into two sub-castes viz. Sad Kamar and Lohara proper. The Sad Kamar have mostly left their caste occupation and have involved themselves in agricultural activities. They are in good number in the ·Panch Paraganas Plains. The Lohara proper are still engaged in their traditional occupation. Their main works are concerned with manufacturing of agricultural implements and domestic articles like sickle, hoe, axe, iron-head of the arrow and parts of plough etc. In addition to iron associated works they also involve themselves in stone cutting and agriculture for livelihood. Since work of the Loharas give more importance to their traditional occupation they do not take much care of their fields. They try to get the agricultural works in association with others. Usually their agricultural output remains poor. they also have not developed aptitude towards milching cattle, fowls, etc.

They speak Sadri and Mundari and in some villages follow the Munda custom of burial in a Sansandiri. Most of them avoid taking meat other than of fowls or goats. But some Loharas do not follow much restrictions in terms of non-vegetarian food and drink. They even eat the carcases of dead animals and accept cooked food both from Mundas and Oraons. Loharas and Sd Kamars do not have matrimonial relation because of their differential social development. Loharas are considered much lower in the social scale than the Sad Kamars.

Loharas have adopted Hindu way of life, about one-third still have tribal traits. Only a few adopted Christianity and other religious.

It is important to note that Loharas do not have exclusive villages of their own but live together with other communities in most of the villages because their professional work is needed in every village. They construct mud walled and tile roofed houses and stone walled and tiled roofed houses. But these days educated and resourceful Loharas have got rural dwellings made of other raw materials based on borrowed designs. Their houses contain separate space for their professional works. They have a nuclear type of family. It consists of father, mother and their unmarried children.

By and large they are monogamous. Their marriages are arranged by parents. The father of the bridegroom pays Dalitka or bride price together with saris. The marriage is solemnised at the place of the bride's father in the month of Magh, Phalgun, Chaith, Baisakh and Jeth.

Loharas have also clan system. They are totemistic. The share of the father in the property is equal to that of his sons. The females have no right in the property. Certainly they have claim for maintenance.

These people do not have Akhara like other tribes. They, however, go to Akhara occasionally and dance and sing alongwith other tribals. They have also developed strong liking for drink. They also celebrate birth and death ceremonies. They do not have liking for distinct dress and ornaments. They use these things prevailing in respective areas.

Bhumij :

The Bhumij are also known as Matkums. This nomenclature is based on tribal language in which it means the children of the soil. Some people even club Bhuian with Bhumij. They are 2,09,448 in number and consists of 2.42% of tribal population of the state as per 2011 census. Previously they were in great number in districts of Manbhum and Dhalbhum. These days they are mostly concentrated in the Dhalbhum area of Purbi Singhbhum, mainly found in the areas lying between the Kasai and the Subranrekha rivers. These people also live in small number in districts of Hazaribag and Giridih. Their habitations are worked either on the banks of the Subarnarekha or the hill ranges. Those who have settled in the densely forested areas were previously known as Chuars and often come to open revolts in the part. Some of the scholars even considered this type as part of Munda because their origin is almost the same. This opinion is based on the fact that a branch of the Bhumij used the Munda language and call themselves Muras. Like all Kolarian tribes they build no temple but worship Mountains or Buru in the form of a stone smeared with vermillion which is set up in a Sarna of Sahira, the sacred grave near the village. Now these things still continue amongst the Bhumij west of Ayodhaya range. On the east of Ayodhaya range the title of Mura has been replaced by the title Sardar and almost all the Bhumij speak Bengali.

The Bhumij have kept good concern with the land which is attested by their keen interest in agricultural activities. They are cultivators and agricultural labourers. They have tried to use the land in a better way. They hardly leave the culturable as fallows. With the help of hard labour they have constructed weils and water bodies for irrigating cash crops, vegetables, gardens associated with their settlements. They also try to produce various crops not only for domestic requirements but also for marketing purposes. Most of them have also started using chemical fertilizer, improved agricultural implements, high yielding seeds and other modern agricultural practices. Some of them depend upon hunting, fishing and food gathering and other forest products.

The rural settlements of Bhumij consist of smaller nucleations having 30 to 40 houses. Their settlements are elongated. They do not have separate settlement of their other tribes. They construct dwellings alognwith other tribes. The dwelling of Bhumij contain two to four rooms with cattle shed. Normally they do not keep cattle feed on the ground because of the problem of termite. They develop platforms for this purpose. They have mostiy mud walled and tile roofed houses. But resourceful Bhumij have also constructed their pucca houses.

The Bhumij also have their own tribal organisation. They maintain their own tribal Panchayat consisting of a headman known as Pardhan. This post descend to the son after the death of Pradhan. People of the village have to obey the decision of the Pradhan.

It is to be noted that this tribe has considerably adopted Hinduism and is regulating the tribal deities to their female folks. They are trying to claim Kshatriya status. They follow the rituals concerning birth, death and other social ceremonies. They also believe in God and Goddess similar to Hindus. They celebrate religious functions like Holi, Diwali, Dussehera, etc. They celebrate birth and death as like Hindus.

Mahli :

The Mahlis are considered as Dravidian caste of labourers, planaquin-bearers and workers in bamboo. Their population is 1,52,663 persons (1.77% of the tribal population of the state as per 2011 census). They are concentrated in the district of Ranchi, Gumla, Lohardaga, Hazaribag. They are also found in some districts of Bihar like Munger, Bhagalpur, Banka and parts of West Bengal, etc. They live mostly in rural areas. They are divided into five sub-castes like Bansphor Mahli who make baskets and do all kinds of bamboo works. Pahar Mahli i.e. basket makers and cultivators, Sulunkhi Mahli, who are cultivators and labourers, Tanti Mahli, who are palanquins and Mahli Munda, a small sub-caste confined to Lohardaga. Social scientists like Risley Conjectures that the main body of the Mahlis consisting of the groups Bansphor, Sulunkhi and Tanti Mahlis are more allied to the Santhals. He, however, mentions that the Santhals are allergic to basket making and carrying palanquins which are the main occupations of the Mahlis. He hold the view that the Mahli Munda group of the Mahlis, probably parked company with the Munda tribe for similar reasons. But his view has not been widely accepted.

Risley holds the view that similar to Munda, Mahlis have also sub-septs related to a tree (Charber), wild pig (Dhilki, Dumria or Dungri), Grain forbidden (Goondhi), Jackfruit tree (Kathargach), a kind of bird (Korketta) etc. The Mahlis do not agree with these sub-septs. However they are aware of septs like Kantiar, Dumaria, Tundwar, Kendriya, Hansdagia and Kesriyar.

The Mahlis have to a considerable extent accepted Hinduism and practice Hindu religious customs and beliefs but some of them still have retained their original occupation of basket making and bamboo works. They have comparatively less inclination towards agricultural activities. But during recent year they have friend to have culturable land and have started agricultural activities. They are growing food crops, vegetables and seasonal fruits. They have also followed the agricultural techniques and modern changes in agriculture adopted by others.

The impact of Hinduism on cultural and social activities of Mahli is obvious by the fact that Mahlis not only take keen interest in Durga Puja but also perform it in their villages. On this occasion they sacrifice a he-goat and one or two pigeons. Certainly they do not engage any Brahmin to perform the Puja. They also worship Ber Pahari (the 'Mountain God of Mundas), Mansa, the Snake Goddess and other Godlings of the animistic tribes. During Karma festival both men and women freely take part in dance and music. Their dance and music do not differ substantially from those of the Mundas and Oraons. The dead bodies are both buried and cremated. The cremation is a later idea adopted due to the impact of Hinduism.

They eat beef, fowls and pork. They are fond of music and dance.

The Korwas :

The Korwas are one of the Kolarian tribes and their total no. is 35,606 (0.41% of the tribal population of the state). They consider themselves as the original inhabitants of Palamu district of Jharkhand, Surguja and Jaspur districts of Chattisgarh, Mirjapur in U.P. However, social scientists have speculated their origin from SriLanka but how or when or under what circumstances this tribe came from Sri Lanka is not very clear. The Paharia and Danr Korwas have a common legend that they have descended from the scare crows set upto frighten wild animals. The Dih Karwas, however, are not familiar with this legend.

The Korwas are short statured people. They are dark brown in complexion and strongly built and active with good muscular development. Their foreheads are narrow. Some of them have black skin, flat faces and projecting lower jaws. In height they are just about 5 feet and 6 inches and the women seldom having higher than 5 feet. They wear beard and moustache and keep long hair.

The Korwas have seven clans like Raj korwa, Manraji, Samat, Edgi, Murung, Birjia and Birhor. Each sub-caste is obliged to marry among themselves. They do not eat and drink with outside their sept. The Korwas living in the villages are called Dih korwas while those living on the hills are known as Paharia Korwas. The influence of totem is strong and a totems are tiger, snake, parrot, wild goose, fish, mango, etc. They are extremely good hunters and bowmen. They still have the nomadic habit and change their homestead frequently. They have no language of their own and speak a mixture of several dialects.

The agricultural method of this tribe has recorded phenomenal changes. Ancestors of this tribe used to practise shifting cultivation which was known Bioda Agriculture but this type of cultivation caused considerable loss to forest resources and it was discouraged by the government. These people also realised the problem and preferred permanent cultivation. They grow maize, paddy, ragi, til, beans, etc. Their agricultural products hardly serve one month food. As such they depend on forest gatherings, hunt, honey collection, labour work and cattle rearing. But lack of money unables them to purchase cattle for rearing and always like to have government support.

Houses are generally built facing northwards. Walls of huts are of Sal posts plastered with clay and cow-dung. The frame of the roof is made of Bamboo and is thatched with paddy straw. Before occupying a new house the owner performs Puja by making offering to Devi, Rascal, Muchukrani and Durpar, the deities of Korwas. They construct their houses close to the forest and hills for forest hunt and herbs and water supply. Their houses remain of one room with verandah. There is only one opening in the room. They keep pigs in one corner of the verandah.

Almost any thing is eaten by Kormas - beef, goat's meat, venison, sheep, fowls, turtels, fish, cattle that die a natural death, hare tiger, leopards, hanuman, monkeys, milk, butter, ghee, etc. and all edible roots. Meals are served first to the old parents and friends in the house and then eat by all members.

There are two marriage systems - the run away or love marriage and second system is arranged marriage in which the father of the boy sends two man of his own caste to the girl's father and ask for his daughter. After getting consent the marriage is arranged. A Korwa may marry three wives but not more. Polygamy is, however, practised only when the first wife is childless. A man may marry two sisters provided the elder one becomes his wife first. He may marry his elder brother's widow.

Among Korwas men wear a Dhoti. It is supported on the waist by a string called danda. In lieu of Dhoti some men simply wear a Bhagwa which is a piece of cloth. They also keep Pagri on the head. Some also occassionally wear an Angana over the body. These days educated Korwas use paint, shirt. Women wear Sari. Korwas are also fond of ornaments. They wear Kanousi of brass or silver on both ears. They also wear bera (bracelets of silver) on the right hand and occassionally on both hands. Ring is also worn on third finger of left hand. They also use necklace of beads, armlets of brass, anklets of kaba, nose ring of silver, etc.

Usually Korwas avoid dancing and singing but dances and singings are performed on the occasion of marriage, births and other functions.

Ili or Handia made of rice is common drink. They serve Handia to their guests on ceremonial occasions. But these days they have also started using local drink which is easily available in the villages and markets.

They have also social organisation for dealing with social problems. They have a Korwa Panchayat having a leader known as Pardhan. Elderly people assist the Pardhan.

Kormas have strong faith in God and Goddess living on hills and trees. Their deities are Devi, Rascal, Durpar, Manusdeva. All these have to be appeased at different seasons of the year by the offerings. They also worship their ancestors. They consider Singbonga as the Supreme body. They also worship moon, earth, Dihwar, Gonhal, Salbahani, Karma, Sarhul, etc.

In case of birth and death they follow the rituals of Hindus. In some case only they also bury the dead body. Normally the dead bodies are burnt and also burried. They celebrate birth and death.

The Cheros :

The Cheros are comparatively developed aboriginals. They are land holding and cultivating class of people. They are 95,575 in number, 1.11% of the tribal population of the state as per 2011 census. They consider themselves to be Rajputs. Virtually they have history of administrative class as they ruled over some Gangetic provinces and also were rulers of Palamu They are mainly concentrated in Palamu Garhwa districts of Jharkhand. Smaller number of Cheros are also found in Rohtas district of Bihar and Gorakharpur district of Uttar Pradesh. In Palamu and Garhwa they retained their stronghold till the accession of the British administration. They had the Rascal chief, sometimes in the reign of Jehangir. They had invaded this region from Rohtas with the help of Rajput chiefs. The ancestors of the Thakuais of Ranka and Chainpur head driven out Rajput Raja of the rascal family. The later retreated into Surguja and established himself there as Raja. During their administration in the region the Cheros conciliated with Kharwars only and used to avoid other tribes.

The Cheros vary in colour but are usually of a light brown colour. Their physique appears comparatively sound. They have high cheek, small eyes, low broad nose and large mouth with protuberant lips. They are divided into two sub-caste - Barahazar or Barahazaria and Tere-hazar or Birbandhia. The two sub-castes are further sub-divided into seven clans having the distinguishing titles of Mowar, Kuanr, Sonwat, Rautia, Manjhi, Sohanait and Mahto. Among the Mowars and Kuwars there is a Barka Mowar and Chotka Mowar and Barka Kuanr and Chotka Kuanr. The Chotka Kuanr are alleged to have come to this region from Buxar. Colonel Dalton is of the view that the Cheros of this region had been expelled from Shahabad because of their ill habits.

Since the Cheros were tribes having experienced of cultivation they developed culturable land and related uses of land in the northern plain areas of Palamu and Garhwa. They left only limited areas for forest culturable waste and pasture. Even today this area has dominance of culturable land and the lower percentage of pastures, cultivable waste, fallows, etc. are available. Certainly areas under settlement, orchard and other uses are remarkable because of comparatively moderate concentration of population and the development of other infrastructures. They have revealed their experience of early agricultural background in agricultural pursuits in this region. They grow Aghani, Rabi, Bhadai and Garma crops. Rice and Wheat are important cereals. Among Bhadai crops maize is grown everywhere. Production of paddy is a common feature. Wheat production has also increased during recent years. These people have also developed interest in production of cash crops with the help of irrigation facilities and adopting agricultural modernity. They also supplement earnings by working in mines, road construction and the extraction of forest.

The Cheros hardly portray social traits like other tribals of this region. Virtually they have social status of high nature. They live strictly as Rajputs and wear sacred thread (Janau). They do not, however, inter marry with really good Rajput family. They worship Hindu gods but also worship such spirits as Baghaut, Cheori, Darha, Dwarfals and others to which goats, fowls and sweets are offered. In these sacrifices priests or Gurus do not take part. A Baia belonging to one of the aboriginal races perform the sacrifices and Puja. They have also priest called Pahan.

The Cheros reside in hamleted clusters. They have mud walled and tiled roofs houses but resourceful cheros have pucca houses having several rooms for different uses. By the side of the house there lies cattle shed and space for keeping fowls. Their houses also contain Varendah where they provide hospitality to their guests.

The marriage systems are similar to Hindus. Infant marriages are not in vogue. The Cheros and Kharwars do inter-marriage. A Chero may remarry firstly on the decease of his wife or secondly if his wife is childless. Widows are allowed to remarry. She can marry her husbands younger brother.

All Cheros women are tattooed. Tattooing is confined to women and is made by the malarian. Birth and death ceremonies are celebrated like Hindus.

Parhaiyas :

Parhaiyas are different tribal group from that of Paharias, Mal Paharias or Sauria Paharia. These people have also concentration in the districts of Palamu and Garhwa with limited number in Ranchi, Lohardaga, Hazaribag, Chatra, Giridih, Dumka, Deoghar and Gaya district of Bihar, etc. They are supposed to be Dravidian tribe. Their total population is 25,585 persons (0.30% of the tribal population of Jharkhand as per 2011 census).

As regards their origin no definite account is available but it is believed that Parhaiyas are in this region for pretty long time. They allege that originally they were the Duarpujaris or priest of the Maharajas of Palamu. They are chiefly found in the Southern tappas, seldom towards the north of Daltonganj. They are a simple people, hardworking and labourious and are good cultivators.

Physique of this tribe is moderately built. Their height consists of normally 5 feet and 3 inches. They have broad, flat faces with slightly oblique eyes and their colour is dark copper. They are by and large honest people.

They have habitations mostly in hilly areas as such their living depends on both forest gatherings as well as cultivation. These people need enough land for agriculture because they still frequently shift their agricultural land. They follow the traditional method of agriculture. They use plough, hoe and other old agricultural techniques. Most of the Parhaiyas are marginal farmers and landless labourers. Due to poverty and lack of resource they fail to manage agriculture properly. These tribes have also been motivated to have permanent cultivation in different areas by facilitating houses, fire acres land, pair of oxen, one cow, few fowls and agricultural subsidies. They practice cattle rearing also, They supplement their slender resources by collecting honey, lac and other jungle produce and barter them with grains, salt, tobacco. A large size of their land has been waste land and pasture.

They do not have traditional Panchayat system. When they were leading isolated living they had Panchayati system but after the independence this system has been weakened by the Govt. organised village Panchayats. They do not have recorded customs but they have carried their rituals on hearing say. The elderly people of the village settle the issues of dispute.

All Parhaiya women are tattooed (Khodna), which is done by the Ghasin or Malarian. Men are not tattooned. The figures are drawn on the arms, neck, chest and ankles of women.

Women and men both use ornaments. Women wear anklets, ring, bracelets of lac, glass armlet, necklet of beads, necklet of silver, ear ring, round water on forehead (tikli) and nose rings. Men wear ear ring, bangle of silver and occasionally necklace of beads.

The deities are Dharati or Muchukrani who is believed to remain within village under a large tree. She is appeased by the offering of goat and fowl Rascal is believed to keep off sickness, provided an offering of a he-goat is made annually. Duar Pahar is said to be a Dhosad who remains in the village. He is appeased with a Male pig. During the Dashara festival a buffalo is sacrificed. Debi Mai is a deity of goodness. They also worship Singhbonga, Parkaiya Debi, Gonhal, Satbahani, Darha, Goria, Chanari, Sokha, Braham, ancestors, etc. They also believe in ill spirits and its measures.

Settlements of this tribe is mostly sprinkled in nature. They construct huts of 6x12 cubits. Houses are made of mud walls and tile roofs. The houses contain doors facing east.

All edible jungle products are eaten and no one knows better than Parhaiyas where to obtain them. Fish of all kinds, young pig and castrated pig called Meda, fowl, goat's meat, deer, hare, doves, patridge, peacock and quails are eaten by them. They do not eat beef and mutton. Marrow is much relished alongwith curd and butter milk.

They believe in arranged marriage just like advanced communities. A man may marry two and three wives if he is able to support them but polygamy is permitted only when the first wife is childless. A man may marry two sisters provided the elder one has become his wife first. They may separate by mutual consent. Birth and death are celebrated in ways of other advanced communities.

Birjias :

Birjias and Brijias are the same. They are mostly nomadic in habit and a very small percentage has adopted cultivation. They are 6276 in number, 0.07% of tribal population of the state (2011). The bulk of them live in a group of two or three families on the spurs of the high hills. They still do a little cultivation by burning the underwood and by throwing some seeds on the ashes.

They are found chiefly in jungle tracks of the southern part of Palamu and Garhwa where the cultivation is done on jhumming system.

They have physique of moderate size. They are well built, strong and able to bear any amount of hardship. They seldom exceed 5'3" while women are about 5 feet. They are dark skinned and have broad noses, flat faces and small black eyes. They have developed resistance capacity to live in climatic regions and rough terrain.

Birjias speak a dialect called Birjia which is a mixture of Mundari and other tribal dialects. Their manners, customs and other ways of life are aking to those of korwas, kharias other tribes.

Birjias are divided into two groups on the basis of social organisation. These are Sinduria and Tiliya. Besides these there are other two sub-castes - Dudh and Ras Birjia. Dudh Birjias do not take beef but Ras Birjias may take beef. Birjias believe in polygamy. They do inter marriage but rituals of male side are followed in the marriage. In case of marriage out side caste Birjas penalise the concerned persons for offering seven times community dinner.

Their settlements contain small number of houses. Settlements are scattered in nature. Settlement contain a few rooms. Houses are made of stone walls and thatch roofs.

Property descends to the son after the death of father. If there is no son the property go to the next male relative and wife-half to each. Daughters do not get the property except given by the father during the life time.

Birjias eat beef, port, deer, goat, buffalo, sheep, Gaur, Dhamin snake, lizard, rat, bull, frog, figers, leopard, bear, peacock and all birds except vulture. They drink milk and take milk products.

Men wear a Barki Gendra which is a sheet or cloak for covering the body. It is in two long pieces stitched together. It is supported on the waist by a cord called Dandador. Women wear Sari and Gendra. These days males have started wearing pant, shirt, Dhoti and females have started using Sari, blouse etc. Male Birjias wear Bera (bracelet) of silver on the right hand. They wear Hisir (necklace of beads). A brass hair pin called Saluka is worn on the head to support the hair. Kanausi of brass are worn on both sides. Women wear necklace, bracelets, armlets, rings, etc.

Birth and death ceremonies are celebrated like other developed castes. The dead bodies are burnt. Feasts, drinks are organised on the occasion of birth and death ceremonies.

Asur :

Asurs are Proto-Austroloid and their language concerns Austro-Asiatic language. They are divided into three sects namely, the Bir Asur, the Birjia and the Agaria. In this region the Birjia has been treated as a separate tribe and the Agaria is a tribe found in Madhya Pradesh. Birjias are Jhum cultivators. Agorias mean for smelting iron and Bir Asur is living in jungle. They are concentrated in Ranchi and Palamu districts. Important areas are Netarhat groups of plateaus and Lohardaga Khunti, etc. Asurs are 22459 in number and consists of 0.26% tribal population of Jharkhand (2011).

As regards origin of this tribe scholar hold different views. There are references to the Asurs in the Rigveda describing them as great builders. They made fortified palaces called Pur. There are old sites of Asur habitation in the form of a garh said to be one of an AsurRaja at village Bhelwadag, six km. south-west of Ranchi. Another old site lie at villages Bihna, 9 kms. from khunti and at Lapundi, Digri, Ridari and Saridhkel in Khunti subdivision.

Physically the Asurs are quite distinctive with high cheek bones and broad nose. These people are stout muscular and of medium height. The complexion of their skin is brown. The Asur women have not that comely physique as that of a Ho or an Oraon women.

Three fourth Asurs are cultivators and one fourth are agricultural labourers. 50% Asur families have less than 5 acres land, 30% have 5 to 10 acres land and 20% Asurs have more than 10 acres land. They do not manage agriculture in proper pay because of poverty. They are highly indebted and exploited by money lenders. They move out in good number for Assam, West Bengal and other provinces as agricultural labourers. Some of them remain engaged in traditional occupation of iron smelting with the help of charcoal. But due to restriction on forest cutting they feel difficulties in managing their traditional work. They grow rice, maize, Marua, vegetables, etc. Asurs also depend on forest produce for their subsistence. They collect fruits, tubers and roots.

They also hunt wild animals for food. They are frequently engaged as wage labourers either in forest coupes where they find employment in the forest cutting or in earth work. These days they have developed interest in growing vegetables like potato, cabbage, cauli flower, etc. which enable them to fetch good money.

Their houses are scattered and in the centre of the village there is an Akhara or dancing Arena. This is used for assembling of Asurs for solving village problems. Their houses are made of mud walls and thatched roofs.

Asur society is divided into a number of killis or clans. A killi is exogamous and totemistic. A marriage in some totemic clan is considered a sin. After clan the family is considered the most important social unit. The major type of Asur family are nuclear, joint and extended families. Their community functions through the community Panchayat. The office bearers of the Asur Panchayat are Mahto, Baiga, Pujar and Gorait. As soon as any objectionable matter arises the Mahto asks the Gorait to inform the whole community to assemble at the village Akhara on a particular date and time. The decision of Panchayat is honoured. The defaulter has otherwise to leave the village for ever.

Asurs like other tribes follow taboos in respect of birth and death. They believe in SinghBonga and Morang Bonga as well as all the local Bongas known to the Baigas.

This festivals are Sohrai, Sarhul, Phagua, Kathdeli and Sarhikutasi. The Sarhikutasi festival is observed with sacrifice of a chicken for the prosperity of their iron smelting industries. Other festivals are common as with other tribals. The Mati and Ojha are also consulted in periods of unusual stress and strain.

The Asurs men usually wear loin cloth called Batoi and use a piece of cloth as wrapper for the upper portion of the body known as Pichauri. These days younger generation wear Kurta, pant, shirt etc. The women wear a long piece of cloth called Paria Sari around their waist. These days they use blouse and other Saris. They are also fond of ornaments like bracelets, anklets, earring and other cheap ornaments of silver, brass, lac and some coloured glass beads are used by the Asur girls. The poor women wear in the ears Tar-saken, a rolled bit of a palm leaf dyed red with black and rings called Dhela in their toe. The women do not tattoo at all as they consider it annoying to Gods.

Their usual diet consists of rice, maize, millets, vegetables and meat. Rice bear and Mahua liquor are their important drinks.

Asurs generally remain confined to their own tribe in terms of marriage but they do not marry in same clan or sect. They have dowry system for a pretty long time. Polygamy is allowed but the second wife is allowed only when the first wife is issueless. The marriage in this tribe is an arranged marriage.

The disposal of the dead body by cremation is rare and only resorted to in case of old people. Some piece are put into the grave and one piece is put into the mouth of the deceased. They bury their dead with the feet to the south. A dinner is arranged after the necessary rituals.

The Bedias :

The Bedia, Bedea or Bejea, are a small Dravidian tribe supposed to be cousins of the Kurmis through maternal aunt. Their number is 100161 persons, 1.16% of tribal population of the state as per 2011 census. Formerly Bedias and Kurmis had marriage relations and used to inter marry but a split occurred when it was discovered that Bedias eat beef. They do not claim any connection with the Mundas nor do the Mundas recognise them as cognates. Their septs are totemistic. In their marriages a barber officiates as priest. The Bedias take Kachchi food from Kurmis only and Kangar Mundas take similar food from them. The twelfth sept of the Santhals, which is supposed to have been left behind in Shampa and has long been separated from the parent tribe, bears the name of Bedia and it seems not improbable that the Bedias may be actually a branch of the Santhal who did not follow the main tribe in their wandering eastwards.

The Bedias are divided into totemistic clan groups. Each group is named after a tree or an animal and cutting or eating of that tree or animal is taboo for the group concern. Marriage outside the community is not allowed. They generally follow traditional rules. They are Patrilineal and Patrilocal, the inheritance and succession going through the main live. Their community Panchayat consists of Ohdar (Sardar) Mahto, Pahar and Gorait (Chowkidar) who decide the community problems.

The marriage is settled after the parties attend maturity. A bride can be arranged only after paying the bride price which is called Dhalitakka. It consists of some rupees, Saris, sweets, rice, oil, Gur etc. The Bedias have a peculiar custom, i.e. the bride price items must be equal to the items at the Chatti of the bride. Marriage is solemnised and feast is arranged.

The Bedias are at the cross road of their tribal religious creed and Hinduism. They worship their natural deities at the traditional Sarna and also the idol at the Hindu temple of Bhagwati. The important tribal festivals of the Bedias are Sohrai, Sarhul, Jitra, Karma, etc. They also observe Dipawali, Dussehra, Makar Sankranti, etc. When Bhatwati is worshipped, they fast on Sundays and Tuesdays which they think to be auspicious and then they offer milk and flowers and sacrifice goat. In the Sarna worship fowl is sacrificed. Though they themselves drink Handia they do not offer it on religious occasions except on the occasion of Sarhul in the Sarna.

The Bedias dispose of their dead by burial as well as cremation. Cremation involves more expenditure and only the old and respected Bedias are cremated. The personal belongings of the deceased go with him and are deposited in the grave or on the funeral pyre alongwith the corpse. Five bones one each from the head, chest and both knees of the dead are collected and put in the water of the nearby river. Resourceful Bedias immerse the bones of the deceased either in river Damodar or in the river Ganga. A dinner is arranged on this occasion on 13th day.

Gonds :

The Gonds are Dravidian tribe. They are found in south western bordering areas of the state. They have come from Chattisgarh and Madhya Pradesh where lies the main concentration of Gonds. Gonds are 53,676 in number and comprises 0.62% of the tribal population of the state (2011). In this region they are mainly concentrated in Simdega subdivision of Gumla district, a few pockets in Lohardaga district, Gumla Sadar subdivision, Palamu and Pashchimi Singhbhum districts. These people are divided into four sub-septs, Gond, Rajgond, Dhokhar Gond, Doroa Gond or Naik. Out of these Raj Gonds constitutes the bulk of the population. They claim presidence over other members of the tribe and have undoubtedly become more influenced by Hindus than other co-septs. Dhokhar Gonds are even today nomadic in character. The Dorao Gonds are adventurists and are mostly fond in Pashchimi Singhbhum.

Different septs of Gonds have developed distinctive relationships with environmental conditions. The Raj Gonds have developed more interest in agriculture than the Dhokhar Gonds who still lack permanent settlement. The Gonds living on hills practise shifting cultivation called Dippa. They supplement their diet with fruits, flowers, roots, leaves and hunting materials. Usually their holdings are small and not enough for their sustinance. But those Gonds who have settled in homogenous areas have permanent cultivation. Among Gonds most cultivators have limited land and are in good number cultivators cum agricultural labourers. They have also engaged themselves as industrial labourers by irrigating to industrial areas of Subarnarekha basin and to the industrial areas of Birmitrapur and Rourkela in Orissa.

The Gonds have their own social organisation. Their council is called the gond Mahasabha. It settles all disputes connected with partition of property, cases of divorce, extra marital relations and crimes against society and religion. The Mahasabha convened by the president. The Mahasabha formulates the policy and propounds the rules of conduct of the Gond society. The ruling of the president becomes a binding for the concerned parties. The secretary assists the president sometimes they call larger Mahasabha consisting of several Mahasabhas in case of grave offences against their society. Besides Mahasabha they have village Assembly convened by the elderly people of the village to decide the social offences committed in the village. Statutory Panchayat have not gained popularity as the traditional village Panchayat. The caste Panchayat still entertains most of the cases.

The religion of Gonds reveals a mixture of Animism and Hinduism. Eminent social anthropologist Risley is of the view that they worship the Gods of Hindus with the assistance of Brahmans. These Gonds who have embraced Hinduism also worship the deities of the Non-Hinduised Gods like Bar Deo and Dulha Deo. They have a whole set of Gods and Goddess called pen. They observe both tribal and Hindu festivals and offer sacrifices in the village Sarana. Gonds celebrate Phagua, Chait Ramnavmi, Janmasthmi, Jitia, Durgapuja, etc. Alongwith scheduled caste they worship Goreya and Bandi. They are devotees of Lord Shiva and and have a peculiar dance known as the Gonr Nach and they believe that they have got their musical instrument Damru from Lord Shiva himself. The Gonds are famous for their song and dance but the female folk do not join the dance.

Gonds have both infant and adult marriage. The rule of exogamy is rigidly enforced and a man must not marry a girl belonging to his own clan, nor should she be one of the groups which worship the same God. For example, the Gonds belonging to Besra Gotra will not marry a girl of the same Gotra. Cross cousin marriage i.e., union between a brother's daughter and sister's daughter is common. Infact cross cousin marriage is preferred. When a Gond wishes to marry of the children he looks first to his sister children upon whom he has his first claim. When a Gond takes a sister's daughter as a wife of his son, such an alliance is known as Dudhlautana (return of the milk). Such marriages are common because it is considered a sort of repayment of obligation of the first generation, when one family gets the daughter of the other family as wife to some male member. In the next generation the family who gives the daughter receives the daughter of other family as a wife to a son. The bride price is called Dangani. But whatever to the bridegroom is known as Dahej. The bride price is avoided in cross cousin marriage. Divorce is allowed on certain recognised grounds, easy adultery laziness in house hold duties, bareness and quarrelsome nature.

The dead bodies are either cremated or buried though cremation is more common. The dead body is disposed of at the earliest. A community feast is given on the third of the death when offerings are made to the deceased.

Gorait :

Goraits are Dravidian tribe. They are found in south-eastern portion of the Ranchi Plateau. Their total number is 4975 and consists of 0.06% of the tribal population of the state (2011).

Since these people are less in number compared with other tribes of the state they are unable to have separate settlement where they can preserve tribal traits and other socio-cultural features.

Virtually their living alongwith other tribes has amalgamated tribal specialities and socio-economic activities. They do not have even separate social organisation dealing with the social problems of the tribe. However they have own social organisation to deal with some important problems.

Goraits are also practising agriculture like other tribes of this region. They have copied the method of developing land for growing cereals, vegetables, seasonal fruits, etc. These days they have also started giving due importance to money fetching crops. Certainly their ancestral inclination towards professional works like musical instruments, making house hold goods and other items still persist and distinguish them from other tribes. Some of them also work as village watchman and messengers. At least one family of Gorait is found in every village. Some of them have even moved out for non-agricultural works in industries or urban areas. They have comparatively better inclination towards education. Certainly female education is still very low.

Goraits still prefer child marriage. Certainly these days they have copied other tribes of the region and also avoid child marriage. Polygamy is permitted in their society. A widow may remarry but she is not obliged to marry her husband's younger brother. A divorce may be granted by the caste panchayat if she frequently runs away to her father's home without the permission of her husband. Proceedings may also be initiated by the wife on the ground that the husband is too old for her or is a habituated drunkard. Divorced women may remarry in Sagai form.

Goraits worship Hindu Gods and Goddesses. They worship Devi Mai and a tribal spirit called Purubia for whom a goat is sacrificed once a year. In case of illness an exorcist (Mati-Baiga) is called to detect the demon or witch who is giving trouble, and if he ascribes the visitation to the wrath of the tribal God, the Pahan is sent for and a goat, pig, sheep or fowl is sacrificed to appease the evil spirit.

Karmalis :

The Karmalis are also considered Dravidian tribe. They are also said to be associated with Santhals because they speak Karmali dialect which is a sub-dialect of Santhali. They usually understand

Mundari. They are 64,154 in number and comprises 0.74% tribal population of the state as per 2011 census. They are found in north-eastern parts of Ranchi Plateau.

The Karmalis are divided into a number of exogamous clans or septs such as Tirki, Samadwar, Kaithwar, etc. Their class are also totemistic. They follow Hindu religion. The Karmali youngmen are given scar marks on the left arm as a sign of their eligibility for attaining a social status, but this practice is on decline.

The Karmalis are basically cultivating tribe. They have developed agriculture like other cultivating tribes of the Ranchi Plateau. Their habitation lies in relatively less uneven areas and is comparatively more suitable for agriculture than the hilly areas. As such they have developed terraced paddy fields on the gentle sloping land and lower areas of the valley, orchards settlements, thrashing ground, burning ground on the Tanr lands. They grow rice, maize, kulthi, ragi, til etc. During recent years they have also started growing vegetables, seasonal fruits for marketing purposes. Some of these people are also engaged in mining, quarrying, livestock and other activities. They have household industries like blacksmithy work. They manufacture sickle, axe, iron head of arrows, etc.

The Karmalis are educationally backward. Certainly awareness towards education has also increased. They have started sending children to school, and literacy among Karmalis has also increased as par with other tribes of this region.

The Karmalis follow to a considerable extent Hindu marriage of tradition. Monogamy is the rule and some are polygamous also. They marry outside their clan or sept which are exogamous clans. They are guided by good or bad omens in selection of a bride. Among good omens are a bear, leopard, a corpse, a deer, a tiger, a pot full of water, a cow or a bullock and among evil ones a sheep, an empty pot or a jackal running across the path in front of the party, an animal being thrown away and a iry branch falling from a tree. If any object of a bad omen comes across their path, the journey for bride selection is postponed. But the negotiation are not abandoned. A Karmali can marry his wife's younger sister or the widow of the deceased elder brother.

Similar to other Hindu tribes Karmali also believe in Gods and Goddesses. They also believe in tribal Gods. They have their own community Panchayat to settle social matters. They celebrate festivals like Karma, Sarhul, Nawakhani, Shohrai, Durga Puja, Diwali, etc. Burial of the dead body is common.

Chik Baraiks :

Chik Baraiks are comparatively developed tribe. They claim to be Aryans and are largely Hindus. Virtually this tribe has been considered as sub caste of Pans in the book of Risley's tribes and castes of Bengal. But apart from a common profession, namely weaving any further connection between them is doubtful. In appearance Chik Baraiks look more fairer than the Pans. They are 54,163 in number and contains 0.63% of the tribal population of the state (2011). They are mostly found in a scattered manner in the districts of Ranchi, Lohardaga, Gumla, etc. There has been difference of opinion concerning the roots of Chik Baraik. There is little evidence to prove that there was an Aryan civilization in the forest clad region of the Chotanagpur before the incursion of the Dravidian races but it may be the Chik Baraik who are considered locally to be the social superiors of the Pans are the descents of some Aryan weaving caste who settled in this region at an early date. It may be noted that in the beginning of the twentieth century Chik Baraik and Pans were not fully classified as animist because of most of them were living like non-tribals. In the census of 1961 the Pans (Sawasis) were enlisted as scheduled caste. Chik Baraik however, are shown as a tribe throughout the state.

The main occupation of Chik Baraik is agriculture. They have full awareness towards the agricultural uses of land like the other Aryans races. They have developed field terraces in a better way and their settlements, orchards and other uses of land reveal slightly different features. They grow important Bhadai, Aghani and Rabi crops depending upon the suitability of the land and other favourable conditions. These people have also developed keen interest in growing cash crops and vegetables. They also supplement their livelihood by weaving of cloths. Perhaps every household may have an old loom to remind one of this prosperous cottage industry in the past. These days a good number of Chik Baraiks are marginal farmers and agricultural labourers. Some of them even go to Punjab, Haryana, Assam for earning more money.

The inheritance of property is patrilinear. After the death of father the property is equally distributed among his sons. The daughters are not given any share in the immovable property.

The family of a Chik Baraik is nuclear in composition. Their family consist of husband, wife and unmarried children. As soon as children are married, they form their separate establishments Joint family system is uncommon.

They are divided into class like Tatia, Khamba, Tanja, etc. Marriage within the same gotra is not permissible. If any boy and girl does marriage in the same gotra that is regarded as an offence. The marriage is generally performed at bride's home under a mandap or canopy. It is solemnised with Sindurdan Divorce is permissible to either party on genuine grounds like adultery, bareness of wife or impotency of husband etc. Polygamy is not preferred. It is a man and women develops extra marital relation with other they are severely penalised.

Chik Baraiks do not have separate habitation and live an integrated life with other communities in the village. The indebtedness is common among them. The incidence of literacy is low, They do not have their own Panchayat because they lack concentration in any village. Their social problems are solved in the assembly of the elders of the community. In terms of alcoholic use they are akin to other tribes but they do not participate much in dancing.

Chik Baraiks bury their dead bodies in the Masan with the head of the corpse towards the north as the holy river Ganga flows to the north of the region. Other rituals concerning death ceremonies are performed like other Hindus. They worship gods like other Hindus. They worship Devi Mai, Durga and Shiva. They perform ancestor worship but do not do Shradh at Gaya. They celebrate a Puja namely, Bar Pahari Surjahi in the month of Phagun.

Kisans

The Kisans are Dravidian tribe. They are 37,265 in number, 0.43% of the tribal population of Jharkhand as per 2011 census. They are also known as Nagesar and Nagesia. The Kisans have clan groups with totemistic characteristic and each member of a clan considers himself as related to the other.

These people have two important clans Sinduria and Telia. The Sinduria has Gasur, Induar, Bagh, Kashyap, Nag and Kasi gotra. These two clans live together and have also marriage relations. Certainly marriage relation between these two is restricted in most areas. The basic difference is that Sinduria's use Sindur at the time of marriage and while Telia can use oil instead of Sindur.

The Kisans have physique of medium to short stature. They have dark complexion and ugly features. They are strong and hard working.

They are mostly cultivators. Most of them have culturable land or chas. It probably refers to their agricultural profesities to distinguish them from some portion of their tribe who did not cultivate. The size of their agriculture holding is small. The decline in lac cultivation has adversely affected their economy. Tamrin, Mahua and Kendu leaves however, supplement their income. Non-agricultural pursuits have little attraction. They take pride to call themselves Nagbansi. All of them practice Hindu way of life. The Kisan rears poultry. They use metal pots and pans.

The male wears Dhoti and female wears Sari. They do not tattoo themselves but the Sinduria women do so.

The Kisans have their own community Panchayat. Its sitting are called whenever a matter is reported to it. The office bearers of the Panchayat are Mahat, Mahant and the Sardar of Kotwar. The offices are hereditary. The part which invites the sitting of the Panchayat bears the cost of the sitting.

By and large the Kisans follow the local religious beliefs and customs. They propitiase certain malevolent and benevolent spirits which are known as different Bongas. They worship both Tribal Bongas as well as Hindu God and Goddesses. They invoke the tribal festivals as well as Hindu festivals. They worship their village deity and also the Sun God. Their marriage customs are substantially the same as those of the Hindus. Polygamous marriage depends upon the economic condition of the persons concerned. The boy's father goes to the girl's parents for marriage negotiation. Widows are allowed to remarry. Both the males and females have right to divorce.

Burial is common, but important persons are generally cremated. All personal belongings of the dead are deposited in the grave alongwith corpse. A community dinner is given at the end of the death rituals.

Binjhia :

The Binjhias are also developed Dravidian tribe but smaller in number. They are 14,404 in number and comprises 0.17% of the tribal population of the state as per 2011 census. They hold the view that they have come from Vindhaya Valley. They describe themselves as Vindhyanivasi. They say that the name Vindhaya was given to them by others when they had settled in this region. They claim to be Rajputs and use the title of Singh or Pardhan. The Binghias of Targa preserve a genealogy which reveals that their ancestors settle in SamVat Sixteen (1657 A.D.), the original settler being one Jasmatideo of Kashyap gotra. The arrival of Jasmati Deo in this region has a romantic background. It is said that on account of internal feuds between the ruling chief Surat Deo and his three brothers the later were ultimately forced to leave the valley of the Vindhayas and come to this region. They are also found in Madhya Pradesh, Orissa.

The Binghias are divided into two sub-tribes, the Paharia Binghias and Dand Binghias. Paharias Binghias live in the hills the Dand Binghias reside in the plains. Dand Binghias are further sub-divided into four exogamous septs. They observe a period of twenty one days as one of the impurities from a date of a birth in a family. They also use sacred threads.

The marriage is arranged by the parents of the boy through the middle men. Child marriage is not in vogue. The sight of Jackel, a snake and oil pot, cry of a peacock, animal fight, falling of trees or carrying of dry logs on shoulders are regarded as bad omen for the proposed union and the middle men are to report on them to the parties concerned lest the bride should die after the marriage. They practice most marriage rituals like the Hindus. There is no ban against polygamy but divorce can be demanded by either party on certain recognised grounds such as desertion, sexual immorality of wife etc. Remarriage is prevalent.

The Binghias are an agricultural land holding community. They are rather very backward in education. They have two hereditary leaders in their community Panchayat namely Madi and Goddi, with their headquarters at Trga and Parba respectively. The punishment usually consists of undergoing some prescribed ritual ceremonies at religious places, compensation to the aggrieved party and community dinner. They have a representative assembly consisting of delegates from every household to decide questions of caste uses. The president of the Panchayat is called Karta and his office is hereditary. The statutory Gram Panchayat setup by the government does not interfere in the matters related social organisations of the Binghias.

They worship Hindu Gods and Goddesses. They follow the rituals concerning birth and death like the Hindus. They cremate the dead body of the elderly persons but the dead body of the children are burried. They celebrate birth and death with a community dinner.

Rich people of this tribe have also adopted considerable changes in terms of dress, food habits, tribal traits, standard of living, etc.

Savar :

Savar is a small tribal community living in Eastern Singhbhum district. They are 9,688 in number and consists of 0.11% tribal population of the state as per 2011 census. This community has diverse opinions concerning sources of their origin. Russel is of the view that these people's ancestors were initally known as Bhill. They have come from Orissa and Madhya Pradesh. S. C. Roy holds the view that the Savars follow the traditions of Hill Kharias. According to him they have also come from Orissa and M.P. They have four sub-castes. Jahna living in Dhalbhum and the remaining three sub-castes residing in Orissa and Madhya Pradesh. They do not have Gotras.

The Savars are marginal farmers and agricultural labourers. They grow food crops like paddy, maize, millets, etc. on small piece of lands near their houses. They lack agricultural infrastructure like bullock, plough, etc. They borrow agricultural implements from the other farmers. Most of them are landless labourers and depend upon wages and forest gatherings. They also produce local wine and do hunting of hare, birds, bear, etc.

The Savars also have traditional Panchayat system headed by Pradhan. The post of Pradhan is hereditary. Pradhan is assisted by Goraits. This society considers witchcraft, sexual relations with other, theft and disobedience of social norms as crimes and Panchayat takes action against these crimes. These panchayats also deals with economic, social and political aspects concerning this tribe.

The Savars confine marriage in their own tribe. They avoid marriage in their relations upto three generation. They have customs of sororate and levirate. Dowry system is prevalent. Middle men play important role in fixing the dowry and the marriage. Parents of bridegroom extend proposal for marriage to girl's parents. Child marriage is still not totally avoided. In case of child marriage the bride and bridegroom are allowed to lead family life only after second marriage when they become adult. They follow the Hindu rituals. Divorce is permissible on valid grounds like barenness, bad character, etc.

The house of Savars are rectangular. The houses are mud walled and thatch roofed. They use logs of Sal and Mahua in their houses. Most of the houses are of smaller size having one room being used for sleeping, entertaining guests, keeping household items, cooking etc. The houses lack windows and have only one door. Certainly mud walls contain space for keeping smaller items.

Males use Dhoti, Shirt and females use Sari, Blouse, etc. Educated persons are using pants, shirts and other modern dresses. They are also fond of ornaments but poor people use ornaments made of silver, glass and forest materials. Brass bangles are used by females are known as Kaharu. They also decorate themselves with flowers, sal leaves on occasions.

The Savars worship most Hindu Gods and Goddesses. They worship Kali and Mansa Devi with special emphasis. Music and Dance are part and parcel of their social life. On the occasions of birth, marriage and religious functions, dances and music are performed. They also use Handia. They perform dances known as Domkach and Pantasalaya.

They have patriarchal society. Property descends to sons after the death of father. They prefer nuclear family. After marriage adult couple lead separate family life. Monogamy prevails in this society but one can remarry in case of bareness, bad character etc. Widow marriage is permissible.

Cremation of dead body is prevalent. Some of them bury the dead body. They provide a community feast after necessary rituals.

During these days these people have also recorded substantial changes in tribal character and adopted modernity.

The Mal Paharias :

The Mal Paharias are the 'hill men'. They are mainly found in Rajmahal hilly areas specially in Damin-i-Koh area. In addition to this region they are also found in smaller number in Giridih and Palamu districts of the region and Banka District of Bihar. They are 135797 in number and comprises 1.57% tribal population of the state as per 2011 census. There are different opinions regarding the origin of Mal Paharias some scholars consider Mal Paharias different from Malers on the ground that Mal Paharias are usually more taller and robust than the Malers. Malers are considered as Barbarians by the Mal Paharias themselves are darker and also dirtier. Outside Damin-i-Koh specially in Deoghar district the Mal Paharias prefer to call themselves Naiya or the reformed race and Pujahar or worshippers. Risley considers Mal Paharias as Dravidian Tribe belonging to Proto-Austroloid Group. Hamilton has expressed relationships between Mal Paharias and Maler. Dalton is of the opinion that Mal Paharias have less connection with tribes of Santhal Parganas. Bal is of the opinion that Mal Paharias reveal considerably different physique, customs, dialect from other tribes of Rajmahal. The Mal Paharias speak Malto language which is considered language of Dravidians.

They do not have clan systems but they are divided on the basis of family or hereditary background like Dehri, Garhi, Manjhia, Laya, Pujar, etc. They avoid marriage among themselves.

The Mal Paharias exhibit landuse and agriculture of varied nature in the Damin-i-Koh area and the less undulating areas in the Deoghar district. In Damin-i-Koh area they cultivate the hill summits. They have clung to the steep hill sides with their Jhum, though they also use the plough in the flatter portions of the hill tops. In the less undulating areas of Deoghar district they form the chief labouring class but many of them have small agricultural holdings and they often hold the post of village watchman. Their cultivation in the plains differ in no respect from that of their neighbours, though they are still addicted to Jhuming, which they call 'Kurao' or 'Kurma-bari'. They grow paddy, maize, millet, bean etc. In the hilly areas they supplement their livelihood from forest gatherings and hunting. But in the plains they have adopted the cultivation of cash crops, vegetables, dairy and poultry activities. Some of them have also moved out for labour works in the neighbouring districts or elsewhere. They have also developed small scale industries making small cots, mats, hunting equipments, local wine, etc. They are also fond of fishing. They also prepare charcoal out of wood for local market. Wood and Sabai grasses are also used for making a number of items for use.

A village is governed by a village council consisting of elderly people of the village. It is presided over by the Manghi with the Gorait as his assistant. These posts are hereditary. The Manghi and Gorait together with the elderly people solve the social problems and sometimes penalise the culprits.

They are mostly Hindus. They worship Hindu God and Goddesses. They worship Sun. Next in honour to the Sun are Dharti Mai, Garami and Singhbahini who bear rule over tiger, snake, scorpions and all manner of noxious beasts. To the earth goats, pigs, fowls, etc. are offered in Asarh and Magh and buffaloes or goats are sacrificed at the the time of Durga Puja to the Goddess Singhbahini. They solemnise karma festival. Besides these deities the Mal Paharias recognise a number of vaguely defined animist power like Chordanu, a malevolent spirit needing to be appeased at certain intervals with sacrifices. They also worship Kali and Laxmi Mai. Ancestor worship is also prevalent. Every household is guarded by its ancestral Gods.

The Mal Paharias villages are linear dispersed. The houses are rectangular in shape and mud walled and bamboo splits walled. The roofs are generally of thatch materials like grass, paddy straw, tree leaves, etc. The villages occupy the summit of high range of hills.

Marriage is either infant of adult. Girls are rarely married before the age of 10 or 11 and usually not until they are fully grown up. A professional mat-maker (Sithu) is usually engaged by the bridegroom's people to search for a suitable bride. When the selection is made, a visit or inspection is paid by the bridegroom parents and if the proposed bride is approved the price to be paid for her is settled by discussion. Thereafter the marriage function is performed with necessary rituals. Polygamy is permitted. There is no limit of having wives but poverty has enabled them to confine with one wife. The widow marriage is permissible.

The dead bodies are usually cremated and a piece of bone is saved from the flames to be thrown away into a river or a deep tanks, the water of which does not dry. After all necessary rituals a community feast is organised. During recent years the Mal Paharias have also witnessed changes in their social habit, cultural traits and way of living.

The Sauria Paharias :

The Sauria Paharias are branch of the Paharias. It may be noted that the Paharias are divided into two branches as the Male Paharia or Sauria Paharia who are found in the north of the Rajmahal hills and the Mal Paharias who are found in the south of the hills and wooded country in the south and west of the Rajmahal. They are 46,222 in number and consists of 0.53% of the tribal population of Jharkhand as per 2011 census. The origin of the name Sauria is doubtful, it has been suggested that it originates in the term Savala, Pahar being applied by Hindus to the Rajmahal hills. The Santhal call them Maler Mundas and the Hindus call them Paharias. A clue to their origin is found in the tradition of the Oraons. Their original home was in the Karnataka, whence they went up the Narmada river and settled in Bihar along the bank of Son river. From the son valley they were driven by the Muslim and they split into two groups : one of these now represented by the Oraons, ascended the Son into the Palamu, and turning eastward along the road, took possession of the north-western parts of Chota Nagpur. The other branch, following the course of the Ganga, settled in the Rajmahal hills and were the progenitors of the Mal Paharia and Sauria Paharias. In these hills, hemmed in by the Ganga on the north and east and shut off from the outer world by thick forest on the south and west these Paharias remained almost untouched by outside influence. They have no characteristics of language in common with the races which surround them from which too they differ in physiognomy, in their social habit, in the way of forming their villages and in their methods of cultivation.

The Sauria Paharias are inhabitants of the northern of the Damin-i-koh where they occupy the hill sides and tops of the hills. They have been driven from the richer valleys by the most enterprising and industrious Santhals. They live in village communities each of which claim as its property certain hills, the boundaries of which are not well defined. The Sauria Paharias are short stature people having long limbs in proportion to their low stature. Their complex is light brown. Their nose is not prominent but they remain flat and broad at the base. Their eyes have peculiar look of the Dravidian. Their hair is long and ringlesed and they keep it well combed and oiled. These people remain beardless. Labour is the hillman's abhorrence but necessity compels them to cultivate a small portion of the land for their actual existence. Beyond this trifling labour they never exert themselves. They are, on the whole lazy and poor.

The Sauria Paharias or Malers support themselves by the Jhum or Kurao Method of cultivation. A patch of land is cleared with axe and fire. The soil is hard and seeds are dibbled in among the ashes. The site thus cleared and cultivated is known as Jhum or Kurao. This process is repeated at intervals of five years, with the result that in some long ranges practically all vegetations have disappeared for miles. The slopes looking as if they had been scored by land slips. These people supplement their crops by the product of the chase but they are not expert hunters. They grow rice, maize, millets, beans, etc. During recent years they have started growing vegetables and seasonal fruits. Besides they also roam over miles in search of honey combs, wild yams and other edible roots.

The religion of these people is animism of the type common among Dravidian tribes. The deities worshipped by them are Ber or Berm Gosain, Bilp Gosain, Laihu Gosain, Darmare Gosain, Jarmatre Gosain. These Gods are not represented by idols nor special form of worship is prescribed. At the same time there is no fixed day for the worship. They have no priests and sacrifices are not offered. There are twenty three godlings, besides several devils and evil spirits who have to be duly propitiated. In case of illness demno advises a puja to Gumonaddu. If recovery takes place sacrifice is offered.

The Saurias settlements are developed on the hill tops and forest covered slopes. Their settlements lack in the valleys and the plains. Their houses are made of mud walls and thatch materials. Dwellings are small size having rectangular shape. Their small houses contain the facilities of cooking space in one corner, storage of food grains and other uses.

They do not have clans. They are divided on the basis of family and hereditary background and occupations. They avoid marriage among their family groups upto three generations.

Dresses of these people have got modernised. The traditional dress is Dhoti, Ganji for males and Sari and topbare for females. These days Pant, Shirt, Lungi and other modern dresses are commonly used. They are also fond of ornaments. Female Sauria Paharias use ornaments made of brass. They use ring, necklace, Hansli, bangles, beeds of coins, etc. They are also tattooed.

Polygamy marriage system prevails in this tribe. One can keep several wives but poverty prevents them to do so. Divorce is permissible on valid grounds. Marriage negotiations are settled by the middlemen and after getting all the formalities the marriage is performed.

Their festivals concern agricultural activities like preparation of the land, planting, transplanting, seedling, harvesting, etc. They offer newly harvested grains to their God and Goddesses. They celebrate Gangi Aria on the harvest of Maize, Osra Aria on the harvest of Ghanghra crops and Punu Aria on the harvest of millets. On these occasions they sacrifice fowls to the deities and ancestors.

They celebrate birth and death. The dead bodies are burried. After necessary rituals a community feast alongwith drink is organised. But during recent years they have also recorded considerable changes in their social and cultural characteristics.

The Koras :

The Koras consist of traits akin to the aboriginals but they have now adopted some of the customs and habits of the non-tribals. They are 32,786 in number and comprises 0.38% of the tribal population of the state as per 2011 census. They have adopted customs, manners, rituals etc. of the Hindus. They worship the God and Goddesses like the Hindus. Certainly they have some Demi Gods such as Panch-Bahinis, Kudra, Rangahari, etc. These Gods are represented by trees, stones and the like other objects of nature which they paint with vermilion and attribute to them the virtues of deities.

The Karas have their traditional occupations like earth cutting but some of them have become cultivators and agricultural labourers. They are growing food crops, vegetables and seasonal fruits like other tribals because of the increasing demand during recent years. People having negligible land or having no agricultural land are working as agricultural labourers. They remain in good demand because of their specialisation in earth cutting. The females of their society sing and dance along with music of drum beating and flute blowing on the occasion of marriages. They follow the birth and death rituals like the Hindus and at the end of the function a community feast is organised.

14

Rural Settlements

The state of Jharkhand having variation in elevation and the presence of scarps, valleys, ridges, hills, forests, plains of varied characteristics has given rise to rural settlements differing in distributional pattern, size and spacing and morphological features. Certainly in areas having relativity less unevenness has encouraged the growth of settlements slightly different from the areas having high dissection or dense forest cover or less fertile exposed rocky areas and steep sloped areas. As such even in the valleys or plateaus or even plains depending upon ruggedness, intensity of slope, accessibility by transport or the percentage of forest cover size of settlements as well as other characteristics considerably differ. The main characteristics of the rural settlements of this region concern dominance of rural habitat despite having considerable urban development based on exploitation of minerals and industrial growth. As such except a few patches of coal belt of Damodar Basin, iron ore, copper mines of Singhbhum, mica mining areas of Giridih, Koderma, etc. most areas even have rural dominance. These settlements are found in highland of Rajmahal especially in the districts of Rajmahal, Godda, Pakur, Dumka. There are considerable number of uninhabited villages in the districts of Garhwa, Palama, Chatra, Lohardaga, etc. Most of the areas have relatively dispersed settlements especially over areas having poor accessibility, forested landscape and rough terrain. The distributional pattern of rural settlement is also quite different from the neighbouring plain areas because this region lack consolidation of dwellings. The difference in a land due to dissected topography and diversity of soils, variation in amount of rainfall and accessibility have resulted diffusion of settlements. Most of the settlements are dotted on the spurs ridge tops, divides and slopes above the valley. As the region has dissected terrain settlements do not have the opportunity to develop into large settlements at any one site. On the other hand small settlements have developed at much convienent sites. These settlements are commonly hamleted. In the forested areas isolated hamlets dot the landscape and show a clean correspondence with patches of arable land in the forest. Houses are generally more spacious and wide inter house space presents a picture strikingly different from that of the consolidated settlements of the plain.

Specialities of the settlement pattern also concern the tribal habitation which is tradition oriented. There are several tribes inhabiting in the region and their way of life differ. The Hos of Kolhan area like to live in large clustered type villages. Their settlements are builts usually on small, barren and round mountains or domes or on fingure of spurs projectiving between valleys. The mixed tribal zone of Panch Pargana Plains possesses hamleted type of settlements where each hamlet may be given to a different tribe or to outsider (Dikkus). The Kharias and Birhors who are more backward build their houses in hilly areas and near jungles and their villages wear a more primitive and disorderly look. The settlements of the Munadas and the Santhal show little variation from the settlements of the Kharias and Birhors. The Mundas generally prefer tile as a roofing material while a Santhals prefers thatch for the roof of their houses. The Oraon prefers to build houses near small streams and rice field which have been cleared in the forest. The settlements are ususally quite small and scattered, but large centres are not exceptional particularly near fortest. The Oraons and Mundas generally do not paint thier houses which are therefore not so good looking as the houses of better class Hos and Santhals.

The typical rural settlement is found away from the main thoroughfares. Those settlements located on the main road have been considerably modified by outside culture such as Pucca building, windows, tin roofs and large wooden doors etc. particularly shops or building of traders and officers. Villages are usually self served. Any need of outside commodity or sale of local goods is looked after by numerous Hats or weekly market dotted all over the region. This market has a very important place in rural life and almost everything of daily requirements is sold there and people from the neighbouring villages regularly visit it.

The main problems which the tribal settlements face are inadequacy of transport and proper water supply. Despite the construction of several new roads, rural transport remains an acute problem due to the hilly nature of the terrain particularly during the rainy season as the main source of water for drinking purposes is the small 'Danri' or hole dug at the foot of the slopes, and the streams are purely seasonal, there is great scarcity of drinking water during the first five months of the year.

Changing Historical Background and its Impact on Settlements :

As the rural settlement is mainly an agricultural workshop any change in agricultural landscape brings modification in the rural habitat. During pre-historic period people were mainly hunters and food gatheres and used to live in rock-cut-caves and busnes. Their residential colony started in the domestication of animals and the practice of cultivation. The information about this change is meagre and speculation but some ideas abouot it have been gained from the evidences available through ancient sites and archeological remains, distribution of religions centres such as temples, churches, mosque, etc., the distribution of tribal communities contemporary historical accounts and literature and nomenlature of settlements. For the convenience of analysis this changing historical background has been subdivied into the following.

(I) The pre-historic settlements : The pre-historic stage of human habitation in this region is shrouded in mystery and it is impossible to give a definite accounts. Except a few lazy legends of the aboriginal races and a few ancient sites there is little material even for speculation. These legends refer to the race of the Asur and indicate the triumph of that race over the original inhabitants of the region, possibly the ancestors of the small iron smelting tribes of iron age of that name, which still speaks a language akin to Mundari. But no clear evidence is available to fix the date at which the Mundas, Oraons, Santhals, Hos, etc. settled in this region, though it seems probable that they were the first to come and enter the region from west and the north-west. The Mundas belong to the South eastern portion of the state. The aboriginal races such as Kharwars, Oraons, Cheros, Hos, etc. practically rule over this track. The Oraons according to their legend came from the strong hold of Rohtas which had probably at an earlier age also been a resting place of the Munda race. This race was probably driven here by the more powerful tribe called Cheros or Kharwars, The Oraon race being driven from Rohtas Plateau entered the region from the north-west. As rugged terrain, dense forests and lack of transport accounts, made the area difficult to access, the advancing non-aryans (pre-Dravidians) encroached only on the more accessible path. It is however difficult to locate the exact occupance unit which was under the possession of the non aryans such as encestors of Kolarians, Marhattas, Chero, Maler (Sauria Paharia), etc. Marhattas settle in the area after the Cheros. Similarly the Malers are also early inhabitants who chiefly occupied Rajmahal highland.

The first ancient sites found in the region also attest to its early habitation. The best example has been found in two villages - Bichna and Bundu (via Ranchi District)) noted for Asurs site. In the village Bichna iron arrows, polished stone slabs, store celts have been found while in the villge Bundu burial ground and ancient site of Asurs have been found.

The earliest settlers have also been classified by historians of the region into two groups - Paleolithic (old stone) and Neolithic (new stone). Inhabitants of both these ages were also expert in constructing soil pots, and toys and stone swords. They lived isolated like island dwellings in a sea of alien races specially in the hilly region. In the Rigveda there are frequent references to the non-aryan people whom the arynas out of feeling of superior race contemptuously called Dasya or Asuras.

After a lapse of time this region received Dravidian settlers who are supposed to have come from the north-western parts of India. They were expert in the contruction of metal sources more particularly of copper. They spread in the entire region. The Kharia is supposed to be a Dravidian tribe, classed on linguistic ground as a Munda tribe. The Kharias are divided into four sub-tribes such as Dhilki-Kharia, Dudh Kharia, Erenga-Kharia and Munda -Kharia, found almost every where in the state.

(II) Aryan settlements : At the time of aryanisation of the Ganga Plain this region also experienced a few habitation in the relatively plain areas of the north Koel-basin and the adjoining plains of the Rajmahal highland and of other accessible parts. Those people were more civilized and left their mark upon the earlier settlers. As a result aboriginals of neighbouring areas were slowly assimilated and where that was resisted, they were pressed further into the inaccessible hilly areas. The Aryan villages are more neat and the densely forested areas remain unaffected by the aryans incursions.

(III) Budhish settlements : Similar to the Aryan hibitation during Budhish period the northern area adjoining the plains experienced some Budhist habitation such as in foothill areas of northern scarps of Hazaribag Plateau and lower Koel Basin. During the regime of Ashoka (C.273 - C.232 B.C.), the atavi or forest Estates acknowledeged Magadh supremacy and several preaching sites were established. At persent there are several villages in Palamu district and surrounding areas of Parasnath hills named after Budhist place names and look similar to Budhist villages of Magadh. In the villages of these areas Budhist relics are found such as Murtia village near Latehar town, Mahobudhi nearly 30 kms. south-west of Netarhat, Champanagra nearly 8 kms. east of Daltongang etc. These villages still possess the characteristics of Budhist settlements.

(IV) Post Budhist settlements : With the downfall of Budhism the Budhist village, of this region underwent some modifications under the influence of Sasank and Pal rules. It is probable that during 700 century A.D.Sasank ruled over this territory as mentioned in the copper plate records Sasaka's reign found at Benu Sagar in south eastern portion of this state. Near by this village there are several antiquaria remains at these places also belong to the Pal period. But due to hilly terrain many rural settlements remained unaffected by such historical changes in sites can altogether be ruled out. People preferred to settle in fortified villages or clustered hamleted settlements.

(V) Settlements of Muslim period : During this period Muslim habitation did not spread more because of poor accessibility. Towards the end of 10th Century A.D. this region received some Rajput immigrants. Even after Muslim conquest of the area thier influence was confined to the more accessible areas and nodal points and rural habitations for the most part remained unaffected. The aboriginal occupants retreated from the extensive plain areas and took shelter in accessible tracts. Consequently there are a few Muslim villages in the interior country side.

During the reign of Shershah a few settlements came into existence in the areas surrounding the Palamu fort near Daltonganj and other accessible points. A ruined mosque constructed by him is also present the Palamu fort of Cheroraja. During the rule of

Akbar and his comandants Sahabaz Turbati, the size of Chotanagpur Raj was further reduced and several Muslim settlements came into existence particularly in the north eastern portion of Santhal Parganas such as Akbarnagar, Kasimchak, Sahibganj (now urban centers) and north western section of Palamu district such as Hussainabad, Amirganj, Alipur, etc.

(VI) Settlements of the British : With the arrival of the British ower in the region in the middle of the 16th century, this part of the country also experienced some changes into settlements. The British constructed several rules to conact inaccessible places for administration and subdivided the area into administrative units like division, districts, subdivisions, revenue thana, police station, etc. They also constructed Dak Bungalow and circles at almost all administrative headquarters. Several rural habitations developed around them which in course of time became permanent. Such permanent habitation also restricted the shifting character of cultivation. Further changes in the distribution of rural settlements came due to increase in communication facilities, as people from the plain areas encroached the inaccessible areas. The British settlement operations demarcated the village boundaries with flag stone at various places after adequate cadastral survey. The department of revenue administration prepared detailed large scale maps of rural settlements in the first decade of the present century, showing individual agricultural plots and inhabited sites, etc.

The rich mineral resources of the region attracted the Britishers but they were not inclined to take any hasty measures against different Rajas of Chotanagpur, although they were generally creating obstruction in matters of revenue collection. During early stage of the present century development of coal mining, gold mining, copper mining, etc. led to the establishment of several mining settlements.

The establishment of police stations enabled administration in a better way. It also reduced the chance of internal dispute between different tribes and also between tribal and non-tribals. People started settling wherever they got facilities of culturable land and pasture. The British always paid much attention towards the tribes especially with a view to win them to Christianity. Several institutions to teach English were opened and free-coaching was given but their influence remained only localised.

(VII) Settlements during post independence period : After the independence settlements of this region also experienced considerable changes due to the developments. The growth of educational institutions of various grades, medical services, roads, development block offices and other commercial and other institutions enabled the region to experience some modification in settlement patterns. In the Damodar valley region the control over flood and soil erosion and development of hydel plants and other infrastructure brought remarkable changes in the neighbouring areas. The region has also witnessed remarkable increase in exploitation of minerals, industrial centres and the mineralised and industrializ belts have given rise to a good number of habitations.

After independence, significant changes have come due to abolition of Zamindari system that made the people free from the clutuhes of the traditional feudal lords who had terrorised the people and did whatever they liked in the past. The tribal people have now selected those areas to settle where ever they felt sufficient resources. The tribal people were also given support in the construction of buildings, wells, schools, etc.

Distribution of rural settlements : It is important to note that there has been variaton in physical landscape based on elevation, dissection and other topographical features and on the basis of these the area has been subdivided into physical units. These physical units reveal contrasting features associated with settlements. The following are the physical units having settlements characterised by its topographic features.

1. **Settlements in the Ranchi Plateau** : The Ranchi plateau characterised by rough terrain on the west and almost flat land on its rest part presents contrasts in the distribution of settlements. The western portion having rough terrain has uneven distribution of settlements while the relatively flat plateau areas have balanced distribution of settlements marked with nucleations of smaller size. The southern extremity of the plateau is undulating area intersected by numerous streams and studded with rocky hills and isolated peaks are marked by relatively dispersed settlements.

The South Koel basin and the southern scarp of the plateau have mostly sprinkled settlements. The hilly areas of the western part of the plateau have widely spaced small villages. The Khunti Plateau inhabited by the Munda and Oraons has hamleted type of settlements. The Oraons have preferred to built houses near streams and rice fields. The settlements are usually small and scattered but some large clusture are also found particularly near towns.

2. Settlements in the Hazaribag Plateau : This plateau is relatively flat in some areas and dissected and forested coupled with poor accessibility in remaining areas. The impact of flatness and ruggedness has encouraged obviously two types of settlements. The relatively flat areas have hamleted clusters but the dissected areas have isolated homesteads. The southern scarps of this plateau have tiny halmets because of steep slope and dense forest. The south - western portion of this plateau is characterised by rolling landscape with a number of peaks of varying shape and size breaking the monotony of the region has semi sprinkled settlements. Growth of settlements on divide is common feature.

3. Settements in the Lower Hazaribag Plateau : The lower Hazaribag Plateau marked by relatively even terrain has more agricultural land than the Hazaribag Plateau. It has comparatively more plain land also. As such it has hamleted clusters and semi - sprinkled settlement. A few of such settlements have turned into agricultural. This region also has mica and coal mines based mining settlements : villages due to exhaustion of minerals. Settlements based on coal mining are more notable in the lower Barakar Basin.

The western parts of the lower Hazaribag Plateau merges im perceptibly with the North Koel basin divided by the Chatra Plateau. This portion is relatively rough and is dotted with widely spaced sprinkled type of settlements. As this part is also more dissected containg a broken country of inter-locking spurs, high ridges and intricate valleys, it presents a distinct distributional pattern of settlements. The upper portion of the Baraker Basin is characterised by semi-sprinkled settlements while the lower portion has hamleted clusters and semi sprinkled settlements. Settlements occur mainly on inter fluvial swells, valleys and raviens. The northern scarps usually have isolated tiny hamlets.

4. Settlements in the Damodar Basin : The structural basin of Damodar running through the middle of the state between the fractural edges of the Ranchi Plateau in the south and the Hazaribag Plateau in the north presents distincts feature for habitation. As the elevation varies between 650 m. near the source and 150 m. at the point where it leaves the state, the basin is characterised by different erosional features and steep and gentle scarps. The distribution of settlements is relatively sparse in the upper section of the basin where erosional features and uneven terrain are predominant. The steep scarps are dotted with widely spaced tiny hamlets.

The lower section of the basin which is marked by even terrain and mining centres has an even distribution of settlements. The settlements are closely distributed in the lower basin.

5. **The Pats** : The highly dissected and relatively higher land areas of the western Ranchi Plateau and the southern Garhwa and Palamu districts are characterised by distinct settlement types. The highland of the Pat is dissected by the tributaries of the river North Koel and the Burha river into many isolated Mesas and steep escarpaorants. These are marked by isolated and tiny hamlets. Almost all parts or flat crested upland namely Netarhat Pat, Daswan Pat, Pakri Pat, Joradumar Pahar, Gari Tanr Pahar, etc. have a few tiny settlements foci, while the southern section of this sub - rigion drained by the river Sankh, has comparatively more settlements foci. Here settlements are very isolated. Only small clearings in the forests where patches of relativerly flat land and along the stream courses are dotted with semi-sprinkled settlements. As the region consists of several rocky spurs of varying relief alternating with deep gorges and valleys, it has many spur point settlements and scarp settlements along the scarps. The valleys are settled with generally even distribution of settlement. The large basin of Chhechhri has relatively even distribution of settlement while its northern most portion where it has a deep gorge, is devoid of settlements The steep scarp of the Pats are also almost devoid of settlements because of difficulty in climbing.

6. **The North Koel Basin** : The North Koel Basin has diversified topographical features and as a result the distribution of settlement type also varies. Dissection is more pronounced in this region than in most other parts of the state as a result of alternately belted arrangement of rock of different hardness from east to west.

The lower parts of the basin where it merges with the Son valley has fragmented or hamleted clusters because the area has elevation below 150 m. The Kanhar river flowing on the south-western border of the region drains areas of highly forested and uneven relief and has isolated tiny hamlets associated with the small patches of agricultural land. The longitudinal ridges arranged in an east-west direction are pearced through by the Koel roughly in the middle creating settlement zones. The poor economy of the peasants and poor accessibility have greatly affected the region in its settlements distribution. The railways and roadways follow the valleys only where settlements are evenly distributed. The forested and dissected uplands are much isolated in respect of communication facilities and contain very sparsely settled areas.

7. **The Rajmahal Highlands** : The north eastern portion of the state known as Rajmahal highland has distinctive physiographic and socio - cultural features which result in the evolution of typical settlements. It has an elevation between 150 and 500 m. and is characterised by rugged topography, isolated conical hill and inter - montane valleys. There are numerous divides which are generally unproductive and give rise to linear settlements upon them as people do not want to misuse the productive valley lands. The linear ridge or divide top settlements are found through the highland. The small valleys of the rivers Ajay, Mor, Bansloi, etc. are inhabited by the Santhals. The conical hills as Tin Pahar are habitable sites of Malpaharias who do not like to settle in valleys and plains associated with the Santhals. The top of the ridges are also occupied by the Malpaharias. In the valleys the small ridges are occuped by linear settlements which generally follow village foot paths running along the ridges and river divides.

The construction of dwellings on the barren uplands ensures maximum utilisation of the cultivable land in the river valleys. The western dissected portion of this highland is noted for scattered or dispersed settlements. The absence of extensive agricultural land precludes nucleations. The south western part of the region which is relatively flat possesses a semi sprinkled settlements. Here large nucleations are found at the flat land while the high regions occupy isolated hamlets. Mineral resoures and communication do not exert significant influence on the distribution of settlement as these are undeveloped.

8. **The Adjoining Plain Land** : There is a narrow zone of low land between the Rajmahal highland and the river Ganga. The alluvial track in the north-east resembles Gangetic valley but it has a few hilly undulations or isolated small hills. This flood plain adjoining the Rajmahal highland has sprinkled settlements while the levee land has elongated compact settlements. Some notable levee settlements are Kasim Chak, Ranipur, Khaspura, Sukhpara, Nulaim, Belpahar, Ghat Salempur near Rajmahal town. Also there are a few hillocks in plain adjoining the highland which have Paharia settlement.

9. **The Panch Pargana Plain** : This physiographic unit lying east of the Ranchi Plateau has relatively gentle slope and even terrain except in the river valleys Due to incission of the channels river hanks have been subjected to intense gully erosion which has laid to the formation of extensive bad lands unfit for settlements. In the western parts adjoining the Ranchi Plateau the surface is broken and bears a dissected appearance having semi- sprinkled settlements. The contact zone or gentle scarps of the west have mostly contact point settlements along the foothills and isolated settlements on the upper portion of the scarps.

10. **The Porahat - Dalma Highland** : The undulating landscape of Porahat Dalma Highland caused by the differential erosion the lavas, is characterised by differential distribution of settlements and their types. The highland stand like a wall dividing the Chaibasa Plains from Panch Parganas Plains. Agricultural land is poor and limited and is confined to small valleys which have tiny hamlets on the adjoining uplands. Along the foot of the highland there lies cluster of hill base settlements. The northern slope of Porahat-Dalma Range is slightly more gentle than the southern slope and has therefore, some habitations.

11. **The Dhanjuri Range** : This highland consisting of Narwa, Rangamati and Dhanjuri ranges has elevation over 500 m. The entire region except a few intermittent valleys has sprinkled settlements. The valley of the river Sankh is dotted with hamleted clusters. The average distance between the settlements in the valley is about 1 km. as against 4 to 5 kms. on the highland. The whole range has been completely dissected by various small tributaries of the river Sankh and is dotted with sprinkled settlements.

12. The Dhalbhum Plains : The fairly feritle alluvial plain of Dhalbhum is marked by semi- compact to compact settlements. This region has a high density of rural population due to better agricultual condition. The presence of rich mineral resources and suitable cultivable land have increased the number of habitation. Good accessibility caused by metalled roads and railways such as the Mumbai - Calcutta Railway has provided further inducement. Along the eastern portion of the Dhalbhum Plain there lies compact settlement and water point settlement but along the western portion there lies hamleted and semi - sprinkled settlements. The settlements are evenly distributed and lie at an average distance of about 1 km. The size of settlement is also relatively bigger than in the other parts of states.

13. The Singhbhum Plains : These plains consisting of Chaibasa plain and Seraikela plain are marked by flat topography with isolated hillocks. These plains have encouraged the growth of evenly distributed settlements. The alluvial plains of the river Sanjay and Kharkai are characterised by semi-compact and hamleted clusters. The adjoining plains of the Kolhan upland and Dalma Dhanjuri Ranges are mostly dotted with semi-sprinkled settlements. The divides of the river are suitable for settlements. Sprinkled settlements are confined to the few isolated hills. There are several mining centres which have attracted rural dwellers to particular sites and have changed the rural landscape. The cultural features such a roads, railways and village paths are relatively more in frequency due to flat topography resulting in more or less even distribution of settlements.

14. The Kolhan Upland : This upland consisting of isolated hills, low and undulating ridges and several intermittent rivers present distinct physical features.The settlements of this region and of three types gragmented or hamleted, semi - sprinkled and sprinkled or disparsed. The distribution of each of these type is clearly related to the nature of the terrain. The settlments are built particularly on high and barren land and this ensures maximum utilization of the cultivated land. The open and the gently undulating northern and north - eastern portion of the high land is concerned with numerous prosperous hamleted clusters. The southern portion is almost devoid of hills and is therefore, closely and evenly settled. The south western portion is mountainous and is thickly covered with forest. It is' therefore, sparsely populated. The Trilposi and Tholkabad areas are reserved forests which are completely devoid of cultivable land and are dotted with a few tiny hamlets only. The central part of the upland is relatively less rugged and forested and has some what more cultivable lands. This portion is characterised by even distribution of tiny hamlets.

15. The Simdega Plateau : This highly dissected portion of the state is characterised by steep hill slopes, narrow valleys, rough terrain has different types of settlements and distributional patterns. The basins of the rivers Sankh and Deo are mainly dotted with semi - sprinkled settlements due to more agricultural land but the highly dissected areas having dense forest are devoid of settlements. The areas of rugged terrain have widely spaced tiny settlements where there is relatively flat topography drained by small stream, the divide is selected for habitation and the low land is selected for cultivation. As the region is poor in mineral resources and economic base, the settlements are mainly controlled by the physical features.

Types of Rural Settlement :

The typology of rural settlement concerns characteristic groupings of rural dwellings in that well defined parcel of the ground which is here known as Mauza. These groupings are intimately related to the geographical and cultural settings. As such the arrangements of dwellings and characteristics of groupings give rise to variations in types of settlements. Broadly rural settlements are grouped into types :

1. Compact or clustered or agglomerated or nucleated and (2) dispersed or sprinkled or scattred. In this region, due to diverse physiographic and socio-economic conditions, non of these extreme types are common. It is, however, possible to identify five sub-types which are more or less branches or variation of the above two broad types. These are (i). Compact settlements (ii) semi-compact or hamleted clusters (iii) Hamlets (iv) Semi- sprinkled or dispersed settlement. In addition some other settlements types such as linear type may also be observed in some of the areas.

(i) **Compact Settlements** : Usually compact settlements are rare in this state except in a small part of the plains adjoining Rajmahal highland and in parts of Panch Pargana and Dhalbhm Plains. This type of settlement is characterised with agglomeration of almost all dwellings of the Mauza at one site. The large size compact settlement consists of dwellings ranging from 50 to 100 houses. Generally these houses remain very close to each other lacking inter-house space and these houses face the narrow and zig-zag lanes without having more front and back spaces. The site of this type of settlement occurs on the high banks of the stream or the dry point where uniformity of soil and relief are found. In flood prone areas growth of dwellings at one flood free site enables the inhabitants to avail the facilities of village boat, school, Panchayat Bhawan, village shop and other amenities developed in the main settlement. The fertile lands coupled with high bearing capacity because of multiple cropping and higher yield enable the inhabitants to live in large number on limited land. Agricultural co-operation coming from generations scattered agricultural fields, dominance of the land lords or tribal chiefs social inter, dependence, etc. have also boosted the developments of nucleated growth. In the adjoining plains of the Rajmahal highland as well as the lower portion of the Dhalbhum plain, moust of the nucleating forces noted above have been responsible for concentration of dwellings at flood free sites or along the metalled or unmetalled roads or some other locational advantages..

In addition to these areas smaller patches of flat Ranchi and Hazaribag Plateau, mostly around the urban centres have also, given rise to compact growth of settlements. Certainly these compact settlements hardly reveal such congestion like the compact settlements found in the plains.

(ii) Semi compact settlements :- The semi - compact settlements represent an intermediate stage between the agglomerated and hamleted settlements. These settlements are characterised by the presence of one easily recognisable nucleated site and one or more smaller hamlets closely linked with the main site by footpath or cart-track in the Mauza. These attached satellite or tolas must have grown due to the increasing pressure of population on the main site which compelled inhabitant to seek living outside the main habitation. It also happens due to the concentration of dominating castes on the main site and weaker people in the adjoining areas. Certainly these people living in the satellites remain economically and socially linked with the main settlements. Some times the development of transport or religious sites outside the main habitation encourage construction of dwellings and growth of satellite settlements, considerable areas of the relatively homogenous Ranchi and Hazaribag Plateaus, have hamleted clusters or semi-compact settlements.

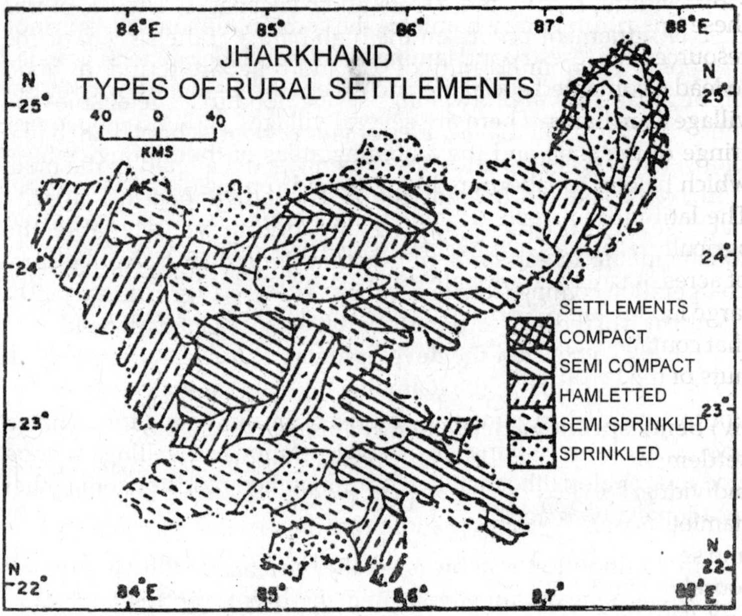

Fig. 20

(iii) Hamleted Settlements :- Hamlets are usually smaller nucleations or tiny foci incapable of exerting any attraction comparable to that which concentrates on cultivated fields. Virtually this type of settlements is determined by both dispersal and nucleating forces. Dispersion forces like lack of large fertile plain, large habitable sites, caste system, etc. cause growth decentralisation of settlements but some nucleating forces like social collision, joint family system, co-operation in agriculture motivate people to live together, the entire area is marked with smaller plains attached to rivulets between divides having capacity to support a few families resulting to growth of smaller nucleatious. In addition these tiny nucleated settlements also come into being on the crossings of small village path or roads or near village amimities.

On the basis of the socio-economic structure, these settlements foci have been sub-divided into three sub-types-- (a) tradional hamlet (b) Desi type and (c) Latifundium type. The traditional hamlets represent the rudimentary habitations of the tribes of the region dotted in the vast forested uneven areas. The main stay of these tiny rudimentary hamlet is the combination of several rural resources like forestry and taming of animals. They generally prefer to lead a contented social life and except the legal authority of the village Panchayat. There are several villages along the northern fringe of the state and the adjoining areas of the mining centres which belong to labourers living to and represent the Desi type. The latifundium type of hamlet is generally associated with the agricultural estates of the region which generally possess hundreds of acres of land. To cultivate this vast agricultural track requires a large labour force. This farm usually carry individual central core that contains the owner together with more numerous and modest huts of the labourers.

(iv) **Semi -sprinkled Settlement** : This characteristic type of rural settlement is marked by the presence of smaller hamlets and individual habitations spread over the entire Mauza. These smaller hamlets generally consist of not more than ten houses at one place and several isolated dwellings adjacent to smaller hamlets. In these loosely connected settlements street pattern is generally lacking and the straggling residences are separated by cultivated fields. Sometimes uneven terrain or patches of forest instead of cultivated fields may intervene the smaller hamlets or individual habitations. These settlements manifest the development of the individualistic tendencies where people have gone out for economic advantages and are indicative of the peaceful atmosphere of locality. Here each occupance unit has its individual cultivated fields nearby orchards, or forests and drinking water facilities like river or well. These settlements are to some extent better than compact dwelling at one place far from their cultivated fields but they are harmful for the development of social life. Here close and warm communal life of the compact village and the feeling of being an integral part of life of a community are generally lacking. Such settlements are mainly confined to the areas having less uneven terrain such as the Panch Pargana Plains, parts of Ranchi and Hazaribag Plateau, etc.

(v) Dispersed Settlements :- This type of rural settlements is marked by the presence of complete diffusion of homestead over the entire village area and relatively long distance between dwellings as compared to the congestion of the plain settlements. In this region, the sprinkled settlements are characterised by the presence of dispersd population into tiny hamlets scattered at an irregular distance and in close association with their individual resources.

The relatively dispersed type of settlement is the characteristics of the settlements of this region. The dissected topography and diversity of soil resulting in a diffusion of surface water and relative security in the part are factors resulting in the growth of this type of settlements. Usually the dissemination of settlements achieves its complete development on fertile soil. The outstanding characteristic of dispersed settlements is the minute size of the settlement unit, viz. the family, and the isolation and the privacy in which it exists. Such isolated living tends to separate people psychologically particularly slope of the region, dissection of the land, lack of large agricultural land at one place, poor living of the tribes requiring no safety of house hold items, etc. become important factor for dispersal of settlements.

In addition to above mentioned types of rural settlements linear dispersed settlements are found in considerable number in the Rajmahal highland. Elongation of these linear dispersed settlements has been caused by the footpath or cart track passing through the region or the existence of ridge top settlements.

Population Size of Village :

In terms of population size of village, the state of Jharkhand reveals considerable special valuation based on nature of land surface, agricultural production and other mining and industrial resources, etc. It becomes obvious by the analysis of the table. 14.1

The above table manifests that districts of Rajmahal Highland have comparatively higher percentage of smaller size of villages but districts located in Hazaribag and Ranchi Plateau and Damodar Basin have more villages having moderate to high size of population.

Table - 14.1
Population size of Villages : Jharkhand (in%) 2011

Sl. No.	Districts	Less than 200	200-499	500-999	1000-1999	2000-4999	5000-9999	more than 10,000
1.	Bokaro	5.40	12.11	22.42	29.13	26.02	4.41	0.49
2.	Chatra	24.33	28.47	24.69	15.61	6.32	0.58	0.00
3.	Deoghar	26.64	36.96	23.62	10.24	2.46	0.08	0.00
4.	Dhanbad	12.93	22.07	27.30	24.35	9.54	3.81	0.00
5.	Dumka	25.52	44.01	22.84	6.10	1.49	0.04	0.00
6.	Garhwa	9.95	13.74	24.53	27.73	20.97	2.73	0.36
7.	Giridih	17.63	27.99	28.07	17.55	7.51	0.98	0.27
8.	Godda	29.27	24.41	21.92	16.23	7.88	0.30	0.00
9.	Gumla	3.82	22.61	37.90	26.22	8.49	0.85	0.11
10.	Hazaribagh	13.58	19.95	24.81	23.97	1.01	0.08	0.00
11.	Jamtara	17.37	33.46	31.61	13.12	4.34	0.09	0.00
12.	Koderma	16.12	23.40	26.34	22.88	10.40	0.69	0.17
13.	Khunti	10.88	41.11	30.24	14.46	3.32	0.00	0.00
14.	Latehar	10.55	28.57	30.57	20.56	9.21	0.53	0.00
15.	Lohardaga	7.39	21.31	27.56	27.27	15.63	0.85	0.00
16.	Pakur	24.89	31.64	24.72	12.27	5.26	1.14	0.09
17.	Palamu	17.30	19.99	27.64	22.79	11.22	0.99	0.06
18.	W. Singhbhum	12.06	29.48	34.23	21.25	4.81	0.00	0.00
19.	E. Singhbhum	21.19	33.00	27.97	14.23	3.48	0.06	0.06
20.	Ramgarh	5.57	11.48	21.64	31.15	25.90	3.93	0.33
21.	Ranchi	5.09	17.82	31.94	27.85	15.28	1.77	0.23
22.	Sahebganj	43.44	20.16	17.57	11.34	5.93	0.82	0.74
23.	Saraikela-Kharsawan	14.57	30.45	35.08	15.71	3.93	0.26	0.00
24.	Simdega	3.56	21.38	29.40	29.84	14.25	1.56	0.00

Source : Primary Census Abstract, 2011.

Rural Settlement Patterns :

While analysis pattern of settlements it becomes necessary to differentiate between distributional pattern of settlements and the pattern of settlements. In case of the former distribution of different types of settlements are analysed. Different areas having different types of settlements are also analysed giving emphasis on geographical conditions. In case of later or the pattern of settlements which also mean the developments of forms of settlements or morphology of settlements are analysed. As such the study of the pattern of settlements concerns the development of varied shapes, changing morphological features - both external layout and internal plan and geographical conditions which determine the morphogenic accounts. The morphogenesis of settlements concerns the development of shape depending of various geographical phenomena. These phenomena include physical as well as cultural condition. Physical conditions influencing external form of the settlement include configuration of the site, surface water like river, canal, tank, pond, well,vegetational cover and the shape of the cultivated fields, socio - cultural condition include pattern of roads, historical events, cultural tradition, etc. The arrangement of dwellings is guided by roads, cart tracks, water facilities and other social amenities. The village cart - tracks and lanes form the skeleton of the internal layout of the settlement. The buildings located in the space within skeleton determine the shape and form of the settlements. Certainly the analysis of shape of the settlement and other morophological features become possible only concerning to settlements having compac growth because the development of shape takes place in case of only compact and nucleated settlements and villages having scattered or other types of settlements other than the compact one do not give rise to any geometrical shape. Thus, the study of pattern involves only these villages which have compact or roughly compact settlements and hamleted clusters or semi sprinkled or sprinkled settlements are left out of this analysis. It is true that different physiographic divisions of this state have some distinct pattern of settlements. In each division the physical elements have obviously influenced the forms to a greater extent than the other social and cultural variables. Areas having natural growth of settlements hardly provide geometrical shape or plans of settlement like the developed nations but mostly provide roughly different geometrical shape.

Important settlement patterns are the following :

(1) Rectangular Pattern : The most common form of the nucleated settlements in the region is rectangular. In this pattern one arm is larger than the other arm of the settlement. The general factors influencing the pattern is the rectangular form of the fields. The aggregation of rectangular or square building in the plots of the same shape produces rectangular layout of the settlement. Most of the houses remain rectangular with their main axis from north to south and east to west so that they may get adequate sunlight. Besides the measurement of the land in the area is generally based on some sort of rectangular or square unit which has resulted similar field pattern. The boundary of the field is sometimes used for footpath and cart tracks. Such fields are easier for measurement and turning of plough. Therefore, people do not want to disturb the ploughable shape of the field adjoining the dwellings, as these fields are the chief assets of the peasants.

The other advantage of rectangular shape lies in maximum accomodation of houses in several rows. People prefer to orient their houses in rectangular form because of certain superstitious belief. Most of the non-tribal and tribal people do not like to construct perfectly square houses as these houses are considered bad from the religious point of view.

Although rectangular shaped settlements are found almost everywhere but areas having more tanks or crossing of roads and cart tracks have usually more such settlements. In the south eatern part of Dhalbhum and in parts of Panch-Parganas Plains tank based rectangular settlements have developed. Certainly such rectangular settlements give rise to Hollow - Rectangular pattern which differs only in terms of open space in the centre of the settlements. Sometimes such open space is also caused by the presence of old site of palace or temple or mosque with a number of trees, etc. In case of a place of worship in the centre villagers do not like to encroach upon the religious sites based on religious believes.

(2) Square Pattern : This pattern is the beginning stage of the rectangular pattern. Square shaped settlements often turn into rectangular after slight more growth in certain direction. This pattern of settlement is also caused by the site having crossing of cart track or roads and other cultural features similar to the rectangular one. In this pattern also hollow space or unbuilt space in the middle of settlements may be caused by the presence of square tank or pond, temple, mosque, Panchayat Bhavan, Educational institutions, etc. hollow square pattern come into being. This pattern is also found almost every where in areas having relatively homogenous land surface and also having good net work of cart-track.

(3) Chess- board Plan : The Chess- board or a gough grid plan is really a by product of the large rectangular settlement. The large rectangular settlement sometimes contain a net of lanes or cart tracks which intersect each other at right angles. The intersection of lanes or cart tracks creates rectangular sub-blocks giving the form of chess board. Certainly such settlement pattern is usually found in planned rural settlement but in exceptional cases only rough chess board pattern can be seen. A suitable example of a rough chess board plan village is provided by **Bindirangroi** in the Subarnarekha Basin.

(4) Elongated Pattern : This is a common pattern usually found in the flood plains where rivers for levese and people prefer to settle on flood free elongated levee site. At the same time metalled roads and cart tracks also encourage elongated growth of settlements along the thorough fare. In such settlements arrangement of dwelling takes place along a line or a series of lines. It prolonging in one direction and restricts in another direction due to certain physical or cultural obstacles. The growth of settlements along the road facilitates inhabitants easy movement as well as the development of shops and small scale industries. The development of levee based settlements provide the advantages of the proximity the river for water for varied uses. There are a few ridge top elongated settlements. Such settlements are also found in areas having development of roads and the flood plains or in the areas having series of ridges.

(5) Circular Pattern : This pattern is also rarely found in areas having unplanned growth but in exceptional cases some villages obtain the circular pattern. This pattern is caused by maximum aggregation for certain facilities like safety and easy accessibility from some central attraction like the temple, pond, mosque, etc. Previously rich people used to construct their house in the centre of the village surrounded by the dwellings of the peasants but example of such pattern has not been identified. Certainly circular shape has developed based on cultural and physical features like the occurrence of temple or mosque or mond in the centre and growth of settlement around these cultural and physical phenomena. Sometimes hollow space existing inside the circular settlement give rise to hollow circular pattern.

(6) Fan Pattern : The fan pattern also rarely occurs in the region. Such settlements have developed near some focal points like temple or ghat on the bank of the river. These features attract the habitation from several directions. Lanes or footpaths converge on such points. Due to convergence of lanes on the focal point to get all sorts of amenities the site becomes just like a fan.

(7) String Pattern : This pattern is more common in the districts of Santhal Parganas and in parts of Panch Parganas. This is more or less a large linear settlement. In this case dwellings develop along cart track, river or nala in a row to utilize the facilities of easy approach. Such habitations possess width of 1 to 2 dwellings. Certainly lengthwise it prolongs to several furlongs. The Rajmahal highland provides a good example of this pattern.

(8) Irregular Pattern : This pattern is more common in the region. It generally consists of larger villages than the average of compact settlements. In this pattern houses are huddled in an irregular fashion following no definite plan. The larger villages generally consist of several tolas having different forms and when combined together do not provide any definite pattern.

In addition to the above mentioned patterns sometimes settlements take the shape of L, T, sickle, triangular, double-nucleated etc. based on differnt local conditions. Some uncommon pattern like semi-circular, amorphous or shoe pattern, herringbone patterns are also seen.

Regional Distribution : It has been discussed earlier that village is determined by various micro geographical conditions such as site characteristics, approach by cart track and internal facilities. These local conditions are not universally found in the entire region and hence the region can not be clearly demarcated in particular settlement pattern. Still predominance of particular settlement pattern can be identified in certain sub-region. The most evenly distributed settlement pattern is the rectangular and square.

These are found in areas having compact and semi-compact types of rural settlements.

The linear pattern is another common pattern. Its occurrence is more marked in the greater part of the Santhal Parganas. The occurrence of circular pattern is not common feature in the region but there are a few circular settlements based on the existence of mosque or temple. The distribution of amorphous pattern, oval pattern, parallelogram pattern is also not the common feature in this region. The irregular pattern is found in most areas of the state. The string pattern is the most common feature in the Rajmahal highland.

Problems of Rural Settlements : Since the development of human habitation has taken place in an unplanned manner lacking scientific consideration of the selection of site, availability of amenities, and other essentials, most rural settlements have problemes related to physical, economic and social conditions. In the adjoining plains of the Rajmahal and the lower parts of the river basins like the Subarnarekha the North Koel, the Damodar, etc. the problems of flood is realised. No doubt, in such areas people have built up their dwellings on high monds artificial or natural but whenever furious flood occurs these settlements get affected. Most of the settlements lack essential amenities like drinking water the facility of bathroom, lavotary and proper approach by roads or cart tracks. The region is marked by dissected landscape and scattered patches of agricultural land which have resulted isolated dwellings or smaller nucleation where the advantages of compact settlements are not available. These isolated settlements only avail the advantages of environment but remain devoid of social and cultural interaction.

Rural Dwellings :

Dwellings, a socio - econmic index, has been the fundamental requirement of human being as it provides shelter and protection from rigours of climate. According to the census of India. A house is defined as a structure or a part of a structure inhabited or vacant or a dwelling or a shop - cum - dwelling or a place of business or work shop or school, etc. with a separate entrance : Virtually rural dwellings concern some shed for the protection of the rigours of climate, the place for rest, keeping household items, agricultural products, intertaining guests, etc. Finch and Trewartha have rightly remarked that the house is a critical focus in the life of any individual since he is forced to return to it for periodic protection. The house then becomes a universal feature of regions permanently occupied by human being and is one of the most fundamental elements of cultural geography. The agglomeration of houses marks the origin of human settlement and reflect the nature of a region since their character is related to environment and to the cultural heritage of the people. This is why the rural dwelling becomes an index of environment, economy and social development. It is true that importance of dwellings has been closely associated with the people when they divorced forest life and cave dwellings and domesticated some animals started pastoral and agricultural socities and constructed houses. Since then houses and its structural design, housing amenities and other features have tremendously changed at par with the increasing civilization. But, the areas where the people are till not much aware of the modernised living, cultural development and economic prosperity, the people still have the rural dwellings not much different from their traditional designs and portray the old culture and poor living.

The rural dwellings of the state also reveal dominating role of environment, economic conditions and socio - cultural conditions. It is true that in most areas of the earth surface the environmental, economic and social conditions determine the elements of house but in areas where people have experienced economic development modernised culture, they have also modernised their dwellings not much dependent upon local conditions like the availability of building material inherited housing designs and other features. In this region most of the houses have been constructed out of locality available raw-materials and they have applied the inherited skills. Certainly settlements located near the urban or industrial centres have recorded a few houses based on materials brought from other areas and also have been constructed on applying modern designs. As such houses of this region reveal manifest spatial variation based on environmental conditions as well as economic development. Areas having considerable industrial and urban growth have houses slightly different from the remote rural areas where almost all houses reveal the dominance of natural control on the basis of differential urbanisation, industrialisation social, awareness, agricultural development climatic condition, houses of this region differ in terms of constructional materials, build in architecture, accomodation, space, sanitation, occupancy rate, etc.

A primary classification can be attempted on the basis of building materials used. A variety of physical products like the mud or soil, brick, stone, tile, timber, grass, leaves, reeds, etc. are used for the construction of dwellings and result in a variety of house types. Accordingly houses are classified into mud wall thatched roof, mud wall tiled roof, brick wall thatched roof, brick wall tiled roof and pucca houses. The houses of this region can also be classified on the basis of build and architecture. There is a marked difference between rural and urban architecture. The rural areas have still traditional art of dwelling construction, as the aboriginals still prefer their traditional designs. A few rich persons in the village have applied modern design from the urban centres but their houses present an incongruous feature in the rural landscape. Similarly urban houses are also classified on the basis of functions such as dwellings , shop-cum-dwellings, industrial, educational, religious houses, etc. In both rural and urban areas houses vary in size from large building to a single room hut showing the residence of the rich and the poor. The size of the dwelling is also determined by the size and type of the family. The caste structure may also play the major role in guiding the size of the dwellings. The houses of the socially depressed communities are small because they are not only poor but there is little parda in their families so that one or two rooms with a verandah become enough. Certainly most of the houses of this region is relatively of smaller size than the plains because of poor economic condition.

As mentioned earlier, there are some poor people living in very wretched condition. They somehow fence the housing area with no regular room. Certainly such houses are negligible. In such houses, the courtyard and shed serve the purpose of the poor persons. Such houses are seen one or two in almost all villages but its number increases in the hilly areas of Santhal Parganas. Certainly, these houses remain non-existent culturally and economically improved mining areas of Damodar Basin and in the neighbourhood of major urban centres.

The area is well marked with houses of smaller size having one or two rooms and verandah. The one roomed house of the poor people represents a rudimentary form of the compact house where human being and animals find shelter under the same roof. Such houses are in considrable number in almost all areas of the state. In this case also the number considerably increases in the hilly areas of Santhal Parganas, this type of dwelling is cheap to construct but unhygenic and uncomfortable as the same room serves as kitchen, store, bed room and other uses. Two roomed houses are also in considerable number because people of moderate economy prefer to have two roomed houses with verandah. Such houses are in large number in the flat plateau of Ranchi and Hazaribag. The ground plan of these houses are roughly rectangular. Medium sized houses consisting of three to four rooms are occupied by people living in joint family. Such dwellings belong to middle class peasant families. The plain areas of Dhalbhum, Panch Parganas and the adjoining plains of Rajmahal have considerable number of houses having three to four rooms. Such houses are also in large number in the Hazaribag Plateau. Houses having five and more rooms are only a few in all areas becuase such houses belong to economically well off people. The occurrence of such dwellings is associated with fertile plain and industrial and mining centres.

House types based on building materials : There are certain materials which are in universal use but some regional differences are observed. On the basis of building materials houses are classified into the following :

(1) Mud wall and thatch roofed houses are usually found in the plains of Santhal Parganas, Dhanbad District and Purbi Singhbhum. Such houses consist of walls made of mud and roof made of grass, reeds and paddy straw. Such houses are oval in plan but the roofs are generally conical in form. Dominance of such houses are in areas having the availability of thatch materials as well as mud. The poor people of Santhal Parganas have considerably good number of mud walled and thatch houses. Sometimes people living in the flood affected areas prefer to construct temporary houses having walls of bamboo split and reeds and roofs of thatch materials becuase such houses have portable character :

(2) Mud wall and tile roofed houses : This house type is dominant in the area. Most areas of Ranchi and Hazaribag Plateau have mud wall and tile roofed houses. Virtually areas including Ranchi

Plateau, Simdega Plateau, Hazaribag Plateau, parts of the North Koel basin, the Pats, etc. form a compact zone having mud wall and tile roofed house. The conspicous occurrence of laterite soil in this portion has facilitated the wide distribution of mud walls while tile roofs are economic. No doubt, other areas of the state like the Dhalbhum plains, the Panch Pargana plains, the Rajmahal Highlands have also houses of mud wall but these regions are not marked by the presence of tiled roof. In these regions, certain geographic and economic condition prevent the people from preferring tile roofs. The comparatively higher amount of rainfall and the easily available paddy straw induce the people to build thatched roof which are cheaper and on which run off rain water is swift.

(3) Brick wall and tile roofed houses : Houses having burnt and unburnt brick wall houses are in considerable number. Certainly houses having burnt brick wall is unimportant almost every where except in the mining areas of Dhanbad where availability of coal enable the burning of bricks. The area is also economically prosperous. In the remaining areas such houses are only a few belonging to the rich persons. Wall built of unburnt bricks is also seen in parts of Singhbhum plains and Hazaribag plateau.

(4) Stone wall and tile roofed houses : This is also a common house type in the state and is the true representation of its regional character. In this house type the physical characteristics of the region put thier special stamp over the cultural landscape. The walls of these houses are made of stone blocks and mud as cementing material. The roof of these houses consist of tiles, iron sheet or straw. This house type is common in the rural but occasionely in urban area. The easy availability of stone from the surrounding hills makes it cheap as building material. The people construct stone wall by piling stone block and cementing them in place with cement in the urban areas and mud in the rural areas. The first precaution in the erection of such walls is that the stone should not be flat surface of each stone block, thus, constitute the two faces of the wall. The precaution necessary for strength is that every stone fixed in the wall should rest on it broad surface at the bottom. The chief characteristics of this house type are durability, cheapness and safety from theft. The stone wall is safe in flood areas as stone is more durable than wood and less subject to conflagration and better than brick for taking shape and strengthening walls. Certainly areas having softer rock have given more number of houses having stone walls.

(5) Brick wall and conerefe roof : The area has also a few pucca houses having burnt bricks and pucca roof. These houses are found in rich villages belonging to a few resource ful persons but the number of such houses conirderably increase in the urban contres or industrial centres. Virtually in urban areas concrete houses have become attraction of the people.

Dwellings have also been classified on the basis of shape. These are the following :-

(1) Rectangular House Types : The rectangular shape of the house is common feature in this region as the overwhelming majority of Indian houses are of rectangular type. Most of the buildings of schools, hostels and planned public houses are of such shape consisting of several rooms in one line joined together with an open courtyard in front of the buildings. Such building also have verandah which serve different purposes. They have windows for ventilation. Most of the dwellings of the aboriginals have rectangular ground plan with sizes varying from about six metres to seven metres in length and four metres to five metres in width. They contain only one or two rooms. The kitchen is divided by a partition wall making one secret seat of ancestral spirits locally known as Bhitharghar.

(2) Square Types : This house type is generally associated with the brick built and concrete houses which are a few in number in rural areas. These square buildings with several rooms belong to the more resourceful cultivators. They have a courtyard in the middle locally know as Angan, the size of which varies according to the size of the building. The door of each room opens in the courtyard. The windowless walls face outside and the front of the room point inward to the courtyard from which only the sky can be seen besides providing sunlight during winter and moonlight nights during summer where people prefer to sleep in the open sky. The coutyard is also used for winnowing and draying of grains and also used for keeping the agricultural implements and food grains. There is a open passage towards the Dalan (sitting place) in front of the house. It accomodates a large family and provides sufficient privacy and protection from theft.

(3) Bee-hive Types : This type of building is a typical house type in the Rajmahal highland. It is similar to the circular house type found in the Kolhan Lake island. Only Birhor tribe of the Rajmahal highlands construct such houses with boughs and twigs of trees. They do not have solidwalls and only shallow gutter surrounds them. Paddy straw is used as thatched material. These houses have single opening consisting of a door of only a few feel high and two to three feet wide. The door is made of split bamboo. There are little round thatched ovals with a low door way through which one can just crawl. Sometimes walls of these houses are also made of wooden logs placed side by side. In the centre of such houses there is central post of about three metres to four metres high from which the structure of the roof slope down in circular form. The wall is raised to a hieght of two to three metres. The roof extends beyond the wall by one foot and serves as a coverage for the raised platform around the wall on its outer side. The shape of some of the religious houses is also circular but in design differs from other houses.

(4) Planned House Types : No doubt planned houses are only constructed in urban areas but a few govt.houses remain in rural areas based on certain plan. In addition to the modern architectral design there are ancestural designs suggested by the Pradhan or Chief of the village. As such people follow the ancestral design and plan of the houses which are being carried out from generations. The modern planned houses consist of enough frontyard, backyard space, separate bed room, store, kitchen, guest room, provision of water supply, proper ventilation, etc. The windows of the planned houses are usually large enough to provide sufficient sunlight and air but the houses based on ancestral plan lack proper arrangement for sunlight for air, water, bathroom, separate kitchen ,store, etc.

Factors Determining House Types

Houses of different areas are influenced by different geographical conditions. In some areas physical conditions become more dominant while in other areas socia-economic conditions become more forceful. Somehow house types of all areas are the result of physical, socio-economic and cultural factors.

Physical Factors : Man taps the local resources and builds houses out of local building materials. The chief building materials available in the region are laterite soil, stones, grass, reeds, timber, etc. The lateritic soil containing high percentage of ferric oxide and bauxite is suitable for the erection of mud walls made of this soil are more durable on account of the registence to weathering. Timber, grass and reeds are also important building materials. When soft stones are easily available, houses are built with stones near the base of the hills. Such houses are more expensive than mud houses. Only the rich people, therefore, avail this opportunity. Brick and tile also constitute the basic requirements for the house. The other building materials associated with houses are grass, leaves reeds, bamboo, etc. which are mainly used for roofing purposes.

Houses reflect the climate in the protection. They offer against cold winds and heavy rain. The weather play a significant role in the variation of house types. Virtually houses are build considering the amount of rainfall, intensity of temperature, direction of hand and the angle of the sun, good ventilation, protection from demp and adequate sunlight are essential for a good house. The intensity of heat and the amount of rainfall also determine the slope and nature of the roof. The fact that the climate expresses itself through the form of the roof is forcefully brought out by the houses of the region. On account of moderate rainfall and easy availability of thatching materials as well as poor economic condition of the masses, tile and thatched roofs are preferred. The slope of the roof is usually between forty five and sixty degree so that rainwater may drain quickly. In the construction of rural dwelling the consideration of courtyard and open space is also essential as the people use the courtyard for various purposes.

Socio - economic Factors : The socio-economic conditions also determine the regional variation in house type. There is a marked contrast between the houses of the upper and the lower classes of people in the whole state. The economically advanced rural areas have good rural dwellings but less developed rural areas have traditional houses. The houses of resource persons remain of relatively larger size and are also built of burnt brick and tiles but the houses of the poor remain of small size usually built of mud and tile or thatch materials. In urban areas houses reveal obvious impact of urban culture and houses are also built of different building materials which are not available in near by region but in case of rural areas houses reveal dominance of local raw material. The rural-urban gulf presents an attractive phenomenon for the study of rural and urban dwellings. The houses of this region have also considerable impact of cultural environment because of tribai concentration. There are marked difference in house type of different tribes. Some tribes like ornamental painting on walls. The Oraons and Munda generally make sketches of different animals on the walls. While Santhal do not like such painting. In certain localities there are some socially more depressed communities like Mushhar and Doms whose culture differ from their neighbours. These people practically prefer to live away from the main village and have only small dingy hut indicative of their poverty and extreme backwardness.

Construction of Dwellings : In the rural areas of this state the houses are usually built up in the months of April and May and most of the labour is provided by the family. The renewal of the old roofs takes place in the early part of the season. A few semi-skilled labours are sometimes engaged. The construction of the Bee-hive type of houses of Birhors of Santhal shows special features. The walls are commonly built with mud, grass, leaves, reeds etc. The surface of the wall is plastered on both sides with a paste of cow-dung. The roof is extended beyond the wall to serve, as coverage for the raised platform in front. It provides safety to walls from rainfall and prolongs it life. The people of the region prefer the construction of pucca houses generally of square type. The flat roof is used for drying grains and enjoying summer nights.

House Ownership and Tenure Status : In rural areas, people construct their own houses and rented houses are found only adjacent to urban centres. Rural dwellers have for generations lived more or less at the same place and possess ancestral houses. The rented dwellings are also non-existent in rural areas but in urban areas the number of rented houses remarkably increases.

Rural Service Centres :

In developing nations where most of the people live in villages and urban centres exist at relatively larger distances the rural service centres play important role in providing services to the rural mass and establishing relationship between the rural habitat and urban centres. These centres may be small town or village performing secondary and tertiary function and having different service amenities like the urban centres on smaller scale. As such they are considered the middle stage between village and town also urban centres. As such they are much regional variation in terms of industrial growth urbanisation mineral resources and its exploitation, transport network etc. has also given rise to a large number of rural service centres scattered in the entire area. Virtually the settlements which have some administrative, or trade, or educational or some other non-agricultural functions have become service centres because they serve the surrounding villages because of their non-agricultural services. These centres portray demographicaly small towns or large village with locational centrality and have some market facilities or, service facilities or other amenities which characterise the settlement for attracting the surrounding rural mass. These service centres not only attract the people of rural areas for availing services but play important role in rural transformation. These centres establish trade relation between the village and the urban centres and collects samples agricultural products of rural areas for other regions and also bring those things which are not produced in the region. Similarly these service centres enable the rural people to avail the facilities of education so that education and technical know how can enrich and rural development can take place. The health services enable them to have the advantage of sound health. The marketing facilities of fertilizer, pesticides, agricultural implements, improved seeds and other requirements enable cultivators to develop agriculture and to grow crops of commercial value. The facility of rural market and its linkage with the urban centres has motivated the people to grow vegetables, seasonal fruits, dairy products and other things which enable farmers to earn more money. Virtually, the products of villages are supplied to major urban centres where demands are high only with the help of the functioning of rural service centres.

Rural service centres of this state are not only working as trade centres or educational, recreational, health but are also growing in terms of industries based on local raw-materials, the repairing shops and gathering of rural masses to these centres is establishing social-contact, providing not only the transaction of the products of rural areas with that of the products of urban areas but also exchange culture, ideology and socio-economic transformation. In the areas having rough terrain and forest clad regions lacking mineralised and industrialised centres there are only a few widely spaced urban centres. In such areas the role of rural market or service centres becomes of prime importance because only people of neighbouring undulated areas can approach the rural service centres and avail the services which are usually found in urban areas but in smaller scale in rural service centres. In the highly industrialised pockets there has been the growth of bunches of towns and in such areas, the importance of rural service centres has not been much remarkable. It is only the areas where less urban or industrial growth has taken place mushrooming of rural service centres has taken place.

Demographically, the rural service centres consist of different population size, varying percentage of secondary and territary workers literacy percentage, etc. Its role increases in accelerating the percentage of literacy in the surrounding villages by catering educational facilities, motivating the people living in the village to involve some of them in non-agricultural occupation or to have occupational mobility and also, aware the innocent workers who are low paid in the rural areas to move out for other destinations in the form of labour migration for earning more money in short period.

Economically, the rural service centres play important role in accelerating agricultural production with the supply of fertilizer, pesticides, high yielding varities of seeds and improved agricultural implements. These centres also motivate cultivators to grow such crops which can enable them to earn more money. The dependency of cultivators on food crops has to some extent minimised by the interest of cultivators to grow cash crops, vegetables and fruits which are highly needed in the markets and cultivators have started growing more and more such crops for earning more money. Some of them have even involved themselves in small scale industries in the rural service centres. The banking facilities in rural service centres have enabled poor farmers to remain free from the clutches of money lenders and exploiters. Percapita income gets, improved and living conditions also get favourably influenced.

Socially and culturally the role of service centres has been much appreciable in less developed rural areas where people of different cultural traits assemble together and exchange their views with each other. In this process, people learn each others social and cultural specialities and social development takes place. Some of these service centres have cultural facilities like means of recreation or religious organisations by which socio-cultural transformation take place.

Similar to the problems of rural settlements, the rurban or rural service centres have also in-numerous problems. These problems concern lack of proper transport linkage, facilities of the developmental service amenities, power supply or public utility services, banking facility, good educational, health and other services. Most of these service centres are connected with the metalled and unmatlled road in bad conditions. Traders coming from outside hesitate to come to these centres. Virtually there should be connection of service centres with the towns of higher order with good transport facilities and service centres shoud be properly linked with the surrounding villages upto the level that the services shoud be made available to the neighbouring villager in proper way. Not all service centres have all the functions and services required for the surrounding area. There should be marketing facilities of food grains, clothing, other essential items, medicine, agricultural implements, fertilizer, pesticides, seeds, etc. which are rarely found in all service centres. At the same time public utility services like health centres, educational centres, telephone and postal services, electricity, water supply, banking, etc. should be available. Attempts should be also made to develop educational institutions of well repute so that movement of students from rural areas to towns should be minimised. Similarly there should be occasional visit of specialised doctors and social reformers for making people socially aware.

15

Urban Settlements

Although most people of the state reside in rural areas, a good number of urban settlements have developed in the state. Majority of these urban settlements have come into being during post-independence years based on the exploitation of minerals, development of mineral and other resource based industries and development of transport and communication and service amenities. The rapid growth of urban centres during post-independence years becomes obvious by the analysis of histogenic accounts of urban settlement. Virtually, the entire region remained aloof from the main stream of civilized life during ancient and medieval periods except little patch. As a result, no exact or connected account is available about the few urban centres that developed earlier. Most of the urban centres owe their origin to later developments particularly after the British conquest. Prior to the British rule, Hindu Rajas and Mughal Kings considered this region to be real Jharkhand (Land of Forest) and they did not establish their headquarter here. Even in the beginning of their rule here the British did not like complete occupation and preferred to keep the region under their influence through Jagirdars. They first of all occupied only the easily accessible areas where they located their administrative headquarter which later developed into urban settlements. The rich mineral resources such as gold, copper, zinc, mica, aluminium, etc. also attracted their attention.

The only town which has ancient origin is Deoghar, about the origin of which we have little authentic information. The history of this town dates back, according to the Puranas, to the Treta yug. According to Shivpuran it was in the Treta yug that the Demon Ravan, King of Lanka (Sri Lanka) tried his best to remove Kailashpati (Shiva) to Lanka but he had to leave God Shiva at Deoghar on his way back. Hence it may be presumed that during the ancient period people might had occupied the lands adjoining the temple and a town or settlement might had existed.

During the early Hindu period Aryan settlement existed along the northern fringe of the Plateau but they were only rural in character. Similar is the case with the Budhist period when only certain parts were in touch with the ideas of Budha and Mahavir and some preaching centres or Viharas and Mathas were founded. Parasnath, Benusagar, Chiyanki near Daltonganj and Garhwa where several idols of Lord Budha and Mahavir are found, might have been such centres. But at that time also, this region must have been such centres. But at that time also, this region must have been economically poor, as the scope for agriculture was limited, the land was mostly forested and the huge reserves of minerals were unknown and unexploited.

During the early phase of Mughal period also, this region remained aloof from urban influence, as the Mughals occupied only those accessible areas which were relatively rich in agriculture and minerals. The north-eastern part of the Rajmahal highland and its adjoining plains where the first areas to be settled by the Mughals and this region was very strategically located. It became a suitable place for the seat of the Government and a camping ground for the marching or retreating arteries of Mughals and Pathans and the Bengal kings (Sultan). It was the battle field of Akbar where many battles were fought between the Mughal forces under Hussain Qulikhan and Dand Khan Afghan (1536) the last independent ruler of Bengal and Bihar. An urban centre had developed named Akbar-nagar.[1] Its original site is known as Agmahal. It was the capital of early Medieval Bengal; Agmahal on the right back of the river Ganga was used as camping ground. Raja Mansingh, the Rajput General of Emperor Akbar selected a vast area lying to the west of Agmahal for the construction of Palace and Jamamasjid, the symbolic of a muslim city. Except for this urban centre, there was no evidence of any other urban centres having developed during the Muslim period.

As soon as the British became acquainted with this region, tremendous changes started taking place which brought in their asking of a number of urban centres. First of all they occupied that most accessible and the mineral rich areas and made their administrative headquarters at Ramgarh, Ranchi, Daltonganj, Hazaribag, Lohardaga, etc. the Purana Ranchi is situated at the foot of the Ranchi Hill. This urban centre came into existence due to the creation of the south-west frontier agency in 1834 by captain

Wilkinson, the first agent. He selected Kishanpur as his headquarter and to avoid confusion with other places of the same name, the station after a few years was designated Ranchi after the hamlet Purana Ranchi. Another important urban centre namely Ramgarh came into existence in the year 1780 as district headquarters of Ramgarh which continued upto 1834. It served as an important commercial centre of lac timber and grains. It was connected with rural and urban centres like Gaya, Chandwa, Champaran, Hazaribag, etc. The most important centre, Daltonganj was found in 1860 by Colonel Dalton, commissioner of Chotanagpur after his name. This urban centre was the nodal point of several important roads and flourished due to trade in forest products such as wood, bamboo and biri leaves.

Remarkable growth in urban centres is seen during post-independence years as the nation experienced considerable growth in agriculture industries, mineral exploitation, etc. This region also attracted the attention of Govt. due to its huge mineral resources, forest products, water power, potentials, industrial capacity and agricultural possibility with the commissioning of the Damodar Valley Projects, the minearlised area of Damodar Basin assumed a new importance. Several industries, coal washeries and other mineral processing industries developed and urban centres like Sindri, Bokaro, Katras, Kargali, Jasidih, Bermo, etc. came into being. Similarly, iron ore and copper mines gave rise to Jamshedpur, Ghatshila, Mosabani, Gua, etc.

In the year 1872 the state of Bihar had 48 towns out of which 8 towns were located in the state of Jharkhand. These towns were Ranchi (12086), Hazaribag (11050), Chaibasa (4641), Daltonganj (1113) were district headquarters and the remaining four towns Chatra (8818), Deoghar (4861), Sahebganj (3251) and Garhwa (3133) were subdivisional headquarters. Most of these towns were engaged in administrative functions. Besides administrative functions, they had small Mandis and Bazars, where trading communities mostly from the plains collected the local produce and supplied the surrounding villages with their requirements of cloth, salt and tobacco. No town was engaged in modern functions of mining and manufacturing. Trade and commerce in the modern sense did not exist in the absence of transportation facilities. Ranchi, at that time was the biggest town of the region and

Hazaribag was next in size of population. Daltonganj, despite being a district headquarter, was the smallest town of the region.

In the year 1881 the region had an urban population of 83116, the excess of population over the last decade living in its mining towns. Between 1872 and 1881 the urban population of the region increased by 69.8%. A part (7.1%) of increase was due to the entry of Lohardaga (3461) in the list of towns. The rest of it was due to the growth of the population of the existing towns through natural increase and immigration. Such a high growth, however, appears to be remarkable because nothing very unusual happened during the period which may explain this phenomena. Therefore, at least a part of this growth may be attributed to the improved method of counting adopted by the census of 1881. During the decade 1872-1881 the growth of the individual town varied between 29.4% (Chaibasa) and 568.3 (Daltonganj). Ranchi and Hazaribag continued to be the biggest and second biggest towns of the region but Daltonganj on account of its tremendous growth improved its position from the 8th to the 5th place in hierarchy. Lohardaga became the smallest town in the region.

Table - 15.1
Urbanisation in Jharkhand (1872-2011)

Census Years	Total Population	Urban Population	% Urban Population	%Decennial Growth	Total No. of UA/Towns
1872	42,99,388	48,953	1.13	-	-
1881	58,96,032	83,116	1.40	69.79	-
1891	65,08,033	86,878	1.33	04.53	-
1901	60,68,233	1,17,975	1.94	35.79	13
1911	67,47,122	1,58,827	2.35	34.63	16
1921	67,67,770	2,44,010	3.61	53.63	17
1931	79,08,737	3,22,475	4.08	32.16	18
1941	88,68,069	5,08,252	5.73	57.61	26
1951	96,97,254	7,60,350	7.84	49.60	35
1961	1,16,06,489	13,33,342	11.49	75.36	7/65
1971	1,42,27,133	22,77,632	16.01	70.82	12/95
1981	1,76,12,069	35,74,045	20.29	56.92	10/101
1991	2,18,43,911	46,41,227	21.25	29.86	14/133
2001	2,69,09,428	59,86,697	22.25	28.99	11/152
2011	3,29,88,134	79,33,061	24.05	32.36	228

In the census 1891 the region had an urban population of 86,878 showing a meagre rise of only 4.5% over the urban population of 1881. This meagre increase of urban population is accounted for partly (-7.3%) by the fact that Garhwa was declassified in 1891 and partly by the decline in population of two more towns namely Chatra (-9.4%) and Daltonganj (-30.2%) the remaining towns reveal slow growth with an exception of Sahibganj which registered a growth of 73.5% during this decade. This general set back in the process of urbanisation was consequent upon the general deterioration of the health conditions in the region following oubbreak of several epidemic diseases. As usual Ranchi and Hazaribag continued to be the two topmost towns. Sahibganj and Lohardaga improved their position in the hierarchy of urban centres considerably but Daltonganj which had improved its position during the previous decade again declined to its original position of the smallest town of the region. This may be attributed to famine conditions in the district.

The census of 1901 recorded an urban population of 1,17,975. The increase of urban population in the region as a whole comes to 35.79%, all of which, however, was not because of the growth of towns which existed during the census of 1891. In 1901 three new towns, Giridih (9,433), Madhupur (6,840) and Bundu (5,469) were added to the list of towns and Garhwa (3,610) which had been declassified in 1891 regained the category of an urban centre. The urban population of the region gained 35.79% on this account. The relatively high growth of urban population during the decade 1891-1901 was due to increase in urban population of Ranchi and Chaibasa. Ranchi and Chaibasa reveal growth rate exceeding 25%. The relative position of Ranchi and Hazaribag again remained unchanged but Sahibganj and Lohardaga declined in importance.

In the census of 1911 the region had 16 urban centres of which three Jamshedpur (5,672), Dumka (5,629) and Rajmahal (5,327) were classified as towns for the first time in the census. The growth of urban population during the decade 1901-11 amounted to 34.63% which raised urban population in the region to 1,58,827. About 14% of the total increase of urban population was an account of the addition of the population of new towns to the total urban population. The rest of the increase (19.4%) was on account of the growth of population in the towns already existing during previous census. During the decade all the towns of the region recorded an increase of population between 2.3% (Madhupur), 95.6% (Sahibganj) only, one town Chatra (-13.0%) declined. Ranchi and Hazaribag maintained their ranks but Chatra was thrown down from third to 6th portion in the hierarchy and Sahibganj jumped up from 7th to 3rd position.

In the census of 1921 there were 17 towns in the region with a population of 2,41,010. Two of these which had been recognised as urban centres in this census namely Dhanbad (11,973) and Chakardharpur (7,944) accounted for 12.9% of the growth of urban population. The rest of the growth 45.7% was due to the growth of population in towns which existed in 1911. Leaving aside the addition of urban population on account of the entry of new towns the actual growth of urban population during the decade 1911-21, and it was higher even than the decade 1951-61. It shows that something remarkable event took place in the region during this decade. The most conspicuous event was perhaps the establishment of the Tata Iron and Steel works in Jamshedpur on account of which the town registered a growth 9111.3% and relegated Ranchi, the biggest town till now to the second position and usurped for itself the position of the largest town of the region. Two other towns Garhwa (+129.3%) and Giridih (+76.9%) registered high growth rate. The cause of the rapid growth of population of Garhwa is not apparent, but that of Giridih may be attributed to the stepping up of the mining of coal in its vicinity. Besides, Ranchi (+2.01%), Daltonganj (26.7%) and Dumka (31.4%) showed higher rates of growth mainly on account of natural increase and rural-urban migration of the rest 5 towns increased by less than 10% and four declined by 10 to 30%. Hazaribag on account of its slow growth stepped down from the second to fourth position and the mining town of Giridih went upto third position.

The decade 1921-31 was one of the rapid growth of population in all the towns. The growth rates of the individual town were uniformly high. The growth of Chakradharpur (+41.87%) may be attributed to the expansion of the town as an administrative centre of the railways. Other towns which grow relatively rapidly were Dhanbad (+36.61%), Jamshedpur (+46.0%) on account of mining and manufacturing activity respectively and Sahibganj for administrative functions. It is important to note that no town declined in population. The old town together accounted for 28.5% of the total growth of urban population of 32.1% leaving only 3.6% which is accounted for by the entry of Jugslai (8,721) to the category of town Jamshedpur, Ranchi and Giridih emerged as the three biggest towns of the region. Jharia (18,037), Bermo (5,674), Kargali (10,127), Bokaro (7,509), Noamundi (6,389), Musabani (8,270) and Saraikela as mining centres were classified as towns during the census 1941. They together accounted for 19.3% of the increase of the urban population in the region. This was an account of stepping up of coal mining in the lower Damodar Basin and iron-ore mining in Singhbhum. The all ready existing towns grew by 36.9% giving a total increase of urban population of the region crossed the 5 lakhs mark in 1941. Garhwa (-27.3%) and Lohardaga (-2.34%) were the only urban centres which declined in population during the decade 1931-41 on account of their decline in trade and administrative function. Bundu had grown by about 10%. All the rest registered growth rates between 10-30%. During the decade the highest growth had been associated with Jugslai (+31.3%) and Jamshedpur (+76.6%) for obvious reasons. Jamshedpur, Ranchi and Giridih continued to remain three biggest towns of the region.

Remarkable growth in urban population took place in the census of 1951 due to development in mining activities. Seven more towns were added namely Gua, Moubhandar and Kharsawn in the districts of Singhbhum, Jhumri-Tilaya and Ramgarh in the district of Hazaribag and Sindri in Dhanbad. All these towns sprung up adjacent to iron ore, mica and coal mines. An increase of 70% in urban population is seen in Ranchi, Kargali and Dhanbad while the population of Hazaribag, Lohardaga, Jamshedpur and Bermo increased by 36%, 43%, 47% and 57% respectively. All these urban centres except Hazaribag witnessed rapid rise in population due to the growth of industrial enterprises. In case of Ranchi, increase in trade and service activities also took place. The establishment of the eastern command at Ranchi and its growth as a regional centre of the region due to improved transport connections with the headquarter of the adjoining district also helped much in swelling the population of the town. In 1951, this urban centre rose to the rank of city with a population of 1,06,849.

In the decade 1951-61, the number of towns as well as urban population further increased. In 1961 the region recorded an urban population 13,33,342 showing an absolute increase of 5,72,992 persons or 75.36% over 1951. This increase is more than the total urban population which the region had in the year 1941. Of this slightly less than 1/3 (31.5%) increase was on account of the emergence of 09 new towns. These towns together claim over 21% of the total population of the region of these Jasidih Bazar (31,605), Kathras (17,244), Kumardhubi (16,542) and Jorapokhar (15,595) were the first four largest new towns. The remaining towns except Panchal (4,700), Muri (1,651) and Netrahat (2,085) had population between 5000 or 15,000 each. The growth of Maithon and Panchet was due to the construction of dam and hydro-electric power plants of Jhinkpani and Khelari for cement industries of Kumardhubi due to their importance as railway junction of Latehar for limestone quarrying of Gumla, Simdega, Khunti, Godda and Jamtara for their being sub-divisional headquarter of Muri for the aluminium industry of Ghatshila for copper mining, smelting and trade and almost all the rest except Netrahat for their being coal mining centres. Netrahat did not fit into the definition of town adopted by the census but on account of its urban look and its being a place of tourist interest it was given the special status of a town.

The older towns recorded high growth rate during the decade 1951-61 amounting to 40.3% of the 22.3, nearly half may be attributed to natural growth and the remaining half to rural-urban migration. In this decade the highest growth of 216.7% was recorded by Sindri on account of the expansion of the fertilizer industry and the establishment of a thermal power station and a cement factory. The opening of the Bihar Institute of Technology was also important in adding to the total population of the town. The second highest growth of 139.6% was recorded by Jhumri-Tilaiya, where the growth was attributed to the expansion of Mica Mining and splitting industry and trade in mica and the development of other industrial and commercial activities incidental upon the location of the hydel power station below the Tilaiya dam. The stepping up of mining and manufacturing and other developmental activities resulted in large growth of population in Dhanbad (+68.3%), Chakradharpur (54.9%), Bermo (54.4%), Noamundi (82.1%), Musabani (45.6%), Ramgarh (35.6%), Pakur (94.8%), Moubhandar (38.8%) and Mihigam (36.8%). All other towns recorded an increase between 20% and 35%. Only two towns Kargali and Bokaro reveal decline in their population. In the face of rapid growth or urbanisation, Jamshedpur, Ranchi and Dhanbad maintained their position as the first three major urban centres of the region.

During the decade 1961-71 the urban population in the region increased by 70.82% over the urban population of 1961. In this decade the highest increase in urban population was recorded in the district of Hazaribag (118.38%) being followed by the districts of Dhanbad (120.7%), Ranchi (84.39%), Pashchimi Singhbhum (76.97%), Giridih (70.62%), Gumla (41.05%), Purbi Singhbhum (36.94%), Dumka (38.85%), Sahebganj (32.29%), Godda (29.77%), Lohardaga (29.41%) and Palamu (25.62%). Districts scoring very high percentage of urban population recorded considerable mining and industrial development. In this decade Dhanbad urban agglomeration recorded an increase of 98.80% population. Jamshedpur urban agglomeration registered an increase 34.26% population. Ranchi urban agglomeration showed an increase of 90.05%.

In the decade 1971-81 the region recorded an increase in urban population by 56.92%. In this decade the highest increase in urban population is recorded in the district of Ranchi (86.71%). Other districts recording increase in population ranging between 60%-80% included Dumka and Dhanbad. Districts recording increase in urban population between 40%-60% included Pashchimi Singhbhum, Palamu, Hazaribag and Godda and the remaining districts recorded the increase in urban population less than 40%. In this decade Jamshedpur urban agglomeration recorded an increase of 54.41% while Dhanbad urban agglomeration exhibit an increase of 49.61%. Ranchi urban agglomeration showed the increase of 88.63% because of swelling of Ranchi township based industrial and urban growth. Phusro urban agglomeration (consisting of Phusro Bokaro, Jaridih Bazar and Bermo) recorded an increase of 47.06% urban population.

Table - 15.2
Level of Urbanisation : Jharkhand (2001 & 2011)

Sl. No.	District	% Urban Population 2001	Urban Population 2011	% Urban Population 2011
1.	Bokaro	45.26	9,83,644	47.70
2.	Chatra	05.30	62,954	06.04
3.	Deoghar	13.71	2,58,361	17.32
4.	Dhanbad	52.36	15,60,394	58.13
5.	Dumka	05.34	90,178	06.82
6.	Garhwa	04.11	69,670	05.27
7.	Giridih	06.42	2,08,024	08.51
8.	Godda	03.53	64,419	04.90
9.	Gumla	04.77	65,081	06.35
10.	Hazaribagh	23.23	2,75,307	15.87
11.	Jamtara	08.54	75,746	09.58
12.	Koderma	17.37	1,41,246	19.72
13.	Khunti	-	44,982	08.46
14.	Latehar	04.68	51,858	07.13
15.	Lohardga	12.67	57,411	12.43
16.	Pakur	05.13	67,512	07.50
17.	Palamu	06.42	2,26,003	11.65
18.	W. Singhbhum	15.48	2,18,034	14.51
19.	E. Singhbum	55.02	12,74,591	55.56
20.	Ramgarh	-	4,18,955	44.13
21.	Ranchi	35.10	12,57,335	43.14
22.	Sahebganj	10.57	1,59,666	13.88
23.	Seraikela-Kharsawan	18.82	2,58,746	24.29
24.	Simdega	06.60	42,944	07.16
	Jharkhand	**22.25**	**79,33,061**	**24.05**

Source : Census of India, 2001 & 2011.

In the decade of 1981-91 this state recorded an increase of urban population 29.86%. The highest increase was recorded in Godda District (58.45%) because of development of township after the formation of the district. Districts recording growth in urban population more than 30% included Hazaribag, Giridih,

Lohardaga, Deoghar, Dumka, Pashchimi Singhbhum and the remaining districts recorded an increase in urban population less than 30%. In this decade the town of Godda recorded increase of urban population of 58.45%. Jamshedpur urban agglomeration recorded an increase of only 21.93% may be lack of further addition in industries. Even Dhanbad urban agglomeration recorded only an increase of only 26.26%. Ranchi urban agglomeration recorded an increase of only 21.28%. Only Bokaro Steel City urban agglomeration recorded an increase of 50.82%.

In the decade of 1991-2001, this state recorded an increase of urban population by 28.99% and the share of urban population to the toal population is 22.25%. This state is more urbanised than many of the major states because of the development of industrial & mining activities. However, this level of urbanisation is still below to the national level. Most of the districts in Jharkhand have a very low level of urbanisation. Only four of the ditricts in the state are highly urbanised which include E. Singhbhum (55.02%), Dhanbad (52.36%), Bokaro (45.26%) and Ranchi (35.10%). The least urbanised districts are Gooda (3.53%), Garhwa (4.11%), Latehar (4.68%), Gumla (4.77%), Pakur (5.13%), Chatra (5.30%) etc. This is mainly due to lack of industrial, mining & infrastructural development, as a result majority of people in these districts still reside in rural areas or migrate to nearby large urban centres.

During the decade 2001-2011, the urban population in the state increased by 32.36%. In the year, 2011, the total urban population is 79,33,061, which accounts for 24.05% of the total population. However, this level of urbanisation is still below to the national average. There is a variation in the level of urbanisation in the state ranging from 58.13% in Dhanbad which is highest to lowest of 4.90% in Godda. The other districts with high level of urbanisation are E. Singhbhum (55.56%), Bokaro (47.70%), Ramgarh (44.13%) and Ranchi (43.14%). The high levels of urbanisation are the result of high concentration of industrial complexes, mining activities in these districts, availability of basic services like health, education, banking, recreational facilities. The development of transport networks in the last decade also provided the impetus in increasing the urban population. The western and southern Jharkhand which includes Bokaro, Dhanbad, Ranchi, E. Singhbhum, Saraikela-Kharsawan and

Ramgarh are more urbanised than the other parts of the state. There are huge deposits of iron-ore & coal in these areas, as a result establishment of basic & ancillary industries has taken place, in which a large no. of workforce is required. The northern districts of Jharkhand which includes Chatra, Dumka, Godda, Pakur, Garhwa, Giridih are lowly urbanised due to the lack of development of industries and less availability of any minerals. The eastern districts like Palamu, Lohardaga, Gumla, Latehar, Simdega, Khunti are moderately urbanised districts.

Urban Trends in Various Classes :

The table 15.3(a) reveals classwise number of towns in the region. In the year 1901 there were 13 urban centres which has increased to 133 during last ninety years. These urban centres have experienced changes in their population hierarchy. In the year 1941 there were twenty six towns consisting of one class I town, two class II, six class III town, seven class IV town, eight class V town and two class VI town but in the year 1991 the number of urban centres have shoot up 133 towns consisting of five class I, seventeen class II, thirty one class III, thirty seven class IV, thirty four class V and nine class VI. In the year 2001 & 2011, the no. of towns rose to 152 & 228 respectively.

Table - 15.3 (a)
Classwise No. of Towns

Years	I	II	III	IV	V	VI	Total
1901	—	—	1	2	8	2	13
1911	—	—	1	4	9	2	16
1921	—	1	1	5	9	11	17
1931	—	2	2	7	5	2	18
1941	1	2	6	7	8	2	26
1951	2	2	6	8	9	8	35
1961	3	4	12	13	20	8	65
1971	4	6	20	27	29	10	96
1981	5	8	22	28	29	9	101
1991	5	17	31	37	34	9	133

2001	7	18	37	35	45	10	152
2011	10	12	39	48	90	29	228

Table - 15.3 (b)
Classwise Urban Population : Jharkhand (2001 & 2011)

Class I Towns	No. of Towns 2001	2011	Total Urban Population 2011	% of Total Urban Population 2001	2011
Class I	7	10	43,28,014	41.1	54.6
Class II	18	12	8,82,716	23.1	11.1
Class III	37	39	12,82,052	20.5	16.2
Class IV	35	48	6,74,280	09.0	08.5
Class V	45	90	6,34,552	05.6	08.0
Class VI	10	29	1,31,447	00.7	01.7
All Classes	152	228	79,33,061	100.00	100.00

The above table 15.3 (b) reveals share of urban population in different classes of towns. The class I town comprise more than 54% urban population of the state. Class II and Class III have 11.1% and 16.2% urban population respectively. The remaining IV, V and VI classes of towns have 8.5%, 8.0% and 1.7% urban population respectively. It further reveals the fact that major towns are not allowing to grow smaller towns which is not healthy for proper development of the region. Special attention is being given for the development of small towns to release the pressure of major urban centres but it is not happening. Major towns are growing at the cost of small towns.

Table - 15.3 (c)
Urban Classification based on Population Size (2011)

Class of Towns	Population Size	Traditional Designation	No. of Towns (2011)
I	1,00,000 & above	City	10
II	50,000 - 99,999	Sub City	12
III	20,000 - 49,999	Large Town	39
IV	10,000 - 19,999	Small Town	48
V	5,000 - 9,999	Small Town	90
VI	Less than 5,000	Small Town	29

Spatial Pattern of Urbanisation and Urban Growth :

The spatial pattern of urbanisation in Jharkhand depict high intra-state disparity where in the district wise levels of urbanisation range from as hish as 58.13% in Dhanbad to 4.90% in Godda as per the 2011 census. The six highly urbanised districts of Dhanbad (58.13%), E. Singhbhum (55.6%), Bokaro (47.70%), Ramgarh (44.13%), Ranchi (43.14%) and Saraikela-Kharsawan (24.29%) form a distinct and continuous belt where most of Jharkhand's urban population has been concentrated. All these districts have considerable urban population because of mineral & industrial based growth of town. Besides these they have been centres of administrative origin in which all the basic urban amenities like health, education, recreation, electricity, banking etc. has been found to be very good in comparison to other areas. In contrary to these highly urbanised districts, there are districts like Godda (4.90%), Dumka (6.82%), Pakur to the east and Garhwa (5.27%), Chatra (6.04%), Gumla (6.35%), Latehar (7.13%) and Simdega (7.16%) to the west which have very low levels of urbanisation in Jharkhand. These districts are still dominated by rural population because in these districts, there are only a few mines and other resources which encourage the growth of urban population. Other districts having more than 10% urban population and below the state average include Palamu (11.65%), Lohardaga (12.43%), Sahebganj (13.88%), W. Singhbhum (14.51%), Hazaribagh (15.87%), Deoghar (17.32%) and Koderma (19.72%). These districts are growing at a natural pace mainly being headquarter of districts having some industrial, educational and cultural significance.

In 2011 census, there are 228 centres in Jharkhand which have been designated as urban including census towns and cities as compared to 152 urban centres in 2001 census (Table 15.3 (b)). The distribution of the urban population in the state by town size shows a "top heavy" or "Mega Cephalic" urban structure wherin most of the urban population is concentrated in the big cities. More than half of the urban population of the state (54.6%) reside in its 10 major Class I cities. The share of urban population in there cities has increased by 32.6% during 2001-2011 census. On the contrary, the share of urban population in Class II, III & IV have declined remarakably during 2001-2011. The share of medium towns in class V & VI has increased noticebly during 2001-2011.

These small towns are more or less overgrown villages having very little characteristics of a urban centre.

Distributional Pattern of Urban Centres :

A glance on the proportion of the urban population and number of towns in various districts of the distributional pattern of towns make it abundantly clear that the process of urbanisation has not been uniformly active all over the region. Urban centres appear to be concentrated in the Damodar and the Subarnarekha basins. About 50% of the towns claiming nearly one-third of the total urban population are concentrated in the Damodar Valley alone where mining and manufacturing activities are at their maximum. The density of urban centres is closely associated with the development of coal mining, coal washeries and numerous after industries. Next importance is the Subarnarekha valley where the towns are not so closely spaced, but are quite important both in respect of their size and function. Else where the towns are wide apart and few and far between.

In the 2011 census, Dhanbad and Ranchi emerged as the major cities with a whopping share of about 28.18% of Jharkhand's population. These 2 cities emerged as two million cities in 2011 census, thus becoming Jharkhand's maiden entry into the country's list of million plus cities. Now, there are 10 class I cities in the state which accounts for 54.6% urban population of the state. Deoghar, Chas and Giridih emerged as the new class I cities during 2011 census (Table 15.4). Dhanbad (483.40%) registered a massive decadal growth followed by Deoghar (106.5%).

Table - 15.4

Urban Population in Class I cities : Jharkhand (2001 & 2011)

Sl. No.	City Name	City Population		% Urban Population		%Decadal growth rate of urban-isation
		2001	2011	2001	2011	
1.	Dhanbad (MC)	1,99,258	11,62,472	3.32	14.65	483.40
2.	Ranchi (MC)	8,47,093	10,73,427	14.13	13.53	26.72
3.	Jamshedpur (NAC+OG)	6,12,534	6,77,350	10.2	8.54	10.58
4.	Bokaro Steel City (CT)	3,93,805	4,14,820	6.57	5.23	5.34

5.	Mango (NAC)	1,66,125	2,23,805	2.77	2.82	34.72
6.	Doghar (MC)	98,388	2,03,123	1.64	2.56	106.45
7.	Adityapur (NP)	1,19,233	1,74,355	1.99	2.20	46.23
8.	Hazaribagh (NP)	1,27,269	1,42,489	2.12	1.80	11.96
9.	Chas (NP)	97,221	1,41,640	1.62	1.79	45.69
10.	Giridih (NP)	98,989	1,14,533	1.65	1.44	15.70
	Class I Cities	24,65,317	43,28,014	41.13	54.56	75.56
	Jharkhand (Urban)	**59,93,741**	**79,33,061**	**100.00**	**100.00**	**32.36**

Source : Computed from Town Directory, Jharkhand, Census of India, 2011.

The spatial distribution of class I cities shows the heavy concentration in mainly two areas. Dhanbad, Bokaro Steel city and Chas together form a major urban core along with Jamshedpur, Mango and Adityapur constituting the other areas. Such restricted spatial spread of urban population into compact urban areas has given rise to many problems in matters related with housing, traffic, pollution and lack of basic amenities. This has led to the growth of urban slums, squatters in many parts of the urban landscape.

In order to analyse the concentration of urban centres it is possible to divide the area into following four zones.

1. Zone of maximum concentration.
2. Zone of notable concentration.
3. Zone of relatively sparse and widely spaced towns and
4. Negative areas.

1. **Zone of Maximum Concentration** : This zone covers practically the whole of Dhanbad and part of Hazaribag, Giridih in the Damodar basin. It has a several bunches of towns. The district of Dhanbad is the most urbanised district of the region. This district has a little more than half of its population as urban. Almost all towns of this zone are oriented to mining and coal-based industries. A number of small towns have merged near the Collieries of Jharia, Bokaro and are closely spaced. This zone has the facility of dense net of railways and roads. Some urban centres are so closely spaced that they seems to merge into one another

and as a result a few town groups have merged, such as Dhanbad, Jharia, Sindi town group, Kumardhubi town group, Katras town group etc. In this zone per square km density of urban population is abnormally high.

2. **Zone of Notable Concentration** : This zone covers an area of about twenty hundred square kms. in Pashchimi and Purbi Singhbhum. The iron ore mines, copper mines and Tata industrial area have given rise to several towns forming concentration in the Subranarekha basin. Towns are based on mineral resource, forest resource and industrial development. Important towns are Jamshedpur, Chaibasa, Manoharpur, Chakradharpur, Noamundi, Mosabani, Gua, Ghatshila, etc.

3. **Zone of Relatively Sparse and Widely Spaced Towns** : This zone is characterized by widely spaced small urban centres and a few isolated towns of major importance. This part has poor mineral resources but is rich in forest products. A few isolated administrative centres have increased their size such as Ranchi, Daltonganj, Hazaribag, etc. This zone includes the area of the Hazaribag Plateau, the Ranchi Plateau, the Koel Basin, etc. The North Koel Basin has an area of about 2000 square kms. and has only a few urban centres. Similarly the districts of Rajmahal highland have only a few urban centres.

4. **Negative Areas** : This zone is characterised by relatively negligible number of urban settlements. It has a poor accessibility and little resources of minerals or agricultural products. As a result only a few settlements have sprung up in the category of small towns. The Pat Region has only one town of Netrahat because it has very rough terrain and unsuitable for urban growth. Only tourist attraction has encouraged the growth of towns. This region also has little mineral resource and little productive land.

Types of Origin of Urban Centres :

This region had been for long, known as Jharkhand (land of forest) and was aloof of modern developments which give rise to urban centres of different types. As such there are only a few older settlements and most of the towns have got recent origin. At the same time causes of origin of most towns appears to be more identical in nature because of mining and industrial development. Certainly there are urban centres which have also originated due to administrative functions, religious activities, commercial functions, etc. Administrative and political causes have also been important.

Administrative Origin : The township of Rajmahal was established as the site was selected for the palace of Akbar by his Jagirdar, Rajamsingh. This site was quite suitable for attacking the Sultans of Bengal. It had a strategic location on the bank of the river Ganga and at a point the hills almost reached the river Ganga. In addition to this town several urban centres came into being for administrative purposes during British Regime. Ranchi town was based on the selection of the site for headquarters of South-west frontier agency by Captain Wilkinson in 1834. The township of Chaibasa came into existence after being selected in 1872 as the district headquarter of Singhbhum. The selection of the site of Hazaribag town is related to the stationing of Ramgarh Batalion in 1780 at the villages Okmi and Hazari. The township of Dumka came into being in 1855 when it was made headquarter of Santhal Parganas. The township of Daltonganj was founded on the Bank of river north-koel by Colonel Dalton, Commissioner of Chotanagpur, 1862.

Religious Origin : This region has also given rise to towns based on religious functions. Deoghar has religious origin. The word Deoghar signifies the house of God and this urban centre came into being due to the establishment of a Shiva temple. The legendary origin of township of Deoghar dates back to Treta yug but it was recognised as an urban centre in 1901 with 8838 persons. No doubt, this is a district headquarter but importance of town lies in religious functions as this centre is an assemblance of several temples, dharamshals and priests (Pandas).

Village Origin : Some urban centres of this state are overgrown villages. They have developed into urban centres due to the growth of educational institutions, small industries, commercial functions, mining activities, etc. This type of origin is very well illustrated by several villages which have been included in the list of towns during last two decades such as Kharsawan, Barkhi Saraiya, etc. Kharsawan, a few years earlier was a large village containing the Palace of Kharsawan Raj but the connection of metalled road and development of other service facilities it got the status of town in the census of 1961. Similarly the village Barkhi

Saraiya overgrew into town in 1961 due to addition of educational institutions; several small scale industries, saw mills and shops.

Hill Resort Origin : Only example of health resort in this region is Netarhat. It is situated on the Pats at an elevation of about 1300 m. Besides there are two other centres of lesser importance Madhupur and Ranchi. The location of Madhupur is linked up with its salubrious climate but the origin of Ranchi town has been influenced by several other factors. But at both the places people flock from outside for change during the summer season. At Madhupur there is a large number of dwellings of people from West Bengal who came here for change of climate during summer. Some of them also lead their retired life here. Here houses are of Bunglow type with much open space and beautiful gardens.

Industrial Origin : Establishment of manufacturing industries has been an important factor in giving rise to urban centres after the independence. Only one town Jamshedpur had originated before independence which too developed rapidly after the independence. This town came into being due to the Tata Iron and Steel Plant at Sakchi in 1908.

The township of Sindri came into being after the establishment of Sindri fertilizer factory in 1951. The establishment of heavy engineering, corporation of Hatia caused rapid growth of Doranda town. Similarly the establishment of a large Steel Plant at Bokaro has led to the origin of Bokaro Steel City. Besides these urban centres, there are several manufacturing based townships such as Kodarma based on Mica-processing, Ghatshila based on Copper smelting, Muri based on Alumininum industry, etc.

Mining Origin : This region has also given rise to a good number of urban centres based on the exploitation of mineral resources. Most of these towns are concentrated in the coal belt of Damodar Basin and Iron ore and Copper mining areas of Subaranrekha Basin. Virtually, coal mines of Dhanbad, Jharia, Giridih, Bokaro, etc. have given rise to bunches of towns in the coal mining areas.

Commercial Origin : Due to poor accessibility, there were before independence only a few urban centres which got their origin based on trade and commerce. These urban centres include Sahibganj, Chatra, Garhwa, Godda, Khunti, etc. Among these Sahibgaj provides the best example as it came into existence due to good accessibility and facilities for trade between the southern hilly areas and the Gangetic Plain in the north. Manihari Ghat on the river Ganga connects south-eastern parts of North Bihar Plain and north-eastern portion of the state. The ferry service has helped much in the development of commercial pursuits. It was an important rural trade centre even during the early British rule as it was most accessible by the river Ganga as well by metal roads. Chatra developed due to trade of food grains, forest products such as bamboo, timber, kendu leaves, lac, etc. Similarly Garhwa located in the North Koel Basin was an important trade centre even during British rule.

Site and Locational Conditions : Usually the term site means the place where the town stands and the term location concerns with the surroundings in which the town stands and grow. These two parameters of urban centres determine the origin, evolution, functional distributes, morphological features, service area, etc. In the consideration of these parameters the undulating terrain of the state coupled with forest patches, highly inaccessible areas, sparse population, etc. debar compact growth of urban centres which are normally observed in homogeneous plains.

Normally towns are categorised on the basis of locational attributes into coastal and inland towns. Of the inland towns the largest number in any country is found to be located on river banks for obvious reasons like the facilities of higher levee land which provides flood free space for habitation, adequate supply of water, proper drainage, facility of transport, etc. Prior to the development of railways and roadways, the rivers in the plains were the chief artery of transport of commodities for territories up and down streams. In the dissected terrain of the state which consists of rejuvenated peneplains, a river is of geographic significance different from that in the plains. Most of the rivers are unnavigable and have no commercial significance; they are seasonal and subject to freshets but have a steeper gradient and rapid drainage. For most of the year they have dry beds of sand or are fordable without forming any barrier or protection and so could not have the same strategic significance as the rivers of the plains. Certainly rivers of this region are of utmost important for industries.

The highlands of this region have given locational advantages of towns based on nodality of roads or railways or both. The city of Jamshedpur is a remarkable example of nodality where railways from the mineralised and industrialised lower section of Damodar valley, from Mumbai and Madhya Pradesh, from Orissa and Calcutta, from the iron mines of Singhbhum and Orissa and roads from all sides converge on the confluence of the rivers Subaranrekha and Kharkai.

In this region the consideration of suitability of site for compact growth of township or nodality of roads and railways, etc. have played important role in only limited number of towns or cities. Virtually there are numerous urban centres which have developed adjacent to mines in the remote areas and afterwards these centres have been connected with roads and railways and urban amenities. As such only a few urban centres have suitable site conditions and locational advantages for proper development of township and most of the towns lack compact area for compact growth of township rather most of these towns have amorphous growth where obstacles like deep valleys, steep slope, rocky projections, etc. create hindrance in compact growth of township. Similarly most of the towns are not well linked with roads and railways with the surroundings like in the plains.

Functional Attributes : Since most of the urban centres of this region have originated due to several reasons, towns portray diversities in functional dynamic attributes. It is true that at the time of origin of towns, most of them perform only one function on the basis of which they get their origin, but in due course the functional attributes change from mono functional character to multi functional or diversified functional character because of the assemblage of people of varied nature and the requirement of services for urbanised as well as people living in surrounding areas. As such functions performed by the urban centres are naturally influenced by the site, situation and environmental conditions. Almost all functions performed by urban centres are categorised into Basic and Non-Basic functions. The basic part refers to the goods and services produced in the urban area but sold beyond the limit of the urban area. As such basic functions become of more use for the people living in the surrounding areas. While the non-basic functions refer to the goods and services produced within the town and are also consumed with the town.

These non-basic functions remain of use only for the urbanities living in that urban centre. As such the growth and development of the town mainly depends upon non-basic functions.

Almost all towns perform several functions required by the people living in that urban centre as well as the surrounding and as per the requirements of the people of surrounding functions develop in the urban centres. Certainly some functions have limited consumable area and some function may have vast service area. But all towns more or less, perform almost all functions usually perform in urban centres like functions related to education, health, recreation, business, industry, transport and communication and other essential services. In the towns of this region administrative, industrial, educational, recreational and transport are more important. In most towns of the mineralised pockets mining functions are of prime importance. Since most of the towns are overgrown villages they are not completely aloof of agricultural activities. Virtually most of the towns of the region have also some percentage of workers engaged in agriculture. There are only a few towns which have originated due to industrial establishments and even today most of the people have relationship with industrial activities like Jamshedpur. Similarly there are urban centres based on mining establishments where most workers have dependency on mining activities like Noa-Mundi, Gua, Jharia, etc. Almost all towns small or large have commercial functions. As in the case of the industrial function so with trade and commerce there appears to be stereotyped pattern and normally it appears that the commercial population is double than the industrial population.

While evaluating the functions and functional specialisation of urban centres several urban geographers and scholars have attempted to suggest different methods for classifying urban functions and evaluating functional specialisation of towns. Here it seems unnecessary to discuss the methodology for the classification of towns based on urban functions as such only important functions performed by urban-centres have been discussed here based on personal observations and qualitative assessment. Important categories of towns based on their functions are the following :-

1. **Agricultural Towns :** It has been noted earlier that most of smaller towns are overgrown villages and have recently been enlisted in the category of towns during the census. As such most of the people living in those villages have not got opportunities to change their occupations and livelihood. Certainly agricultural significance is getting day by day replaced by Secondary and Tertiary functions. These towns do have specialisation in transactions of agricultural products. Some of these towns are Khunti, Simdega, Bundu, Barkisaraiya, Kharsawan, etc.

2. **Mining Towns :** This category of towns is more important in the region specially in the mineralised portion of the Damodar and Subranarekha Basin. The coal field's of Jharia, Bokaro, Giridih, Iron ore mines of Singhbhum and Mica mines of Giridih, etc. have given rise to several mining towns. These towns are Kargali, Bermo, Jaridih, Katras, Noamundi, Gua, Musabani, etc. The chief features associated with mining towns vary so rapidly as to make it impracticable to generalise about the class as a whole. Just as mining towns vary in appearance, depending on particular type of mineral commodity so do they vary greatly socially and economically. The location of these towns is guided mainly by site factor.

3. **Industrial Towns :** In this category of towns also the region reveals some importance. This region has urban centres having household industries and manufacturing industries. Smaller towns still have small scale industries based on local resources like Biri making industries or handloom cloths or agricultural implements etc. These towns having small scale industries are Godda, Pakur, Chakradharpur, Bundu, Hussainabad, etc. In case of manufacturing industries there are a few towns recognised as manufacturing towns. These towns are Jamshedpur, Bokaro, Sindri, Kumardhubi , etc. In addition to these manufacturing towns there are other towns where manufacturing industries have developed. These towns include Ranchi, Jharia, Dhanbad, Doranda, Giridih, Jhumri Tilaiya, etc.

4. **Commercial Towns :** It is true that almost all urban centres do have some workers engaged in the functions related to trade and commerce. Such towns are generally dispersed through out the region as their location seems to be guided by centrality in productive hinterland. Some of these towns are Sahibganj,

Daltonganj, Giridih, Garhwa, etc.

5. **Transport Towns :** This region having rough terrain possesses poor accessibility. Consequently the region contains a few towns specialising in transport functions. Amongst these towns of Gomo, Chakradharpur, Ranchi, Madhupur, Daltonganj are more significant. Most of these towns are important as railway and road centres from which routes diverge in all directions.

Besides above mentioned groups of towns some other functions are also developing in the region and in the future may have their decisive role. The town of Netarhat is growing as health resort, tourist interest and educational centre. Deoghar is noted for religious functions.

Hierarchical Order of Urban Centres :

Similar to urban functions the categorisation of urban centres on the basis of service importance becomes necessary for planning purposes. Ranking of towns is also termed on the basis of population size as well as on the basis of functional and service importance. In true sense the hierarchy of urban settlement denote the ranking of urban centres into successive groups on the basis of size or some other selected Criterion. These groups of urban centres are functionally organised in a nesting pattern and are arranged in such a manner as that lower order centres occur within the influence areas of higher order urban centres. The higher order centres occur at a greater distance than the lower order centres.

In the analysis of hierarchy of urban centres also several scholars have applied varied statistical methods. The well known method of the study of Central places is adopted by Christaller. Other scholars have propounded some other methods. It is true that in all the methods attempts have been made to evaluate the central functions performed at the centre. Virtually hierarchy of towns should be termed as hierarchy of central places. It is also true that almost all towns with a few exceptions are central places because these towns perform basic functions meant for the people living in the surrounding areas. Only negligible number of towns perform non-basic functions which have no service area it is rather exception. As such all towns perform functions meant for urbanites as well as the people living in the surrounding. These towns have service area of different size depending upon specialisation of services and these towns become central places for the surrounding country side. As a matter of fact the development of town or very existence of town, the development of town and functional characteristics of towns etc. depend upon the resourcefulness as well as size of the surrounding upto which urban centres cater services. Such functions or services which an urban centre performs not nearly for itself, but primarily for the surrounding areas or the service area which is dependent for many socio-economic needs on the centre, but not a constituent are known as 'Central functions' and the centres or places where in these functions are performed, known as central places. A central place can be only that where an exchange of central goods or services for population other than the local one exists, or in other words, where exists a certain central institution through which such an exchange is made possible. Thus a central place can be defined as a permanent human establishment, where an exchange of goods, services and necessities socio-economic in nature, primarily and basically for population other than the local one, exists and which therefore commands as its region, the immediately contiguous encircling areas.

As almost all central places perform certain essential functions, the areas they occupy vary in size according to their service capacities. The small market centre, the large market centre, the regional centre, and the large regional centre are all central places, but they vary in character and importance. The inter-relationship of central places or different order shows a series of inter-connections between unlike establishments.

This region also reveals central places or varied hierarchical orders. These orders have been based on services rendered by the urban centres located in the region. The term region implies an area with certain characteristics by virtue of which it is a suitable unit for some particular purpose of business and administration. It is considered as homogeneous area consisting of a set of association whether of land or people, such as industries, farming, distribution of people, commerce or the general sphere of influence of service centre.

A service centre contains numerous social institutions, adequate people to support those central institutions and a residential area grouped around and within easy excess to the centre.

These service centres are the following :

1. **Regional Capital :** Ranchi is the Regional Capital. It has got central location in the region and has developed distinguishable set of tributary areas which overlays only on outer margin with larger cities like Kolkata, Patna, etc. This centre has large service area. It has a concentration of wholesale and retail traders, branches of all important banks, educational institutions of higher learning, health services of repute, etc. It occupies the highest rank in hierarchy of urban centres though its population is relatively smaller than Dhanbad urban agglomeration.

2. **Regional Centres :** There are five regional centres namely Dhanbad, Jamshedpur, Bokaro, Hazaribag and Daltonganj in the state. These regional centres have relatively less service significance than the regional capital. These regional centres have also large service area. Jamshedpur having large workers engaged in industries but Ranchi has more workers engaged in tertiary services. These centres also have good educational institutions, health facilities and other urban amenities.

3. **Sub-Regional Centres :** These centres include Chaibasa, Deoghar, Sahibganj, Dumka, Ramgarh, Jharia and Giridih. These are relatively smaller in comparison to the regional capital and regional centres and occupy small service area. These are comparatively more evenly distributed than the higher order service centres. Most of these centres are district headquarters.

4. **Large Market Centres :** There are several large market centres which serve the surrounding areas higher than the smaller urban centres. These are Pakur, Godda, Madhupur, Bundu, Maithon, Bermo, Lohardaga, Khunti, Gumla, Rajmahal, Jhumri-Tilaiya, Chatra, etc. These are almost evenly distributed all over the region, except those dominated by very rough topography. These centres generally have degree college, hospital, post and telegraph office, a few branches of banks, etc.

5. **Small Market Centres :** There are a good number of small market centres. These are more uniformly distributed over the entire region. These centres include Kharsawan, Saraikela,

Musabani, Moubhandar, Gua, Noamundi, Barkisaraiya, Latehar and Netrahat. These centres do not have all services but they generally possess educational institutions, banking services and health facilities.

Morphological Characteristics : In terms of morphological features urban centres of this region portray varied characteristics. Variation in morphological characteristics are marked by their planned and unplanned growth, site characteristics dominated by uneven land surface and network of roads and streets. The morphology mainly concerns the ground plan; external outline and internal arrangements of streets, forms and materials of buildings and the functional structure of the town. The actual buildings of the town may change over a period of time, but the ground plan remains relatively permanent as regards the layout of streets, the arrangement of buildings and the relation to these to the main cultural domains like temple, church, theatre, market place, etc.

The physical and man made characteristics of site, the historical background, the skyline, the green spots, the water bodies, the physical and cultural dominants and indeed all the elements of urban landscape are important components of urban morphology[2]. Consequently it is to the ground plan that one should turn for the basis of morphological treatment of town rather than to site architectural style. In case of the urban centres of the region, the impact of multiple geographical factors; natural and cultural or man made appear more obvious. Amongst the natural factors rivers and divides appear to have played a very important role. Other such features are ridge tops, flat plateaus, existence of mineral exploiting centres, etc. The cultural features consist of elements like old market places, mosque, temples, churches, roads, railway stations, administrative offices, etc.

Generally there are three systems of urban ground plan viz., the irregular system, the radial concentric system and the rectangular system. The towns of the region show a broad similarity in their ground plan. They are generally straggling in form and belong to irregular system. Large compact sectors are uncommon within the urban centres. This is caused by local topography which is undulating consisting of micro-divides and intervening lowlands. As a result there is absence of large compact sector and predominance of straggling growth.

The topographic irregularities of the region have greatly influenced the morphological characteristics of the urban centres. As such it becomes important to study the topographic feature of the sites which may vary from micro divides, rivulets, undulating land to ridge, scarps and hills, thereby affecting the external layout, internal arrangements of streets as well as the drainage system.

Apart from the physical conditions, the existence of cultural features have also influenced the development of urban patterns. The historical background of some of the urban centres has also played important role in their make up. Such centres grow as socio-politico-economic centres of kingdoms or principalities of local chiefs. For strategic reasons they had commanding situations and were generally built on defensive sites on river banks. Houses were huddled together in a compact manner, in some cases the towns were fortified with walls resulting in greater compactness.

The selection of sites, the planning of network of roads and the designing of houses bear the mark of their historical origin. For example the urban centres of Rajmahal, Deoghar, etc. present various imprints of historical events. The patterning of the Pre-British urban centres is entirely different from those established by the British. Urban centres of the earlier period had the tendency of concentration of all activities in a small area, people flocked within a limited area and created congestion and over crowding. There was hardly any space left for open lawns while towns or portion of towns established by the British consisted of wide roads, sufficient open space and planned houses such as Daltonganj.

The study of geographical zones like Inner zone, Middle zone and Outer zone is an essential aspect of the morphology of urban centres. The Inner zone primarily comprises of the core business area also known as C.B.D. The Middle or Intermediate Zone includes the markets and the residential quarters and is referred to as the integuments. The Outer zone consists mainly of residential quarters and civil offices.

The business sections of the inner zone usually known by the name of chauk. It represents the commercial hub of the town. It comprises of the general bajar, the specialised markets such as the food grain, vegetables and brassware markets. Surrounding the business section are primarily residential zone with scattered and small markets and retail shops. Industrial Areas may not be distinctly marked but modern factories are generally located on the outskirts of the town.

Functions and functional areas also influence morphology of towns because the industrial, commercial, administrative centres differ from each other in structure and appearance.

Huge industrial establishments and glamour and prosperity to the urban centres attract people from outside to settle in such urban centres. Industrially prominent urban centres have clear cut and distinctive morphological features such as industrial development on the fringe, well planned residential colonies and recreational facilities for industrial workers. In the commercial centres there is concentration of business shops, store houses, etc. in the central commercial areas. The administrative centres have imposing office buildings, meeting halls, official quarters and as such presents different morphological look.

The changing transport network within and outside the urban centres has also influenced the form and appearance of urban centres. In the early days people used to cluster within easily accessible points which outright became site of urban settlements. Such centres obviously are incapable of growth in circular pattern because of the location of towns along the road or by the site of river. The introduction of railways has brought about far reaching changes in the development and expansion of urban centres. Outside the confines of the town proper is the railway colony a distinct functional zone in towns which are important railway centres. The railway colonies lie adjacent to the railway station and includes the railway offices and residential quarters for the railways employees. The railway quarters are lined along the rectangular pattern streets and feel the spaces formed by the network of streets. Such feature is more obvious in the morphological look of Madhupur, Chakradharpur, etc.

The horizontal expansion of the urban centres also takes place due to the development of means of communication like railways, roads radiating from the service areas and within the urban boundary. With the construction of metalled roads connecting the surrounding service areas, the growth of settlements, industries and cash crops develop along the roads along the fringe areas. As such the original external forms of the urban centres get distorted with new additions in certain directions. A good example is Daltonganj where two roads, the national highway and the state highway traverses the middle portion of the town and have caused the development of business, administration, educational and other public utility services along such roads.

Civil lines are mainly administrative and residential in functions containing government offices, courts, post and telegraph offices, public gardens, parks, clubs, dakbungalows, etc. In these areas, metalled roads lined with shady trees are found. The extent of civil lines depend upon the size and administrative importance of the urban centres. In these civil lines, most of the buildings are brick built pucca houses, generally constructed in western style with some modifications in the form of addition of verandah and courtyard to suit the local climate. Sometimes industrial establishments, educational institutions, cantonment, administrative offices, jails, railway colonies, etc. serve as nucleus for settlements and help in the horizontal expansion of the urban centres. In the Ranchi city the addition of the towns of Doranda and the industrial areas of Dhurva has considerably expanded the city area. So far as internal structure of urban centres is concerned, it is found closely related to the purely local physical features of the site such as hillocks, scarps, slopes, etc. Some of the urban settlements are built along narrow divides. These divides are traversed by roads and subsidiary lanes which foster human habitation as well as industrial and social institutions. Sometimes intervening lowlands consisting of rivulets or streams, lakes, depressions, etc. exist and affect the compact growth of settlements. Almost of all the towns of the state excepting a few which are located in the homogeneous areas are characterised by convergence of roads following the micro-divides and the resulting town plan is straggling, radial or star like with frequent absence of large congested clusters of built-up area.

General Appearance :

The general appearance of the urban centre is represented by a consolidated arrangements of roads and streets, buildings and building materials, plans of houses, open spaces, etc. Many of the urban centres have developed from the villages and presents unplanned features. Roads of such towns are still jig-jag with narrow bends and houses differently piled with a closed frontage. But some localities which have developed in a more or less in a planned way have good houses such as railway colony of Madhupur, Chakradharpur, Gomo, Ranchi, etc. The main roads are generally metalled, cemented or pitched and often broad but as we proceed outward from the business centre to the residential areas the streets become narrow, unmetalled and crooked though paved with bricks or stones in most cases. This road again becomes wider at the end of the city and on the peripheral parts of the town. The streets of the rural markets or small towns usually consist of unmetalled path.

Dwellings of urban settlements of this region are also influenced by the size and functional features of the urban settlements. Larger towns of the region have majority of the houses built of brick and tiles. During recent years, the number of concrete houses has remarkably increased. Certainly older Mohallas of the urban centres still have some mud walled and tile roofed traditional houses. In the smaller towns there is predominance of brick walled and tile roofed houses in the main area and mud walled and tile roofed houses in the peripheral areas. In the newer urban settlements whose origin is linked with particular industries or mineral exploitation, the buildings are generally pucca and have been constructed in a planned way.

In some towns, expansion has taken place in various phases. These additions reveal contrasting features depending upon the architectural style of the period and the way of life of the people. In such old towns, some old buildings constructed by some business men present diverse features. In such towns streets remain narrow and hardly permit two way traffic of vehicles. In some towns, fragmentary remains of past culture and partial superim position of modern aspects are also observed. Such features are seen in the town of Deoghar.

Patterns of Towns :

Similar to rural settlements urban settlements also exhibit different patterns which have relationship with site and situational conditions. The internal ground plan is related to physical factors of the site such as hill or a ridge or micro-divides, etc. It is also related to historical features such as the existence of religious sites, forts or old buildings of the Zamindars or Tribal Chiefs, etc. The external ground plan or the form is an expression of the past history of the town as well as the present physical factors. It becomes obvious by the analysis of the changing pattern of external shape by the additions in later periods. The undulating areas of this region presents the following urban patterns :

1. **Linear Pattern or Elongated Pattern .** A few towns have elongated growth. Usually settlements get elongated either along the metalled road or due to ridge top settlements or in the valley bottom. The valley bottoms, however, are not generally suitable as town sites because they may be flooded after heavy downpour or they may be subject to landslide. On the other hand ridges are almost free from these two hazards. As soon as we leave the divides of ridge tops, the amount of drainage increases. The headstreams become pronounced and the surface become dissected. Towns having elongated growth are Latehar, Netarhat, Giridih which have developed on the ridge tops. There are a few linear settlements based on the rivers. The river guided elongated settlements are Daltonganj, Kharsawan etc. Because of the appearance of the natural levee due to higher amount of deposition of silts, this zone is immune from floods. It is obvious that such a zone is narrow and linear and the settlements located along the bank are bound to be elongated. The river provides the advantage of adequate water required for urban settlement and better drainage. In addition to these there are elongated settlements based on metalled roads. It is true that highway provides the facilities of easy movement as well as the facility of trade and commerce. The influence of coals is well marked in case of the linear growth of small towns like Vhowrah, Kharsawan, Godda, etc.

2. **Rectangular Pattern :** Many towns of the region appear to have a rectangular shape. Such shape has been the result of composite forces of site. It generally occurs in flat or plain areas where irregular terrain is lacking. It may be stated that the rectangular villages are to a considerable extent responsible for the rectangular pattern of the town.

The urban centre of Sahibganj has a roughly rectangular shape. Its shape has been determined by physical as well as cultural factors. The river Ganga controls its northward expansion and the rough terrain to the south limits the southward expansion. The township has developed well along the Bhagalpur Howrah loop line and the river Ganga. It is also linked with North Bihar by steamer services. Even, at present the township is expanding east and west along the road and railways. Other towns having rectangular shape are Angarpathan, Kerkend, Ghatshila, Madhupur, etc.

3. **The Circular and Semi-Circular Pattern :** The occurrence of such town plan is rare. Only a few towns in the region have roughly circular shape. In such urban centres street pattern remains radial with the convergence in the centre. A good example is Deoghar which has maintained roughly circular shape due to various conditions. Most of the people depend upon the religious activities of Pandas who spend most of their times in the temple and do not like to go far away for habitation from the temple premises and the main market. Most of the main roads also converge in the centre. Its external shape is controlled by the river Darwa on the west and hilly terrain in the north and east.

4. **Triangular Pattern :** There are only a few urban centres like Chatra, Barkisaraiya, Seraikela, etc. which have roughly triangular shape caused by the convergence of roads. Besides the attraction of roads several factors such as depressed lowlands and forest beyond the settlements have restricted its expansion. The triangular shape of Seraikela has been caused by the river and the curved road. The Barkisarai town has triangular shape due to convergence of main road and other approach road to railway station.

5. **Irregular Pattern :** There are several urban centres in the region which have irregular forms. Their shape does not fall in any of the above mentioned categories. Principal cause of their irregular shape is the irregular growth of the town from the villages. This form comes into existence due to interaction of various physical instructions like hills, valleys, lowlands, nalas, swamps, tanks and cultural features like railways, roadways, churches, temple and other urban amenities. Such urban form present an irregular plan having a labyrinth of streets, irregular of both width and alignment with no main thorough routes or dominant focal point and with houses irregularly dispersed in

relation to one another. A few example are Ramgarh, Loyabad, Gomoh, etc.

6. **Street Pattern :** The study of road net is an important aspect of urban morphology as it provides a skeleton of urban settlement. It influences both the growth and alignment of settlements. Only a few towns have planned street patterns. Streets in the civil line, railway colonies and cantonment are broad metalled and lined with shady trees while streets in the older parts of urban settlements vary according to the size and importance of the place. Outward from the business section of the urban settlements streets become narrow and crooked in the poorer residential areas. In the section bordering the civil lines streets widen again and are sometimes much better kept that business section. The condition of streets deteriorate as the town decreases in size and importance, so that in smaller towns streets outside the main bazar are usually unpaved.

On account of rough terrain, most towns have irregular growth of roads. The street pattern is consequently irregular while planned areas of these settlements may show one or more of the street patterns which are discussed below.

Rectangular Street Pattern : The rectangular street pattern comes into origin when several streets cut each other at right angle. The buildings also occur in rectangular or square blocks. It provides a short cut inter-connections between various parts of the city. Because of the rectangular mesh, the city is divided into equal areas and this is helpful in administration and daily urban life. This pattern is more common in newly developed parts of town such as planned areas of Sindri and Hazaribag.

Grid-Iron Pattern or Recti-Linear Pattern or Net of Square Pattern : Checkerboard or grind iron pattern is so wide spread a feature of the great phases of town building, it has come to be regarded as the norm for the plan towns. It consists of Grid Plan, with straight, parallel streets intersecting each others at right angles. It is found in Jamshedpur, parts of Ranchi, Bokaro, Dhanbad, etc.

Radial Pattern : Radial Street patterns occur when several roads or lanes radiate from one centre or converge at one point. In the region there is lack of such road pattern. But rough radial plan occurs in the township of Daltonganj where several roads have converged at a point. The most striking example of radio-centric street pattern is to be found in the planned areas of Sindri where free roads have formed perfect circle.

Winding Linear Pattern : Such street pattern usually occurs in mountainous areas where construction of straight roads become impossible. Consequently several roads following the terrain are necessary. An example of such street pattern is provided by Chakradharpur where two roads run in the same direction having the same curvature.

Irregular Street Pattern : It is the most common feature in the unplanned portions of the towns of the region. Topographic irregularities are chiefly responsible for irregular street pattern. This is commonly found in the old parts of the towns of Deoghar, Ranchi, Jamtara, etc.

Build : The building is the components out of which the urban settlements are made and thus becomes a significant aspect of morphological study. The dimension of the building, the plan of the building, the number of storeys, building materials, slope of the roof, architectural style, standard of the house, the provision of open space, sunlight, ventilation, etc. become essential. Except the planned portions of the new towns like Bokaro, Jamshedpur, Sindri, etc., the urban centres of the region are mostly unplanned. Consequently most of the urban houses consist of traditional designs having still mud or brick walled and tile roofs. Brick houses predominate in all urban centres of the state. The difference arises only in their size and structure.

As regards roofing materials, there is not much diversity in urban areas. Even today tiles continue to be important materials. Certainly recent development in the cities of Ranchi, Hazaribag, Khunti, Bokaro, etc. have pucca houses. Architectural style of the building is the material expression of the wants, the sentiments of the age in which it is created. Style in architecture is the peculiar form which takes place under the influence of climate and materials at command. Even today more people have houses based on traditional design. Certainly resourceful persons and elites are applying modern architectural style in their houses ignoring the social and environmental conditions.

Morphological Zones : On the basis of the layout of woods, types, density and distribution of houses, distribution of open spaces and water bodies and marshy areas, three broad zones can be delimited. These are :-

1. Inner Zone
2. Middle Zone or
3. Outer Zone

The Inner Zone : consists mainly of the older areas of a town. This zone geographically occupies the higher ground in the centre and is fully built up. It represents two types of clear cut sections — a. Business areas along the road and b. Residential areas in the interior. The ribbon shaped business areas lie along the main thorough fare of the town. Retail shops occupy the ground floor while upper floors in most cases are used as banks, offices and residences. The street system diverse from this zone to the various sector of the town. The older parts of Ranchi, Deoghar, Dumka, etc. present winding lanes and bye lanes which indicate their emergence without a broad plan. These lanes have developed entirely in response to the local needs, and in order to avoid the smallest obstacles such as small tank or private lanes become winding in character. The houses are mostly of better class having one or two storeyed of pucca structure. From the congested central parts the buildings extend radially along the main roads towards less congested ranges where they show a linear pattern of distribution. In the denser parts of the zone open spaces are rare except an occasional garden attached to individual house. Functional character of the inner zone is very complex and diversified. It rarely specialises in a particular function. In contrast to newer civil lines, railway colonies and military cantonment which were laid out in planned fashion to serve definite functions, the contiguous older parts of the town have chequered history with various cultural and commercial foci at different periods. The domes and minarets or mosques and pinnacles, temples are prominent in their sky lines.

Middle Zone is predominantly residential in character with some overlapping of business functions. This zone is well linked with the city properly by numerous roads which are broader than the lanes of the inner zone. In place of crooked lanes wide and straight roads crossing one another at right angles are found. This zone specialises in residences, public buildings, small cottage industries and administrative units, etc.

Outer Zone is rather rural or semi-urban in character and generally possesses more open landscape dotted with isolated groups of houses and settlement. The chief characteristics of the outer zone are the broader and straight roads showing better planning in their construction. It is mostly composed of lower class 'kucha' houses.

This zone also includes the cantonment which has vast open spaces dotted with officer's bungalows, barracks, offices, etc. all beyond the approach of general public. A number of recently developed planned colonies other than civil lines also lie in this zone. Several unplanned older rural settlements exist with old markets catering to their daily requirements. Moreover, most of the big educational and medical or health centres lie in the outer zone of these cities, because they require more space recreational centres like regional parks, stadium and picnic spots also lie in this zone along the roads.

Sub-urban Zone : Besides the above mentioned zones another zone usually developes in the fringe area beyond the municipal limits. This zone, though under the influence of the urban centre shows a strikingly rural features. In this zone over 50% of the land remains under cultivation but the nature of cultivation changes from foodgrains to the production of vegetables and fruits. Sometimes heavy industrial establishments develop in this area due to the availability of cheap land near roads.

Morphological Characteristics of Ranchi :

Ranchi, the second largest urban agglomeration of the region, is the capital of the state. It comprises of Ranchi and Kanke. It is important in terms of administrative, industrial and commercial functions. It is situated almost in the centre of the region ($23^{0}27'$ north and $85^{0}20'$ east) at an elevation of seven hundred metres. Its urban landscape is characterised by Ranchi conical hill, beautiful Ranchi Lake, Christian Mission and churches, irregular pattern of roads and lanes and new emerged planned areas having specialisation in different functions. The morphological characteristics of this urban centre is also growing due to industrial centres and habitation of people of various cultural background.

This is a straggling unplanned urban centre. The layout of roads, streets and buildings and style of dwellings, govt. offices are the result of continuous addition from the pre urban nucleus of Purana Ranchi near the foothill of Ranchi to present vast urban areas. Ranchi consists of several morphological units of various characteristics of different evolutionary phases. The older parts of the town being the Purana Ranchi, Upper Bazar have mostly irregular growth and winding lanes. Another patch of heavy concentration is found in Chutia towards the east of Hazarlbag road. The remarkable development of the town has taken place along the main road also known as Chaibasa road in the north-south direction. The north-south alignment of the town along its axial high street bears witness to the fact that this road plays the most important role in the movement of traffic of the town. The hub of city lies at the junction of three important roads radiating to Hazaribag, Chaibasa and Kanke. This morphological unit of the city is characterised by its retail shops, vehicular congestion, multi storeyed buildings, etc. In this area landuse is characterised by offices, consumer services, educational institutions, hotels, banks, post offices, theatres, etc. 'Typically the area has the city's highest buildings, the highest retail shopping and the greatest day time population concentration. It is the transit system as the land value is the highest in this portion.

The upper bazar area acts as collecting and distributing centres of the city and also supplies goods to retail stores. Ranchi hat is situated in the centre of the old nucleus of the town at the foot of Ranchi Hill. This hat is the place where the farm products from the surrounding areas are brought and sold. As substantial part of this transaction takes place through whole salers located near the hat in upper bazar. The Upper Bazar wholesale area is closed to the Central Business District (CBD) and the retail traders find it convenient to their supply of goods from the wholesellers.

During recent years Ranchi has become the seat of immense industrial activities with concomitant development, commercial and cultural activities and has started losing its previous landscape of open and cultivated fields and Adivasis huts interspersed between the Pucca houses, one of the largest industrial concern is the heavy engineering corporation which lies nearly 10 kms. south of Ranchi. A large colony has developed in the area. Besides there are marked concentrations of industrial enterprises near the railway crossing, on the main road, the lower Bazar south of the church road, the lower Bazar near Sardar Hospital , the upper Bazar near Ranchi Lake and old and new Hazaribag road. Most of the units situated on the main road and in the lower Bazar are Auto repair workshops, Automobile servicing centres and small machine shops. In the upper Bazar area various food processing industries predominate as this is the main grain and grocery area. Saw mills and several other industries are located in the peripheral areas such as the railway crossing in the south and Purulia road.

During recent years apart from the addition industries in the peripheral parts there have been several residential colonies like Gandhi Nagar, Ashok Nagar, etc. where a large no. of residential quarters are available. The upgradation of this city as the state capital and related administrative functions are likely to add new feathers to the morphology and functions.

Morphological Features of Jamshedpur :

Jamshedpur urban agglomeration consists of Jamshedpur, Tata Nagar Railway Colony, Mango, Adityapur, Bagbera, Jugsala, Chhota Gobindpur, Haludbani, Gadhra, Sarjamala and Ghorabandaha. This town group forms the largest in the state and is located in the middle of the Subarnarekha valley on the confluence of the rivers Subarnarekha and Kharkai and covers an area of more than 50 sq. kms. The city plan looks like a rough parallel of ram. This agglomeration owes its origin to the establishment of the steel plant as this site was choosen for the Tata Iron and Steel Company Ltd. Foundations were laid further for the building up a planned industrial township which later proved to be spontaneous in character and explosive in growth and development. The spacious well planned city of Jamshedpur which surrounds the steel plant is a city with peaks and paved streets, hospitals, educational institutions, water works and sewage, temples and mosques, a race tract and a big football stadium and a dozen playgrounds.

Morphologically, the city comprises of an irregular shaped settlement consisting of planned and non-planned streets which divide the area into a number of large pockets. These pockets have been filled in by a series of grids which are usually rectangular in the case of older settlements, but spider web pattern in case of newer township like Indranagar, Nildih. The internal structure is a complex mosaic resulting from piecemeal planning. This however was unavoidable as the enormous growth of population was absolutely unforeseen. Thus, despite the recent and planned origin of Jamshedpur its evolution has been akin to organic town.

The residential buildings are mostly on the north and west while the factories and their official quarters are in the south and east. The south-eastern portion is characterised by industrial landscape as the powerhouse, cake ovens, plate mills, calcining plants, furnaces, steel bars, etc. are all located in the part while to the north-west the well planned residential and business areas with extensive parks and recreation grounds attribute the landscape. The settlements, in general, tend to form linear pattern along the roads, but in some cases they are distributed wide apart in the township area of Jugsalai, Sakchi, Nildih, Beldih, Khutadih and Sunari.

Architecturally, the city contains mostly rose of quarters and bungalows, built on modern lines and marked by a zonal diversity in plan and structure. The roads are wide, well built and shaded. The Housing Colonies are built at different elevations leading to diversity of landscape and excellent facilities of drainage. The best in this respect is the picturesque new colony of the Telco, while the worst are the old areas of Kadma, Sakchi etc. A peculiar feature is the mud built and tile roofed houses of the Labour Bastis. The worst areas lie in the north-west near the rivers, north-east in Sakchi and in the industrial zone from Gomuri Road, Kalimati Road to Burma Mines, the Railway Colonies and Sundarnagar. Industrial concerns have been powerful factors in determining the physical structure and landuse pattern. Most of the pucca houses belong to different companies. They usually consist of rose of single storey to two or three roomed quarters with all amenities.

The market area has modern appearance with large show rooms, multi storeyed buildings and a wide thoroughfare. The old narrow lanes of Sakchi on the other hand are comparatively congested, the larger colonies are a rather unique feature of the city. There are labour colonies having ordinary houses which provide ugly look to the city.

The internal street pattern of Jamshedpur is complex and presents a mixture of unplanned old village. Street markedly determined by the site conditions and the planned grid iron-ore radial streets which fall in the intermediate areas. This is due to the lack of an early Master plan. Kali Mati Road has a network of cross roads, and Sakchi exhibits a multiple grid pattern. The defects of early planning and strong influence of the hilly nature of the site upon the street pattern is to be seen in the frequency of curves and bends in the roads as also in the undulations .

References :

1. Karimi, S. M. (1968) : 'Akbarnagar (the forgotten capital city of Bengal)' the Journal of the Bihar Research Society, Special Vol., p. 11.
2. Ahmed, E. : op.cit., p.267.

16

Natural Regions

The river basin of Damodar covering large areas of Gondwana Trough lies roughly in the middle of the state in between the Ranchi and Hazaribag Plateaus. This basin comprises the drainage areas of Damodar and its tributaries. The river Damodar rises in the Khamar Pat (about 2000') in Palamu district and flows eastward. The length of this river is 538 km out of which 290 km lie in the state of Jharkhand and is considered the upper Damodar Basin and the remaining length of 248 km lie in West Bengal and is considered the Lower Damodar Basin. This Basin comprises an area of 12.8 thousands sq. km, out of which 11.1 thousands sq. kms. of area lies in Jharkhand state. In this basin, the Damodar Valley Project was designed on the pattern of Tennessee valley project of U.S.A. This project was planned with a motive to control furious flood caused by the river Damodar especially in the lower portions, soil erosion, production of hydel power, afforestation, development of irrigation, etc. The Damodar valley Corporation came into existence in 1948. The multipurpose project encompasses the entire basin of the Damodar and its important tributaries, the Konar and the Barakar, with four dams and power houses.

The Damodar Basin comprising most areas of the Gondwana trough and adjoining plateaus has geological formations of complex nature. The formation of the area within the trough basin is dominated by the Gondwana formations and subsequent sedimentation, igneous intrusions, upheaval and structural events. The adjoining areas have complex geological history because of its relation with the ancient landmass of the Gondwana land and prolonged denudation, igneous intrusion, tilting, warpings, upheaval, etc. Consequently, these two areas the trough basin and the adjacent plateaus provide distinct evolution of landscapes. The faulted basin of Damodar has rock formations of Gondwana period in east west direction from the Hutar coal field in west to Jharia and Raniganj coal fields in the east. The bulk of the Lower Gondwana made up of sandstones, shale and beds of coal belong to the Permian and Triassic Preiods but the subdivision trespasses down into the upper carboniferous period. The Upper Gondwana is only represented in the area of Rajmahal highlands. The formations of Lower Gondwana provide greater evidence of local disturbance, steep dips and faulting than the upper. The Damodar valley region also recorded depositions of some Rajmahal lavas during Jurassic period. The Basin areas mostly consist of sandstone and shale but in the adjoining plateaus harder Gneiss and Schist are dominant. Igneous activities have considerably transformed the characteristics of the parent rocks.

The fault line scarp bordering the valley provides marked breaks in the long profile (Thalweg) of the Damodar, the site of the Konar dam built on such a knick point or minor knicks, e.g. at Bermo. On the successive and post-tertiary rejuvenations there is a series of erosion surfaces in the catchment. The highest is the 1200 metres Lohardaga-Netarhat highland dividing the Damodar and North and South koel where from the basin starts. Afterwards it comes the basin area separating Ranchi and Upper Hazaribag Plateau where the plateaus consist of about 675 metres elevation. In the lower portion this basin passes through the area having less than 200 metres relief. Because of these different levels descending abruptly to the lower levels by receding scarps, the terrain as a whole is upgraded causing rapid drainage, floods and soil erosion. In the upper courses of this basin the rejuvenations have entranched the river fairly deep into the terrain so that even short embankments can impound reservoirs. In these areas there are resistant residuals of hard rocks which have enabled the construction of dams. The catchment area above the Barakai-Damodar confluence is roughly triangular and is subject to high flood frequency in case of heavy downpour. As such major dams have been constructed near the downstream apex of the triangular of Maithon and Panchet hills above the confluence of Barakar and Damodar. The area is composed of ancient hard rocks and is immune from earthquakes which has favoured the construction of dams. Both sides of the basin consisting of Hazaribag Plateau in north and Ranchi Plateau in south have steep scarps. The lower areas are comparatively less undulated. Most parts of the districts of Dhanbad and Bokaro form peneplain. The residual hills appear as monadnocks and are composed of Gondwana sandstone. These hills are Lugo, Jhumra, Panchet, etc. The hills are covered with forests. Rajrappa waterfall is a discordant junction fall below which the imposing gorge and many juvenile topographic features make a classic segment of landscapes. The general slope of the basin is from west to east.

The basin of Damodar has several rivers which join the main river Damodar from northern and southern plateaus of Hazaribag and Ranchi respectively. Important rivers are Damodar, Konar, Bokaro, Jamunia, Garga, Gabai, Safi, Ijari, Maula, Haharo, Barakar, Nalakari, Venera, etc. Altogether the rivers form dendritic pattern. The rivers are rainfed.

The climate conditions vary from one end of the basin to the other. The mean annual temperature is 23^{0}C in the west to 26^{0}C in the east. The basin receives an average rainfall of about 125 cms. most of which is concentrated in the four rainy months. The torrential downpour in the upper basin causes excessive erosion on gneiss rocks, sandstone and shales on account of which lower course of the river is choked. The rain from June to September ranges from about 80 to 90 cms. while during the period of this retreating monsoon the amount of rainfall remains 5 to 10 cms. The winter season remains relatively dry having less than 2 cms. rainfall. The rain from March to May ranges from 5 to 15 cms.

The Damodar Basin having gneissic structure in the upper portion and sedimentary formations in the valley reveal great variations in typology of soils. The soils found in this region are sandy, clayey, loamy and lateritic. The clay and loam are found in mica-schist region and in the low-lying areas. Coarser and lighter soils are found in upland areas. Lateritic soils are located in escarpment and hilly areas. The soil erosion is common feature specially on steep slopes.

Variation in climatic conditions and soils and differential human interference in the form of mining and cultivation forests cover also differ. The hilly areas are more forested and the peneplain areas are almost devoid of dense forest. The steep scarped areas also have reasonably good patches of forest. The forests are of mixed deciduous type. Important trees found in the forest are Sal, Sisam, Semal, Mahua, Kusum, Palas, etc. These forests also have Sabai grasses, bamboo thickets and lac.

The Damodar consisting of rugged terrain in the upper reaches, steep sloping scarped area and the peneplain areas reveal contrasting landuse pattern. Forest areas are confined to the hill and scarps but the plain areas have comparatively smaller vegetative cover. Certainly plain areas have isolated trees. Culturable land is comparatively much more in the plains but in hilly areas small patches of culturable land is found. Waste land caused soil erosion and mining is found in both areas, the hilly areas have waste land due to soil erosion and rocky projection. While the mining areas have due to dumping of waste materials and collapse of the top layer of the mines after use. Certainly afforestation scheme has, to some extent, controlled soil erosion in the upper reaches. Construction of dams, reservoirs, transport, power plants, development of industries and urban centres have also engaged considerable area under non-agricultural uses. The area is noted for concentration of mines specially in the district of Dhanbad, parts of Giridih and Hazaribag and this portion has considerable land under habitation, infrastructure and other uses. In this region rural landscape hardly dominate the landscape having culturable use. In this area plantation has also been developed to minimise industrial pollution. Pastures and Fallow land are found in almost areas.

Similar to variation in landuse agricultural pattern also portrays diversity. Paddy is the dominant crop but in higher areas people grow maize, ragi, kulthi in higher areas but in lower areas paddy and wheat are grown. Certainly some remarkable changes have taken place in terms of the production of cereals as well as vegetables and seasonal fruits. Previously wheat was produced in limited areas but these days homogeneous areas having irrigation facilities are involved in production of wheat crops and vegetables. Due to rapid urban growth in this region people have developed more interest in growing vegetables and seasonal fruits for fetching more money. Some of them even, have developed interest in dairy.

Uneven terrain has been obstructing factor in development of irrigation channels, however, in undulating areas artificial reservoirs on smaller scale have been developed for irrigation purposes. In the Damodar Valley region some major irrigation projects have also been developed. The construction of dams at Tilaiya, Konar, Maithon, Panchet, Tenughat have also enabled irrigation in limited areas. Farmers have also constructed walls for irrigation of small areas particularly for growing vegetables and seasonal fruits.

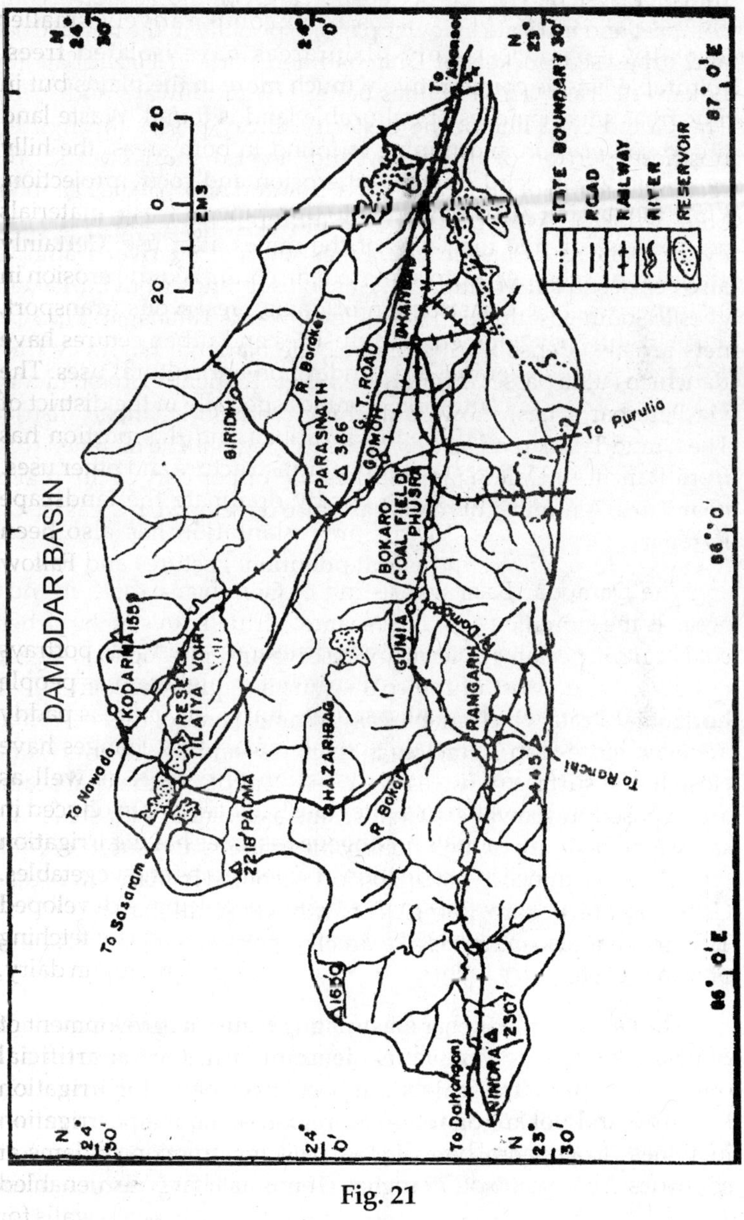

Fig. 21

The Damodar basin having cluster of mines, agglomeration of urban centres and a good number of industries has a better transport system. The nets of transport and communication become compact and intricate in lower parts, especially in the mineralised and urbanised pockets of Dhanbad and parts of Hazaribag and Bokaro districts. This area has better belt of railways and roads. The Grand cord line of the eastern railway passes through the northern portion of the valley, but several short lines fill the rectangles south of the Grand Cord Line and north of the Damodar bounded to the west by Gomo-Chandrapur lines and to the east by the Danbad-Pathardih line. Lines south of the river Damodar also connect from Mahuda to Asansol. A branch from Barkakana goes to south-east through the valley towards Tatanagar. The road nets are also dense in this region. The National Highway from Ranchi to Patna passes through the valley. Ramgarh-Dhanbad road via Paterbar, Chas, shows branching roads in Dhanbad district. The Grand Trunk road also connects this region. The metalled road from Ranchi to Daltonganj touches the upper part of the region near Kuru. A number of roads have also developed in areas east of Ramgarh.

The Damodar Basin consisting of Gondwana rocks in most areas is the store house of coal seams. Virtually this is the richest coal region specially in terms of reserve and production of good quality of coal. Most of the coal seams are quite thick and lie in horizontal strata which enable easy mining. Some of the coal seams are even more than 25 metres in thickness and the layer is found close to the surface of the earth. At several places the coal seams are exposed in the valley. Coal seams associated with shale and sandstone beds are of high quality bituminous and subituminous type. The coal mines of the region is distributed in Jharia, Raniganj, Chandrapura, Bokaro, Ramgarh, North and South Karampura, etc. This area has also minerals like fireclay, chinaclay, inferior quality of iron ore, etc.

The Damodar Basin has encouraged industrial growth because of coal resource, power plants, dense network of transport and other industrial infrastructure. In the lower part of the valley in Dhanbad and Bokaro districts the industrial landscape resembles in fullness to a mineral based industrial landscape of some western countries. There is no other region where such large section of the working population is employed in mining. The number of those engaged in manufacturing and transport is partly associated with mining and manufacturing. In the upper part of the valley mining and industries have developed in patches. The Huge Sindri Fertilizer factory, the Cement Works at Sindri and Khelari, no. of coal washeries, coke ovens, engineering works at Kumardhubi, the refractory works in Dhanbad, Glass factories, the large steel plant at Bokaro, power plants at Chandrapura, Patratu, Bokaro, Sindri, Tenughat, etc. provide industrial look. The way the industries are growing in the region it may emerge a large industrial belt having very high density of population, extent of urbanism, degree of mechanisation and in other respects similar to the industrial zones of highly developed nations.

The area is marked by very uneven distribution of population. There are areas having concentration of mines and associated industries and urban agglomerations where density of population is very high. On the other hand the rugged hilly areas and steep sloping scarped areas portray sparse distribution of population with a few smaller patches of human concentration. For example, Jharia-cum-Jorapokhar-cum-Sindri development block has the density of over 1200 persons per sq. kms.; Dhanbad-cum-Keduadih-cum-Jagata development block has the density of over 917 persons per sq, km., the Bermo development block has the density of 1742 persons per sq. km. etc. On other hand the development block of Kuru, Petarbar, Gumia, Kasmar, etc. have even less than 600 persons per sq. km. This region is inhabited by tribals like Munda, Oraon, Kharia, etc. in rural areas and the urban areas have dominance of non-tribals.

The rural settlement type is relatively dispersed, dispersal being more marked on the steeper slopes and in forested tracts. The rural dwellings are mostly mud walled and tile roofed. The mineral and industrialised areas have given rise to bunches of urban centres more particularly in districts of Dhanbad and Bokaro. The district of Dhanbad alone has 2 statutory towns namely Dhanbad and Chirkunda and 44 census towns. The districts of Bokaro and Hazaribag lying in Damodar Basin have also developed a large number of towns. Important towns are Dhanbad, Bokaro, Phusro, Sindri, Jharia, Kumardhubi, Ramgarh, etc.

The Rajmahal Highlands :

The Rajmahal highlands form north-eastern projection of the Chotanagpur highland as well as the Peninsular highland of India. It contains most areas of the Santhal Parganas except a narrow strip of alluvial plain on north and east. The area lies between 23°40' N and 25°18' N. latitudes and between 86°28' E and 87°57' E longitude. This region has also physical demarcation. The boundary on north and east of this highland is defined for some distance by

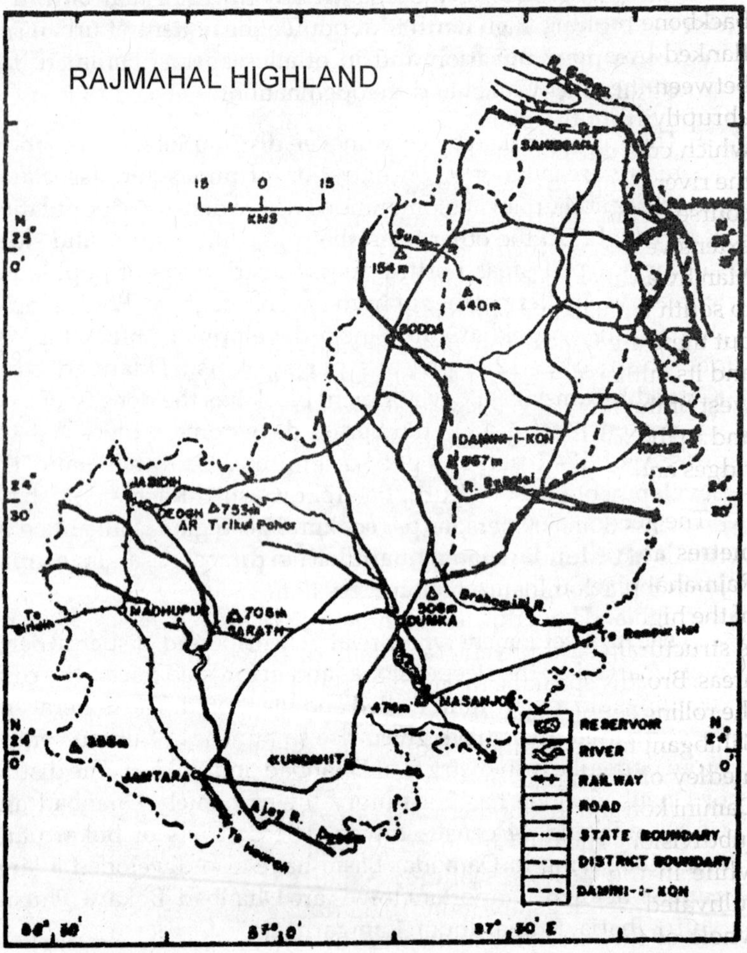

Fig. 22

the river Ganga which separates the region from Katihar and Malda, while portions of the southern boundary coincide with the Barakar and Ajay rivers of Damodar basin, which separate it from Manbhum and Burdwan. The western boundary is roughly determined by the Mica belt and Hazaribag Plateau in the West. It comprises an area of 14189 sq. km and population of 69,69,097 persons in 2011.

The highland of Rajmahal is an upland tract with a hilly backbone running from north to south. To the north and east it is flanked by a long but narrow strip of alluvial tract hemmed in between the river Ganga and the Rajmahal hills. These hills rise abruptly from the plains, forming a wall 320 m. to 640 m. high, which cuts out into the alluvial plain of the Ganga and compels the river Ganga to bend to the east before it takes finally southerly course. From Sahibganj the hilly areas stretch southwards in an extensive range which is divided into two parts by the Burhait or Manjhwa valley. The Burhait valley extends over 36 km from north to south with an average width of 8 km. It is surrounded by hills, but there are five narrow passes leading to the plains. The ranges and its outliers form a central block of hilly country. To the north-west of the range there lies a level fertile tract as Tappa Manihari and to the west and south the hills give place to a series of rolling ridges and undulating uplands.

The general elevation of the region ranges between 160 to 320 metres above sea level, although the northern-eastern edge of Rajmahal plateau themselves are less than 150 metres. The mass of the highland covering about three-fifth of the Santhal Parganas is structurally and physiographically distinct from the adjoining areas. Broadly speaking the region is divided into the hilly portions, the rolling country and the flat country. The hilly parts extend from Sahibganj to the southern limit of the region. It is made up of a medley of hill ranges and valleys and includes the whole of the Damini koh and the southern and eastern portions of Dumka Sadar subdivision. The hills are in many parts still covered with jungle while in the valleys, scattered small villages surrounded by cultivated clearings are found. The rolling country includes the whole of the west and south-west. It contains long ridges with intervening depressions, in places rocky and in places covered with scrub jungle. The flat country lies between the Ganga and the hills which is largely culturable and often liable to inundation.

The general slope of the area is from north-west to south-east except in the small alluvial tracts in the north-west where slopes towards north-west and sends its drainage to the Ganga.

There are a few loftier peaks which attain the elevation of about 640 m. Among these peaks Mahuagarhi (532 m.), Mori, the highest peak (641 m.), etc.

Late Prof. R. P. Singh has divided the region into seven morphological regions like (a) Scarp on the west and north, (b) Rolling peneplanned surface in the south with skeletons of ancient ridges, (c) Dubrajpur sandstone topography presenting a contrast to the flat lava plateau, (d) Scattered Gondwana deposits with coal fields of Hura, Jilbari, Chuparbhita, Pachwara, Brahmani from south to north, (e) Resistant Lava Plateau of Rajmahal, (f) the Rajmahal hills and (g) Gently grading country from west to east along the major tilt axis determining the general trend of the eastward flowing rivers.

The geological formations of the region presents complex and interesting features. This region consists of geological formations of early geological periods as well as later additions in the form of lava capping and features created on account of tectonic events. The basement rocks are granite, gneiss, quartzite, etc. The extremely dissected Rajmahal Plateau consists of Jurassic basalt underlain by gneisses and granites. There is succession of basalting lava flows with inter stratification of basalting lava flows with inter stratification of shales and sandstones. The sedimentary bands are held to have been deposited in the interval of times which elapsed between the volcanic outbursts by the circumstances that the different bonds of shale and sandstone differ from each other in mineral content. The upper surface of the shaly beds has been hardened and altered at places by the contact of the overlying basalts. The sedimentary bands are chiefly composed of hard, white and grey shale, carbonaceous shale, white and grey sandstone and grit. The trap rocks are dark coloured dolorites. They vary in character from a fine grained, very tough and hard rock. Dykes are rarely found. Tectonic movements have given rise to faults, fissures, warping, etc. The surface of the ground is much obscured by superficial deposits.

The region is drained by the Ganga, Gumani, Bansloi, Mor, Ajay, Brahamini, Morang, Barakar rivers. The river Barakar separates the region in the south-west. The river Ajay and Mor with their numerous tributaries carry the drainage of the western half of the region not into the Barakar but into the Bhagirathi. The streams which rise within the Rajmahal hills follow the same general direction from north-west to south-east and issuing through the passes in the hills, join the river Ganga after it has made its great curve southwards. Most of the streams in the hilly areas have well defined channels and high banks. At the time of rainfall they come down in flood and become rapid torrents, impassable owing to the velocity of the current. In the hot season they are reduced to a mere thread of water not more than a few feet deep with a gentle stream trickling through the sand.

The river Ganga enters the region a few kms. west of Teliagarhi and flows eastward as far as Sakrigali where it bends to the south-east. The course of the river frequently changes. The river Gumani rises in the Rajmahal hills in the extreme east of Godda district and makes its way north-east through its scoured gorges. This river is joined by the tributary Morel at Burhail. The river Bansloi rises at a hill called Banspahar in Godda district and flows in easterly direction. The river Brahamini rises in the west of the Dudhua Hills in the north of Dumka district and flows in southerly direction. Its main tributaries are the Gumro and Ero. The Mor rises in the Tiur hills at the extreme north-east corner of Deoghar District and drains the central portion of the region. This river is also known as Morakhi or Mayurakshi. This river is joined by Bhurbhuri, Dhobai, Tipra, Nunbil, Sidh, Dauna, etc. The river Ajay rising in Kiul District drains the south-eastern portion.

The region also has a few waterfalls. The most picturesque water fall is the Moti Jharana, situated about 3 kms. south-west of the Maharajpur Railway Station. This waterfall has 60 feet water drop. There are also two small falls or cascade on the Brahamani and Bansloi rivers at Singhpur and Kuskira respectively. There are several hot springs in Pakur and Dumka districts. The hottest spring is known as Laulandah situated along the bank of the stream Boru, about a km. north-west of Shivpur village in Maheshpur development block.

Owing to the portion of the region near West Bengal and hilly landscape climatic characteristics vary from the neighbouring areas. The alluvial strip has the damp heat while the hilly areas are swept by the hot westerly winds in summer. In this undulating country, the winter months are very cool and the winter rains are not oppressive. During the season the temperature rises from the end of March and becomes severe and the hot westerly winds are extremely disagreeable. In the month of January monthly mean minimum temperature remains 51.1^0 F and monthly mean temperature of the day remains at 63.5^0 F. In the month of June monthly mean minimum temperature remains 78.1^0F and day temperature 85.5^0F. The rainfall hardly exceeds 2 cm between November and April but increases to 8 cm in May owing to the influences of occasional cyclonic storms. In June the rainfall is 25 cm and in July the heaviest fall of 32 cm occurs. August and September are also rainy months with a fall of 30 and 25 cm respectively. The October rainfall, when the monsoon is practically at the end, is important to agriculture, the average rainfall of this month is 9 cm.

Nearly 25% area of the region is under forest cover. Most of the forests are confined to the hills and the steep sloping areas. The adjoining plains have orchards. There is a protected forest apart from the unprotected open jungle. Sal is the predominant tree in the region. Its distribution is almost everywhere except in culturable areas. Other trees are Semal, Palas, Pipal, Kathal, Mahua, Sisam, etc. In plains and valleys the chief associate of Sal is Piar. On the higher slopes and crests with thin and poor soil sal decreases in number and size and is replaced by bamboo and gamihar, etc. Sabai and the thatching grass, khar are the most important grasses as associates of bamboo and needs.

Soils of the region also differ from hilly area to the adjoining plains. The Rajmahal trap soil, eroded scarp soil surrounded by red soils characterise the soils of the hilly areas. The foothill balther soil occurs in eastern fringe. In the alluvial plains, clayey and sandy soils are found,

The region reveals spatial variations in land uses depending upon its relief features. The plains exhibit more culturable use of land, limited fallows, considerable waste land due to shift of river and non-agricultural uses. In the hilly areas having rough terrain more than one-third area is under forest and scrubs and the remaining areas have patches of agriculture alongwith the valleys, waste land in the rocky surfaces, habitations on hills and farms, pastures, orchards surrounding the settlements and other non-agricultural uses. The north-western plains is fertile and has more areas involved in agriculture with some non-agricultural uses of land. The valley of Burhait in the hilly region presents a distinctive land use pattern because of suitability of land for cultivation and other non-agriculturable uses. Cropped area in the region varies from about 40% to 65% and the areas sown more than once to the Net Area Sown varies between 40 to 60%. Important crops of this region are paddy, Maize and Marua. In the hilly areas rice, maize and marua are grown but in the adjoining plain and river valleys paddy and wheat predominate. In the adjoining plains of Ganga maize and jute in Bhadai season and wheat in Rabi season remain important. During recent years production of oil seeds in hilly areas and production of oil seeds in plains have remarkably increased because of more commercial value.

About 30% net cultivated area is irrigated. The chief source of irrigation is tank which is locally known as hirs. People create dam across the valley and store rainwater for irrigation purposes. There is an irrigation project on River Mor. The dam lies at Massanjore near West Bengal border. This dam enables little irrigation in this region also. Well irrigation has also developed in the region for growing vegetables.

The region is not as rich as the Damodar Basin and Subarnarekha basins but there are a few minerals of economic importance. Chief mineral products of the region are coal, building stones, limestone, pottery clays, iron ore, copper and lead ore. The south-western portion of the region contains the small Deoghar coal fields. The boundaries of these coal fields are often faulted. The coal is neither plentiful nor of good quality. In the southern portion there are coal fields named as Hura Coal fields, Chaparbhita coal field and Pachwara coal field and Brahamini coal field. These coal fields are lithologically similar to the Barakar beds of Rajmahal coal fields, consisting of alterations of grit, sandstones and shales with occasional beds of inferior coal. The hilly areas of this region contain a considerable variety of rocks suitable for building purposes. The basaltic trap affords a durable building material.

Besides there are a number of sandstones suitable for building purposes. Laterite is found on the top of some of the hills. China clay has long been known to exist at Lohandia. Fireclay occurs on the western side of the Rajmahal hills in the northern coal fields. Iron ores are found in considerable quantity in the basaltic trap and trappean beds as well as in the sandstones. Copper ore has been found at Behroki and Bodhbank.

The region has not registered phenomenal increase in industries except the cottage handloom industries of the villages and a few rice and oil mills in towns. Virtually industries of this region are of a primitive character. They mostly consist of the exploitation of the natural resources such as manufacture of iron, production of lac and Tusser. Lac is generally raised on the palas tree called in Santhali Muru but in the north and east where palas trees are few in number the bair or plum bush is used for the purpose. The Paharias, Santhals and Khatauris rear Tusser worms on Asan trees. Coarse cotton cloths are woven by village weavers.

The area may be considered less developed in terms of transport and communication except in southern and north-eastern margins. The main line of eastern railway passes through the southern portion connecting Jamtara, Madhupur, etc. and the loop line passes via Sahibganj and Pakur. In between these two rail route, the vast hilly areas are devoid of railways. However roads are numerous. Certainly road network is less in the hilly areas. Important roads link all district headquarters and urban centres with neighbouring districts and states. Important roads are Dumka-Bhagalpur, Dumka-Jamtara, Dumka-Deoghar, Dumka-Sahibganj, etc. Rivers in the plain are navigable. There is an air strip at Dumka.

The highland of Rajmahal reveals uneven distribution of population as well as unbalanced rural-urban ratio. The average density of this region is 504 persons per sq. km, but the district of Deoghar has an average density of 602 Godda has 580, Sahibganj has 580 persons per sq. km. The remaining districts have less density like Pakur (497), Dumka (351) etc. than the regional average. But there are development blocks having more than 1000 persons per sq. km like Pakur (1270), Mahagam (1206), Thakur gangati (1175) and Meharma (1111) etc. This region has 10302 inhabited villages and 1808 inhabited villages. Important urban centres are Deoghar (2,03,123), Sahibganj (88,214), Dumka (47,584), Pakur (45,840) and

Rajmahal (22,514). There are 25 towns in this region out of which 10 are statutory towns and the rest 15 are census towns. Even the tribal and scheduled caste population is unevenly distributed. The district of Dumka has (43.22%) tribal population being followed by Pakur (42.10%) and Jamtara (30.40%) etc. In case of scheduled caste the highest percentage has been in Deoghar district (12.74%), which is followed by Pakur (12.70%) and Jamtara (9.21%) etc. The percentage of urban population is highest in Deoghar (17.32%) which is followed by Sahibganj (13.88%) and Jamtara (9.58%) etc.

The rural settlement is partially dispersed, dispersed being more common in the intensely dissected Rajmahal hills where the frequent site is on the mid slope of the spurs. Some tribal people have developed hill top settlements. Linear dispersed settlement is common in the hilly areas. More villages are of small size. In the district of Sahibganj there are 1349 inhabited villages, out of which 43.44% villages have population less than 200 persons and 20.16% villages have population between 200-499 persons. Similarly in district of Dumka out 2688 inhabited villages 25.52% villages have less than 200 population and 44.01% villages have population between 200-499 persons. In Deoghar district there are 2354 inhabited villages, out of which 26.64% villages have less than 200 population and 36.96% villages have population between 200-499 persons. In Godda district out of 1599 inhabited villages 563 villages have population size less 200 and 24.41% villages have population between 200-499 persons.

The North Koel Basin :

The North Koel Basin lies in the north-western portion of the state of Jharkhand. The area of the basin is demarcated by the plain areas adjoining the river Son and Kaimur Plateau in the north, the Pat region in the South, the Hazaribag Plateau and the Damodar Basin in the east, the water divide between the North Koel River and the Kanhar river in the west. This basin comprises almost areas of the district of Palamu except a narrow strip in the form of homogeneous plain in the north, a portion in the southeast of Palamu District which falls in the jurisdiction of the Damodar Basin and narrow water divide area and valley of Kanhar along the western border of Garhwa. Thus, the North Koel Basin is an approximately equal to the area of Palamu and Garhwa districts which is 12,777 sq. km and consists of 39,89,631 persons.

Structurally the basin area of the River North Koel presents some contrasting features than the main landmass of the chotanagpur highland because it contains geological formations in limited areas of the Vindhayans in the north-western corner. These formations are succeeded in the South by the harder gneisses and granite which are further succeeded to the south by the softer Gondwanas, particularly in the Amanat Basin of this region. Again the areas having gneisses and granites are found and in the extreme south lateritized basalt of the Pat region is found. Thus, the geological formations of different rocks areas having distinctive hardness, erosional resistance and weathering have given rise to

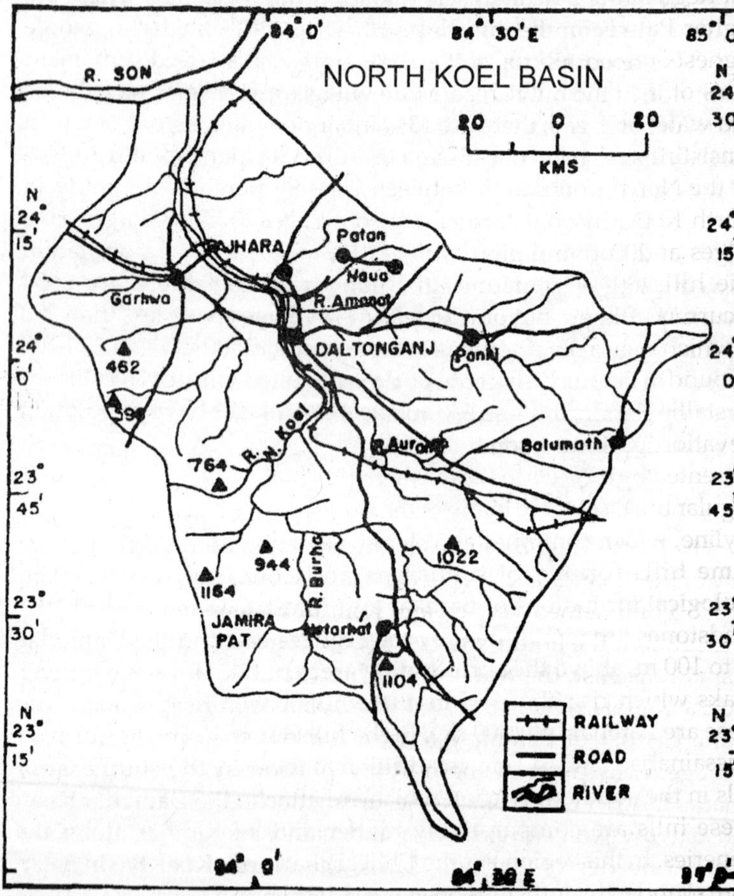

Fig. 23

variation in landscape. Virtually, the basin appears as a saucer shaped elongated valley. Different portions of the basin, thus, exhibit rocks related to marine, fluvial, aerial depositions, igneous intrusions and eruptions. Important rocks are granite, gneiss, schist, sandstone, quarzite, limestone, etc. The bed of the Koel is rocky upto a short distance beyond its confluence with the Auranga from that point it is composed mainly of sand and until the Son is reached there is only one serious obstacle in the form of rocky exposure consisting of a ridge of gneiss rock that crosses the river below the village of Sigsiri.

Phsiographically the area consists of three units, the parts of Pat Region, the Divided Terrain Region and the Valley Areas. The higher Pat region lies in the southern portion and presents the highest erosional surface. The divided terrain areas lie along both sides of the basin which coincide with Hazaribag Plateau in east and water divide in the west. The general outline of the hill system consisting of a series of parallel ridges running east to west pierced by the North Koel has presented varied landscape. The course of North Koel shows great denudation. There are many spurs and ridges and outlying peaks, rising from the general land surface. The hills are conspicuous for their irregularity of form and occurrence. Their contour depends mainly on the nature of rocks of which they are composed but every variety of form and outline is found. The most numerous and ancient are those composed of crystalline and metamorphic rocks. Some of these hills even have elevation upto and some time beyond 1000 metres The outline presented by the hills in the northern half of the region are generally angular but elsewhere many of the ridges present an almost regular skyline, which continues as a steady elevation for a long distance. Some hills formed of sandstone and conglomerates having geological formations of Barakar and Mahadevas are found. The sandstones are represented by long ranges east of koel averaging 65 to 100 m. above the surrounding areas but occasionally having peaks which rise about 65 m. Here hills have flat tops. Some of these are Latehar hill (300 m.), Bigha hill (416 m.), Khaira hill near Hussainabad (545 m.) etc. In addition to these hills there are some hills in the southern portion which are attached to the Pat region. These hills are comparatively higher and provide picturesque sceneries. In this region lies the Chhechchari valley of Saucer shape is surrounded by lofty hills on every sides. This valley is a complete basin with the scrap of the great tableland of Sirguja on the west, on the south the range that overlooks the Barbe valley in Ranchi Plateau, on the east Netarhat and Pakri Pat and on the north the spur crowned by the fort of Tamolgarh and the Burha Pahar, 1000 m. high. The lower basin areas consist of excavated basin in the divided terrain region. The basin contains fertile alluvial soil on the surface but the basement rocks are the igneous, metamorphic and sedimentary. Even in these plain areas some isolated hills and little undulating land surface are found.

The North koel with its tributaries determine the drainage of the most areas of Palamu and Garhwa districts. The general line of drainage is from south to north towards the Son following the general slope of the region. The principal rivers are the koel, the Auranga, the Amanal, the Garu, the Orangai, the Banki, etc. The river North koel rises near Gumla and carves out defined elongated valley in the western higher plateau of Ranchi and enters the Palamu District below Netarhat near Rub after flowing nearly due west for about 30 km it turns north at an almost complete right angle through a gorge at kutku and flows between the districts of Garhwa and Palamu and joins the river Son near Haidarnagar. The river Auranga rises near Soheda in a pass and flowing in winding course in north-westerly direction for about 75 km joins the river koel near Kechaki. This river has tributaries like Tata, Sukari and in lower portion it is also known as Aurangai. The river determines the drainage of South eastern portion of the basin. The tributary Amanal rises in Hazaribag Plateau and flows almost due west and joins the river koel 8 km north of Daltonganj. The tributaries of the river Amanat are Jamuna, Khaira, Jinjoi etc. The river controls the drainage of central eastern portion of the basin. The river eastern Banki rises near Chhatarpur Plateau flowing due west and joins the river koel near Majiawaon. The river western Banki rises near Bhurki and flows in north-east direction and joins river N. koel near Railway Station, Utari Road. This region decides the drainage of north-western portion of the basin. The river Uria flowing in eastward direction joins the river N. koel near Garhwa. Other tributaries of the koel rising in the western water divide are the Araraj, the Bijka, the Saphi, the Satbahini, the Gunga, the Kulhi, etc. In addition to these rivers draining in the region there are a few rivers which have some drainage area in the districts of Palamu and Garhwa. These rivers are Kanhar in extreme south-west, Pandu in the extreme north-west and Jatane in extreme north-east and the Damodar in the extreme south-east. Altogether, the river N. koel with its tributaries provide an appearance of Dendritic Pattern.

The climate of this region on the whole is dry and bracing. Certainly the southern hilly areas present climatic conditions slightly different from the remaining areas in terms of temperature and amount of precipitation. The average annual rainfall of the area is 133 cm. The northern portions of the region get comparatively smaller rainfall than the southern portion. Hussainabad near the Son basin in the extreme north gets an annual rainfall of 108 cm. while the southern hilly areas record 182 cm. The rainy season receives about 85% of the annual rainfall. The cold season starts from November and continues till the middle of March. December and January are the coolest months. In association with cold waves which occasionally pass over the area in the wake of western disturbances, minimum temperature goes down to less than 2°C. In the southern parts frost occur in December and January. In May and early part of June, the maximum temperature goes as high as 46°C on individual days. The daily range of temperature ranges from 16° to 17°C during the winter and summer months, while in monsoon months it is only 6° to 8°C. The relative humidity remains lower in the afternoons than in the mornings except in the monsoon months when there is little difference.

This region also exhibits variation in vegetative cover. The southern portions having rough terrain, high peaks, steep slopes, little human habitation have reasonably higher percentage of forest cover while the remaining areas have comparatively less natural vegetation. The natural vegetation of the basin except the southern hilly areas is of dry deciduous type in which gamhar, sal, sisam, kusum, khair, palas trees are found. In addition to these trees bamboo thickets and orchards have also developed near the settlement sites. In southern areas reserve forests form the main vegetative cover. Sal is the most important tree. Large stretches of forest lands which were once covered with fine forests have since been completely denuded of all vegetation and stand out as desolate, naked wastes. Large parts of such wastes have been gullied and deep ravines have been formed there in. Certainly during recent years some afforestation schemes have been taken to increase vegetative cover and to control the increasing wasteland.

The impact of varied rock formations in different areas, intensity of forest cover and climatic conditions has laid to variation in soil types. The southern portion having thick mantle of lateristic has given rise to red lateritic soil. The divided terrains existing in eastern and western portion yellow brown laterite soils on the higher areas and Domat soil in the lower areas. The lower portion of the basin has clayey and sandy loam. The soils existing in the lower basins are comparatively more fertile where different crops are grown.

The North Koel Basin has more spatial variation in land use pattern based on variation in relief features, forest cover and human habitation. The southern portions having considerable forest cover has more areas under forest. A few isolated dwellings dotted in the hilly region having most sparse population has not disturbed the vegetative cover much except small patches of culturable land and other uses. In the divided terrains the culturable land has comparatively more developed than the southern hilly areas because of increased population pressure of low and comparatively better transport network, more development of culturable land, waste land, fallows and other non-agricultural uses. The lower portion of the basin is dominated by culturable land, residential areas, pasture, play grounds, orchards and other non-agricultural uses. Virtually forest cover in this portion becomes non existent except isolated trees and orchards.

The North Koel Basin reveals variation in cropping pattern. The agricultural pattern is characterised by the dominance of rice. The don areas constituting about 45% of its cultivated land are mainly devoted to rice cultivation in the Kharif season and wheat in Rabi season where irrigation facilities are available. The Tanr areas are devoted to ragi, Maize, Oil seeds and Pulses, Arhar, etc. The next important crops in order of acreage are wheat, maize, gram, arhar, etc. The other considerable crops are oilseeds, sugarcane and vegetables. The area has not recorded considerable development in irrigation facilities. Bunds or Reservoirs or Ahars are distributed in the northern region. There is proposal for Koel-Karo Project for developing power resource and irrigation facilities. There are several minor irrigation schemes in the basin like Phulwaria Irrigation Schemes, Adra Irrigation scheme, Jinjoi

Irrigation scheme, Sone Irrigation scheme on smaller rivers.

The region is well served by transport systems . The railways aligning the North koel river has contributed to the prosperity of the region. The main rail route connects Barkakana to Dehri-on-Son which passes almost along the Koel Basin. Roads follow the rails and serve as trade routes of distinguishing character. Important roads connect Daltonganj to Aurangabad, Daltonganj to Garhwa, Daltonganj to Ranchi via Latehar, Chandwa, Daltonganj to Pankl, Garhwa to Ramanujganj of Sarguja district of M.P., Garhwa to Mirzapur, etc. Besides there are roads connecting development blocks and rivers. North koel is navigable only in limited northern portion especially in rainy season. There is a airstrip near Daltonganj.

The north koel Basin having geological formations of different periods consisting of igneous, metamorphic and sedimentary rocks has mineral resources of considerable importance. Certain areas provide minerals related to Archeans and some areas have minerals related to Gondwana, Vindhyans, etc. The portions of the area having thick mantle of laterite have considerable occurrence of bauxite in which high percentage of alumina with some titamium oxide and a little silica is found. It is confined to three areas namely Netarhat Plateau, Joradumra Pahar, Garitanr Pahar and Jamira Pat. Iron ore is found at Gore, Datum, Biwabathan and Sua. At Gore magnetite occurs associated with tremolite schist on a small hillock. At Datum magnetite occurs in bands and lenses and small pockets. Crystalline dolomite and dolomitic limestone occur at a number of places. These places are Bakoria, Saohawadih, Chauparia, Banpahar, Harhipahar, Kolthia, Pandwa, Balumeth, Majhalio, Dolomites are usually white in colour but in some places due to the presence of serpentine they are greenish. Graphite is found at Sokra, Khamdih, Kui, Ajlatua, etc. Mica of inferior quality is found near Daltonganj, Lesliganj, etc. Fireclays occur a Rajhara and near Latehar. Coal is an important resource of this region. Coal bearing Damuda strata occurs in three separate areas namely Auranga, Hutar and Daltonganj. Auranga and Hutar coal fields lie in belt of Damodar Valley coal field but Daltonganj coal field exists in the valley of North Koel. Daltonganj coal fields cover an area of about 300 sq. kms. The coal is non-coking and is of second grade. Rajhara, Pandwa and Lohda coal mines are important.

Although the area has varied types of minerals and forest resources industrial development has not taken place as per expectation. Only mentionable industrial activities of the basin are the manufacture of lac, shellac at Daltonganj, Garhwa, Latehar, etc. Handloom industry-weaving cotton cloth and coarse woollen blankets, biri industry at Daltonganj, Garhwa, Saw mills, etc. are other industries.

The impact of uneven terrain coupled with unequal agricultural potentialities and other resources is obvious in the uneven distribution of population and its characteristics. The density of population decreases from south to north depending upon the decreasing intensity of ruggedness. Again the divided terrains existing on both sides of the basin provides population characteristics of different nature. The main concentration of population is in valley plains and in the northern homogeneous areas because of suitability of the land for agriculture, better facilities of roads and railways and urban growth. The district of Garhwa has an average density of 323 persons per sq. km. but the district of Palamu has an average density of 442 persons per sq. km. while in Latehar district the average density is only 169 persons per sq. km. There are some development blocks of this region having density of more than 400 per sq. km. like Daltonganj - 442, Garhwa - 596 because of urban centres and development blocks having density between 400-600 persons include Utari (457), Meral (494), Bishrampur (424), Hussainabad, (485), Hariharganj (588), Lesliganj (584) etc. Development blocks having less than 300 density include Garu (132), Mahuadanr (117), Latehar (260), Balumath (253), Chandwa (189) etc. There are urban centres like Daltonganj (78,396), Garhwa (46,059), Hussainabad (29,241), Latehar (26,981) etc. The urban population is only 8.02% in the North Koel basin. Garhwa has 5.23% urban population and Palamu has 11.65% urban population and Latehar has 7.13% urban population.

Important tribes of the region are Kharias, Kharwars, Bhoktas, etc. This region has 23.48% tribal population and 24.38% scheduled caste population to the total population.

The areas have very sparse population in the southern hilly terrain where isolated dwellings are found. The divided terrains of the region have relatively dispersed settlements but the northern homogeneous plains have compact and semi-compact pattern of settlements. This region has 3,304 inhabited villages out of which 459 villages have population less than 200 persons, 672 village have population ranging between 200-499, 909 villages have 500-999 population size, 698 villages have population between 1000-1999 and the remaining villages have population over 2000. Uninhabited villages in the region are 223 and there are sixteen urban centres.

The Ranchi Plateau :

The Ranchi Plateau, a rejuvenated peneplain is located in the heart of the state. It is bounded by the faulted Damodar trough in the north and Simdega Plateau and Porahar highland in the south, the Panch Paragana plain in the east and the Pats in the west. It comprises most areas of the Ranchi district and about half areas of Gumla district. It comprises an area of 4500 sq. miles and has nearly 41 lakh persons.

The geological formations of the area concerns the formations of the oldest geological period to the tertiary upheavals. The geological formations of the area is represented by Dharwar sediments with the basic intrusives. The initial sedimentary formations have got highly metamorphosed into various parts and the remnants of earlier sedimentary and igneous rocks are known from the inclusion of phyllites and schists of varying dimensions in the granite mass and extensive areas of Khondalites. The Ranchi plateau is made of Khondalites which is garnetiferous sillimanite schists decomposing easily into earthy soils. Phyllites are the important rocks in the south-eastern portion. It is finely foliated, silvery grey rock with silky texture and of a varying lithological character. The phyllite is often associated with slaty rock and impersistent bands of quartz schist and quarzite. Granite containing good proportion of felspar and mica are found in large areas. Rocks containing minerals like garnet, kyanite, sillimanite are found. Outcrops of quarzite are seen in many places. They are usually intimately associated with the silicate rocks. Recrystalised quartz veins are also noticed to resemble the quarzite. Some areas have basic rocks having linear inclusions of unaltered dolerite and the more common epidorite and hornblende schists and granulite. They crop out as dark coloured bands, elongated strips and lenticular patches of varying length and width. Most of the areas have enormous intrusive mass having wide variations from fine to coarse grained and porphyritic types and from massive to highly gneissic type. The granite rock is rather fine grained towards its border and looks like micro granite gneiss. In the central portions granite gneiss vary from normal medium grained rock to a porphyritic materials with large crystals of potash feldspar. Vein rocks are found extensively among the older rocks. Coarse pegmatitic intrusive are quite abundant in the south east portions. In the northern portion the region has also Gondwana formations. In the western portions, outcrops of the Deccan Trap lava flows have caused the occurrence of basalt and lateritized rocks. Besides the older igneous and metamorphic rocks predominantly found in the region, it has also wide spread plateau laterite and the less common valley laterite. Actually most of the valley laterite is due to accumulation of the derital materials from the disintegration of plateau laterite. These cappings of Moorumy laterite are spread all over the region. The high level laterite is the residual weathering product of the trap which occur in the western region.

Being the uplifted peneplain the region consists of a

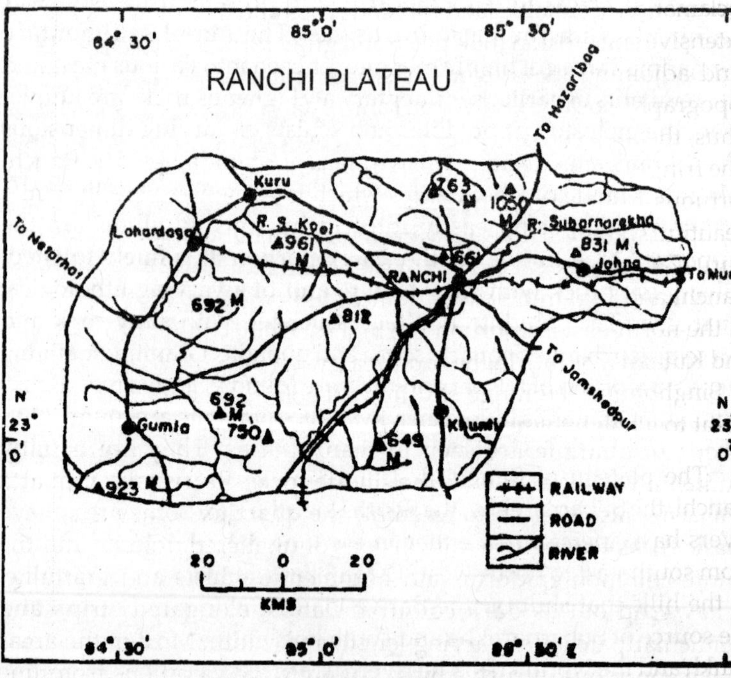

Fig. 24

considerable extent of homogeneous land surface with an average elevation of about 2000. It is gently undulating area intersected with numerous streams and rivers and studded with low wooded rocky hills and isolated peaks which form a conspicuous feature of the landscape. The average slope of 1^0 to 3^0 is indicative of the flat character of this uplifted peneplain being evident from the uniformity of the plateau surface, low relief within the plateau surface, isolated residual low hills, high degree of truncation of structure, the rivers being indifferent to softer Dharwars and harder gneiss and granites[1]. The gneissic hills form typical landscape. They rise abruptly with bare rounded slopes, the result of exploitation of the outer layer which expand by chemical changes of their constituent minerals under humid tropical conditions. Their summits are generally bare and dome like sometimes rugged and irregular. The highest portion of the region is a ridge 12 to 15 km south-west of Ranchi from where the main rivers like Subarnarekha and south koel originate. Because of considerable deforestation, reclamation of suitable land for food crops and other developments extensive soil erosion took place and gully formation by which the land adjoining the streams has been cut into typical bad land topography locally known as Tanr or land of cultivated upland. Thus, the area consists of extensive cover of Tanr and don lands. The fringes of the plateau are mountainous and have given rise to tortuous hill passes. The area locally known as ghats presents beautiful scenery. There are a few important hills known as Marang Buru (780 m.), the sacred hill of the Mundas about 15 km south of Ranchi and Pyramid shaped Ranchi hill crowned with a temple. In the north-eastern portion the highest points are Paina (865 m.) and Kutaim (789 m.). In the south-east where the plateau falls away to Singhbhum, the range shutting off sonapat valley rises at one point to a height of nearly 960 m.

The plateau of Ranchi is drained by several rivers like the Kanchi, the Subarnarekha, the South Koel, the Karkari, etc. the main rivers have their source either in the long ridge of high country from south-east to north-west, 12 to 15 km south west of Ranchi or in the hills that shut off the Barwe Plateau. The former contains the source of Subarnarekha and south koel and in the later rise the Sankh and the North koel. The river Subarnarekha originates south of the village Nagri, 15 km west and south-west of Ranchi urban centre. The river pursues a winding course to the east till it descends the plateau forming the steep fall of Hundrughagh (102 metres). It then turns south. This river has three important tributaries like the Kokro, the Kanchi, the Karkari which flowing in easterly direction drain the south-eastern portion. The river South Koel rises on the northern slope of the same ridge as the Subarnarekha and flows in a north-westerly direction and collects waters of the Bandora River and the Birgora river and flows almost parallel to Ranchi-Lohardaga Road. Again this river passes through the south-western portion. This river has tributaries like the Karo, the Chata and the Bonai. These areas have a number of waterfalls. The most picturesque is the Hundrughagh waterfall, about 36 km north-east of Ranchi urban centre. In the rains the falls present a most impressive spectacular, as waters red with the soil through which they flow, fall thundering over the chift in a cloud of spray. From the top of the falls one can have a magnificent view of the river twisting through a narrow gorge in the forest clad hills to the plains of Hazaribag. The Gautam-Dhera or the Jonha falls near Jonha is comparatively a smaller fall. The Dhashrughagh, 33 km south-east of Ranchi near Taimara is formed by the Kanchi River falling over a edge of a rock in a sheer descent off 36 metres amid rocky and Sylvan Scenery. Panchghagh, about 13 km from Khunti, is confluence point of five rivers. There are a few lakes in the Plateau which characterise the landscape. These lakes are Ranchi lake in the centre of Ranchi and the Kanke lake.

The climate of the Ranchi Plateau is cool and pleasant because of moderate elevation and some vegetation. Certainly due to industrial and urban growth some changes in climatic conditions have been experienced. Relative humidity also remains smaller so summer season is also comfortable. During summer season the temperature goes upto 30°C and it comes down to only 17°C during winter months. December is the coldest month with mean monthly temperature of 22.9°C and the mean daily minimum at 10°C. The temperature in January and February is slightly higher than in December. After February the temperature rises steadily till May and it becomes maximum in the month. The mean daily maximum temperature in May is 37.2°C. The highest maximum temperature recorded at Ranchi is 43°C and the lowest registered is 2.8°C. Ranchi receives rainfall almost throughout the year. The concentration of rainfall, however, occurs in the months of June, July, August and September. During the monsoon months the region receives about

82% of the annual rainfall. July and August are the rainiest months. Average annual rainfall of the district is recorded at 137.5 cms. In the months of December and January the rainfall obtained is only 3.7 cms.

The Ranchi Plateau having rock exposures of different geological formations has soils varying in different areas. Certainly most of the areas have laterite red soil but in some areas of western portion black soil has developed on account of decomposition of parent basaltic rocks. In the south-eastern portion , again in some areas, black soil is found but in remaining areas soil is derived from the decomposition of granite and gneiss. The soil capping is very thin. Normally the thickness of soil at the base of ridges is more in the intervening valley portion. Alluvium derived from granite gneiss is a coarse sandy soil. It is a rich soil quite suitable for agriculture but soils on Tanr regions become less productive because of the erosion of softer soil.

The impact of relief, climate and human interference is significant in case of vegetal cover of the area. Earlier this plateau was covered with dense forest but presently the central and eastern portions have been almost denuded of forest and only small scrubs jungle with few big trees are found on the hills. A few patches of jungle dotted among the cultivated lands are the only traces of the forest. The forests in real sense, is available in scarp areas and hilly areas of the region. Important trees are Sal, Asan, Gamhar, Kendu, Semal, Mahua, Jamun, Kathal, etc.

Relatively homogenous land surface has encouraged considerable use of culturable land. The culturable land of this region is divided into two classes, viz. Don and Tanr. The Don lands are the terraced low patches on which rice is grown and the Tanr are the upland areas which produce a coarse variety of rice, pulses, oilseeds. In south-eastern portion Don and Tanr lands are known as 'Loyong and Piri'. The Don lands are prepared by levelling and embanking the slope so that they may retain water. These lands remain close to the rivulets or water body which also helps in irrigation and retaining moisture. In lower areas, the land is mostly used for cultivation and negligible percentage remain fallow or waste land or under other uses but in the higher areas or Tanr areas land is used for construction of dwellings, orchards, pasture, playground, graveyard and other non-agricultural uses.

During recent years considerable industrial and urban growth have accelerated the percentage of non-agricultural use of land. The crops of the region fall under three main harvests : Aghani, Bhadai and Rabi. The Aghani is the winter crop which is harvested in the month of Aghan and is composed mainly of winter rice. The Bhadai is the Autumn crop harvested in Bhado. It include Bhadai Paddy, Ragi and Maize. The Rabi crops include wheat, gram, pulses, vegetables. Other important crops of the region are Gondli (small millet), Marua, Urid, Oil seeds, etc. Apart from food crops this region is important for horticulture. Almost all villages have mango orchards and other fruit trees like Jamun, Kathal, Sarifa etc. These days seasonal fruits are also grown. The production of vegetable has remarkably increased. Cultivators grow two to three crops of potato, cabbage, couliflower, beans, carrot, tomato.

In terms of irrigation the area is not much resourceful. The hill streams remain almost dry except in the rainy season and do not offer as much scope for irrigation as the perennial rivers are absent. The Ahars (reservoirs), wells and danries (springs) are the only traditional sources of irrigation. The government has also developed some irrigation schemes for conserving rainwater by controlling the rivulets.

There is a radial network of transport which converge from all direction. Two national highways No. 23 and 33 pass through the region. Important national highways and state highways are Ranchi-Hazaribag road, Ranchi-Bahragera, Ranchi-Netarhat, Ranchi-Chaibasa-Jamshedpur, Ranchi-Daltonganj, Ranchi-Purulia, etc. This plateau has also good rail connection. It is linked with Lohardaga with broadgauge, Rourkela with broad gauge, Gaya with broad gauge, Chandrapura to Dehri-on-Son, etc. Rivers are not navigable. There is a good aerodrome at Ranchi.

Although Ranchi Plateau forms heart of the Chotanagpur highland known as store house of minerals, it is not very important. On the Ranchi Plateau itself some poor quality asbestos occurs near Itki. A series of Baryte veins occur in the granite gneiss, some 21 kms east of Ranchi. Important places are Silwai, Bahea, Bongabera etc. Some chromite occurs near Silli in Hotag Hill. Some galena occurs near Silli and Silwai. Beryl occurs in Pegmatite west of Kilkajaratoli. Cassiterite occurs in the southern slope of the hills of north of Paharsingh. The southern scarp is an important zone for steatite and talcschist.

Notable industries of the region include the heavy engineering works near Ranchi, handloom weaving of cotton in the villages, shellac industry distributed over the region, ball bearing factory at Ratu, Aluminim factory at Muri, Cement Industry at Khelari, high tension insulator factory at Namkum, electrical appliances at Tati Silbai, etc. As such the region contains varied land surface, it has also uneven distribution of population. In this region Ranchi district has the average density of 572 persons per sq. km. and Gumla district has the average density of 191 persons. Development blocks having very high density include Kanke (703) because of urban growth, Ratu (733), Mandar (540), Silli (393), etc. Developments blocks having low density include Lapang, Angara, Irki, Rania, Torpa, etc. The Ranchi district of this region has 1296 inhabited villages and 15 uninhabited village and fifteen urban centres. Important urban centres are Ranchi (10,73,427), Churi (24,876), Khelari (20,010) etc. Other towns are Itki, Khunti, Bundu, Muri, etc. Tribal population in Ranchi district is 35.76% and scheduled caste is 5.25%. Gumla district, of which some areas lie in Ranchi Plateau has tribal population 68.94% and scheduled caste 3.17%. Important tribes are Munda, Oraon, Kharia, Chikbraik, Gond, Mahali, Lohara, Kharwar, Bedia, Asur, etc.

The typical rural settlement is a consequence of relatively even land surface and absence of forest cover on the peneplain is less dispersed than in other areas. Virtually there are settlements of considerable size and look like fragments of compact settlements. Certainly nucleations remain so marked in the settlements of this region. The rural dwellings are mud walled and tile roofed. In the Ranchi district of the region there are 1296 inhabited villages out of which villages having less than 499 population are 297, 775 villages have population of 499-1999 persons and 224 villages have population more than 2000 persons.

The Pat Region :

The Pat region is located in the western portion of the state. Its area lies in the southern portion of Palamu and Garhwa districts and north-western parts of Ranchi and Lohardaga districts. Virtually this region consists of more areas of Palamu and Garhwa and smaller areas of Gumla and Lohardaga. This region has obvious physical demarcations in the north by the koel basin, the Sankh basin in the south, scarp of the higher Ranchi Plateau in the east and water divide in the west. Its area stretches between 23^{0}10' N. lat. to 35^{0}35'N. lat. and 83^{0}55' E. longitude to 84^{0}20' E. longitudes and comprises an area about 500 sq. miles and has about 04 lakh persons.

Geological formations of this region is complex. Forming parts of the ancient landmass it is supposed to have rock formations of early geological periods like gneiss, schists in the basement. Since the area got uplifted at the time of Himalayan formation the region along with Ranchi Plateau achieved higher elevation. This higher portion is supposed to have recorded upliftment during third stage of the Himalayan and witnessed lava cappings at the time of Deccan Lava outpours. This region formed outliers of the Deccan lavas. The lava covered land was elevated during the Tertiary period to attend its altitude and diversified relief. In the process of

Fig. 25

lava cappings and other structural disturbances basement rocks experienced acute metamorphism. As this region contains Archean formations as well as formation of later geological periods different type of rocks are found. Important rocks include granite, gneiss, schist, basalt, laterite, etc.

The impact of structure on relief is more obvious in this region. The successive upheaval of this part of the plateau has enabled high elevation as well as lava cappings have protected the rocks from erosion. However, the differential erosion has brought variations in landscape. This region is the highest area in the entire region. This region consisis of several smaller plateaus known as pats. These pats are separated by the tributaries of the North Koel, the Sankh, etc. These pats have maintained elevation ranging between 1000 to 1200 m. Pats enveloped by lavas are generally known as Mesas and Buttes. There are outliers of the Deccan lava and have maintained flat tops.

It has been noted that this region was elevated after lava cappings in tertiary period. In this process 300 m. elevation was achieved in the Early Tertiary Period and another elevation of about 300 m. was recorded in Middle Tertiary Period. The surface lavas have been weathered into laterites. The western part of this plateau has steeps escarpments of Bagru Pahar and Rudani Pat Mesas. The highest waterfall is Burhaghagh on the river Burha. Important higher Pats are Ganeshpur, Netarhat, Mamira which have elevation of about 1250, 1260 and 1225 metres respectively. Important hills are Shanu and Saryu which are dissected by the rivers koel and the Sankh. The fundamental geomorphic features of the region are steep scarps facing the west, narrow valleys, steep slope, etc. The general slope of the Pat region is towards the north and the east.

The Pat region is drained by a number of rivers. The Rivers originating from this region flow towards north, east and the south. Rivers draining towards north are the Nakti, the Bahera, the Chaupat, etc. which become Burha river in the north and after flowing some distance it merges with the River koel. This topographic region has also given rise to tributaries of the South koel and tributaries of the Damodar river wich flow towards south and east respectively. A tributary of Burha river also flows towards west. Thus altogether the river system present in the area provides a radial pattern. As such the slope of the region is steep, most of the rivers exhibit little curved courses. The eastern portion of the Pat region which is attached to western Ranchi Plateau acts as a divide between the north koel and the south koel basins. From its south-eastern loop till Jamira Pat, it forms the watershed between the Sankh and tributaries of the North koel. The rivers of this region forms a number of picturesque waterfalls.

The high elevation and dense forest cover have characterised climatic conditions slightly different from other regions. The area remains cold in the winter season and cool and warm in summer season. The average annual temperature is a little below 23^{0}C but the temperature sometimes comes down to 2^0 to 3^{0}C in the month of January. Similarly during summer season the temperature hardly goes very high due to elevation and vegetative cover. As a matter of fact the summer temperature hardly remains uncomfortable. Because of elevation and vegetation this region also records comparatively high amount of rainfall. The annual rainfall is about 180 cms. The concentration of rainfall however occurs in the month of June, July, August and September. During the monsoon months this region receives about 90% of the annual rainfall. July and August are the rainiest months of the region. In the months of December and January the region receives only 2 to 3 cms. of rainfall.

The impact of decomposition of parent basaltic rock has played dominant role in the formation of soils in the region. Most areas have laterite soils followed by black soils in smaller areas. The layer of soils is thin in most areas. Soil erosion is prominent feature.

The impact of relief, soils, climate and human interference is important in case of vegetal cover of the region. The region is covered with dense forest from Bahar (Outer), Barwe and Bhitar (Inner) Barwe to the other end of the region. The wet deciduous dominate with sal, gamhar, kusum, palas trees and tall bamboo clusters. The forested scarps and winding ghats are cleared by the natives for cultivation.

The area is marked by distinct land use pattern because of hilly terrain inaccessible narrow valleys, steep scarps etc. Even today there are areas which are very difficult in terms of accessibility and these areas have remained aloof of the native habitations. There is wide coverage of forest. The forest scarps and winding ghats have been partly cleared by the Asurs for cultivation. There are pockets of agriculture on the descending slopes but the valleys reflect paddy culture in low lands and dry crop culture on the uplands. Jhum and shifting cultivation is still practised in some areas cultivators abandon the cultivated fields for regaining fertility. Important crops grown here are paddy, maize, kulthi, arhar, beans, etc. Aghani paddy in the valleys but maize and kulthi are grown in the higher areas. These days people have developed interest in fruit trees specially guava, mango and other fruit trees. Production of vegetables has also considerably increased. Asurs prefer growing maize and millets. Apart from agricultural use of land there has been some areas involved in settlement, transport, waste land, etc. Some areas are engaged in bauxite mining in the vicinity of Lohardaga at Bagru, Pakhar and other Pats.

The Pat region is poor in terms of transport network. However metalled roads connect Netarhat with that of Lohardaga. There is a road connecting Mahuadanr, Netarhat, Anksi, Garu, Daltonganj, etc. Besides metalled roads there are unmetalled roads connecting important places, waterfalls, picnic spots, etc. The rivers are not navigable because of hilly terrain. The area lacks air strip also.

This region contains enormous bauxite. There are pats having good reserve of bauxite where mining is done. Important mines are located in the vicinity of Lohardaga at Bagru, Pakhar and other pats. A vertical section reveals laterite, kaoline, bauxite, interrupted bauxites over the denuded granite and gneiss beds.

In terms of industries the region is also extremely backward. The only metallic resource of industrial use is bauxite which is not in industrial use in this region. It is used elsewhere. Other industrial resource for industries are forest products and agricultural products. Forest based industries are also almost non-existent because of poor transport and other conditions. Valuable woods of this region are in industrial use elsewhere. The agricultural products and horticultural products are not developed on large scale which can give rise to the development of agro-industries. Virtually natives of this region have poor size of agriculture which hardly meet domestic requirements. There is however good scope for the development of lumbering and paper industries. There are small scale industries located in villages and rural service centres like making of agricultural implements, furniture and other domestic necessities.

As the region is hilly the pattern of population is sparsely and unevenly distributed. The development block of Mahua danr has population of only 74,732 persons (2011) having per sq. km. density of only 117 persons. This development block consists of only 3.2% scheduled caste population but the area has considerable tribal population i.e. 79.93%. There is only one statutory town named Latehar which has a population of 4017 persons in 2011. Part of Lohardaga lying in this region has the density of 307 persons per sq. km. which is more accessible and developed agriculturally. Bishunpur block of Gumla which lies very close to Netarhat has the density of only 102 persons per sq. km. In this block the scheduled caste population is only 1.64% and the tribal population is 89.14%. The Chainpur block has the density of 120 persons per sq. km. This block contains 1.37% scheduled caste population and 81.85% scheduled tribe population.

The development block of Mahuadanr has 101 inhabited village out of which 8 villages have population size less than 200, 28 villages have population ranging from 200-499, 44 villages have population from 500-999 and only 21 villages have population more than 1000 persons. Similarly the development block of Bishunpur has only 68 inhabited villages and more villages are of smaller size. Most of the villages form scattered type of settlements. Netarhat is the smallest town located on a pat of same name. It is linked with Pakri Pat on the South. It is noted for educational as well as tourist centre. There is a wooden chalet of the Governor (now principal Bungalow) and the Dak Bungalows. Sun rise on its south-eastern face is famous and Maghnolia point on its opposite rim is reputed for sun set.

The Subarnarekha Basin :

The Subarnarekha region lies in the south eastern portion of the state. Although the dimension of the basin is vast as the river Subarnarekha originates south of village Nagari, 15 km west and south-west of Ranchi city and pursuing a winding course to the east till it descends from the plateau forming the steep fall of Hundrughagh and then turns south and enters the district of Singhbhum and forms the main valley. This river forms two distinct basins the upper basin and the lower basin. Most areas of the upper basin lies in the district of Ranchi having its course in the Ranchi plateau and the Lower Ranchi plateau. The lower basin lies mostly in Purbi and Paschimi Singhbhum. The region as a whole is bounded by the Trough Basin of Damodar in north, the Kolhan upland in South, the Panch Pargana Plain and the Dalma Range in East, the Porhat highland and the Ranchi Plateau in the West. This region comprises some areas of Ranchi, Purbi Singhbhum and Paschimi Singhbhum. The region extends roughly between 22°10' N. Lat. to 23°30' N. Lat. and 85°15' E. Long. to 86°55' E. Long.

The river basin covering areas of Ranchi Plateau, Lower Ranchi Plateau, Porhat-Dalma Range and Dhalbhum plain provides complex geological formations, varied landscape, differential drainage, economic activities, industrial development, human habitation, etc. Virtually the river passing through the higher and

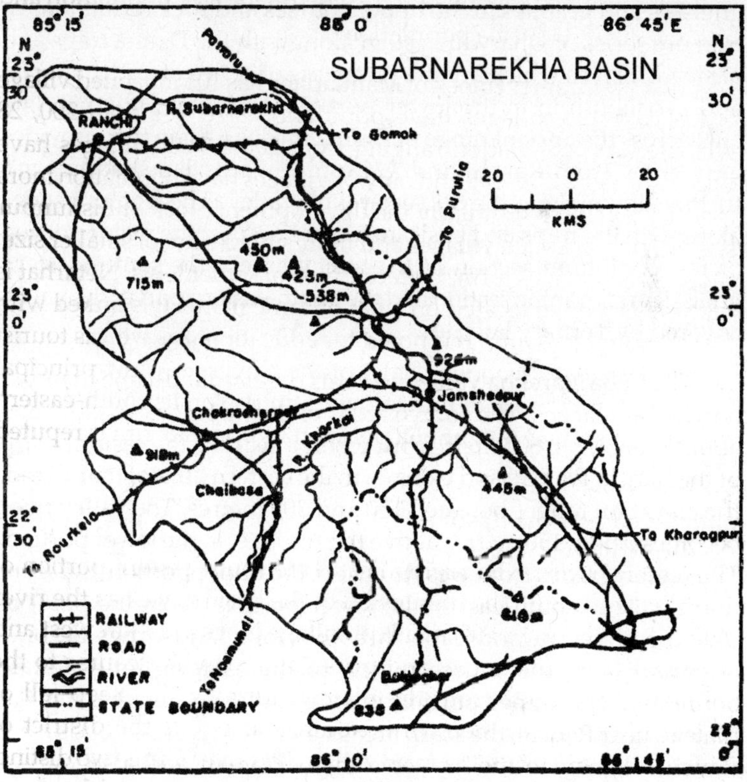

Fig. 26

the Lower Ranchi Plateau, Dalma range, Tertiary gravel region portrays variations in rock formations, occurence of minerals, soils, vegetative cover agricultural activities, growth of industries based on different metallic minerals, settlement of different types etc. The source areas of the basin have crystalline rocks, which have been metamorphosed. Important rocks consist of granites consisting of coarse to medium size crystals of different minerals, gneiss, schist, limestone etc. When the valley descends to the lower portion of the Ranchi Plateau, the region again consists of the same type of rock formations. Near the Dalma range where the river cuts the highland of Dalma to maintain its course consists of rocks of volcanic eruptions. In the lower portion of the basin there lies rock formations of Dharwarian system having the oldest sedimentary rocks which have been highly metamorphosed and contain good reserve of metallic minerals. Virtually the region is composed of gneisses and granites in the upper courses and rock formations of iron ore series of Dharwar system alongwith the Dalma traps and basic intrusives which are the remnants of series of anticlinorium and synclinorium in the lower portion. The river Subarnarekha cuts across the ancient range of Dalma through a gap that provides easy entry from Ranchi and Asansol towards Jamshedpur and further down along the Subarnarekha valley. The iron ore series alongwith the traps and basic intrusives occupy the major portion of the Dhalbhum section of the valley except the south-western gneissic and granitic section and the south-eastern part of the valley covered by Tertiary laterites.

The Subarnarekha valley presents diverse landscape in upper and lower reaches. The source areas of the basin has an elevation about more than 600 metres but when it reaches the eastern scarp of the Ranchi Plateau and extends to the eastern limit of Jharkhand, the elevation ranges between 150 and 300 metres. The valley floor becomes below 150 metres above the sea level in the lower portion. The region grades from west to east in the upper portion and from north west to south-east in the lower portion. From the point of view of physical landscape three obvious landscapes become apparent from the source region to the eastern limit of state boundary. The upper portion is an undulating area having flat plateau near Ranchi, the scarp areas where the river descends from Ranchi Plateau to the Lower Ranchi Plateau and also form a waterfall named Hundrughagh. The Central portion consists of relatively homogenous landscape upto the Dalma range which provides rough terrain and also separates the lower portion from the central part of the basin. The average elevation of Dalma range is more than 300 metres but the river Subarnarekha has made narrow passage across the Dalma range. The lower portion of the basin is considerably homogeneous with isolated hillocks.

This region is drain by the river Subarnarekha and its tributaries. The principal tributary of the Subaranrekha is Kharkeai which is formed by the junction of two mountain streams rising in the eastern Kolhan range of hills, namely Terlo and the Koranjai, of which the latter forms for about 27 km boundary between the Kolhan and the Mayurbhanj. The Terlo joins the Koranjai on the boundary, and some six km lower down at the tri-junction point of Seraikela, Mayurbhanj and the Kolhan, the river is known as the Kharkei. It continues in a north-westerly direction, forming the boundary of the Kolhan and Seraikela with several sharp bends one of which brings it within 8 km to the north-east of Chaibasa. A few kms. further north, at a point opposite the solitary conical Hindu Hill (317 m.) it ceases to be the boundary and enters Seraikela running north and then east till it debouches in the Subaranarekha on the boundary of Kharsawan and Dhalbhum. The river Kharkai is joined by the river Sanjay near Lengtasai about eight km south of Gamharia station on the south eastern railway. The river Kharkai is also fed by several streams from the Kolhan upland among which may be mentioned the Iligara, the Jamiragera and the Roro. The Roro or the Raro is about 54 km in length and is joined by the Jamira just outside Chaibasa, which is situated on its eastern bank. The sanjay rises in the forest clad hills of Porahat, north-west of Sonua. It flows in an easterly direction and falls into the kharkai river near Lengtasai. The Sanjay receives no streams from the south but from the north it receives the Binjai which drains Karaikela, the Sankua from kera and several tributaries from Kharsawan. The tributaries joining the river Subarnarekha from eastern side in the Central plain include Garra Nandi, Loubhang, etc. Teh Garra Nadi breaks through the Dhanjori range separating the Narwa pahar (350 metres) from the Rangamati Pahar (607 metres). It then turns south-east and flows parallel to the Dhanjori range and joins the Subarnarekha opposite Rakha mines, west of Mahulia. The Loubhang joins the Garra Nandi where it turns south-east after flowing out of the gorge. Another tributary of the Garra is the Dudh

Nadi which drains the Dhalbhum plain and joins the Garra before the later enters the gorge between the Narwa Pahar and Rangamati Pahar on the north side of the Subarnarekha river the tributaries are the Gurma Nadi, Bagalduta Nadi and the Dimrajhor Nadi. In the extreme south-east the river Subarnarekha is joined by the river Sankh about 3 km south-east of Musabani.

In terms of climatic conditions the upper and the lower portions of the basin present slight difference because of differential elevation, vegetative cover and proximity to the sea coast. The rainfall decreases from the lower portion to the upper portion blurring summer season as well as during winter season. During winter season the average rainfall of this region ranges between 1.5 to 2.5 cms. and during summer season the rainfall occurs between 10 to 15 cms. During rainy season the rainfall increases from lower to upper portion and it ranges between 80 to 90 cms. The mean temperature ranges between 80^0 F to 88^0F during summer season and it ranges between 80^0 to 84^0 F during rainy season. The temperature ranges between 65^0F to 75^0F during winter season.

The Subarnarekha basin contains soils varying from the upper reaches to the lower based on varied parent materials. The upper portion contains lateritic soil while the lower portion contains the gravely soil. South of Dhalbhumgarh, the gravel country of red soil dominates till the narrow alluvial fringe along the Subarnarekha gorge appears near Baharagora.

The basin portrays spatial variation in vegetative cover. The lesser undulating lands in the lowest portion are almost devoid of forest. In these areas isolated trees and smaller patches of natural vegetation break the monotory of the area. Certainly orchards close to the settlements are frequently found. Valley plants and the hilly terrains have considerable forest cover. The wooded hillocks show sal and mixed forest of deciduous species. The hills in the Dhalbhum, made of sedimentary rocks, have thick forest. Important trees other than Sal are gambhar, mahua, semal, sisam, neem, bamboo thickets, etc.

The Subarnarekha basin presents diversity in landuse pattern. In the upper portion of the basin, especially in the source areas, agricultural land dominates with patches of forest cover and human habitation. In the scarp areas the percentage of land under vegetation increases and culturable land coupled with homestead land decreases. Again the percentage of culturable land, fallow, pasture and non-agricultural use of land increase in the central portion till the Dalmarunge. In the Dalma area the culturable areas and built up areas sharply decline because of the increased percentage of vegetative cover on the hilly areas. In the lower areas, industrial and mining uses of land and dense network of transport, habitation accelerate the non-agricultural uses of land. This area is, however, important for agriculture point of view. There are three kinds of land namely Gora, Bad and Bera. Bera lands are those which are at the bottom of the valley and in depressions which receive the washings of the slope and are naturally irrigated. They are richest of all the lands, yielding good crops of winter rice followed occasionally by linseed, khesari and other pulses. The land just higher up the slopes are called Bad lands and grow early rice, cereals, pulses and miscellaneous crops. The uplands which are composed of light soils are known as Gora lands which are situated close to the villages. Owing to their proximity to the villages they are well manured and cultivated to give two to three crops annually. Crops are Surguja, Mung and Urid. In the lower portion, the lands are divided into Bahal, Kanali and Bad. Bahal means the best rice lands situated at a very low level. Kanali or Nali means rice lands made in beds of small Nala or rivulets. Bad lands are terraced rice lands which are generally located at high level next to the Gora or uplands. Bad Lands also include some lands growing crops other than rice like sugarcane fields and Kalimati which grow vegetables and other valuable crops. Besides these there are Bastu or homstead lands, Udbastu or cultivated bari land or uplands growing cereals and pulses.

The areas of this basin consist of geological formations of Dharwarian system and other older geological periods which contain metallic and non-metallic minerals. The iron ore series formed incourse of ancient sedimentation is noted for huge reserve of superior quality haematite iron ore and other metallic minerals like manganese, copper, etc. The banded haematite, quarzites crop out as ridges, the most important of which are arranged in the form of a narrow horse shoe open to the north and close to the south. The ridge forming the western side of this horse-shoe is known as the iron ore range. In this area there are several mines containing huge deposits of iron ore. Important iron ore mines are Gua, Manoharpur, Noamundi, Jamda, Lukidburu, etc. Iron ore reserves of Gua and Manoharpur have iron ore containing 60% of iron. Iron ore fields of Noamundi contain iron ranging from 59.65% to 66.90%. Much of the haematite found in the Porahat region occurs as a fine powder. The deposits are scattered and impersistent. The iron-ore mines of Lukidburu contain bands of haematite quartz schists inter bedded with phyllites. Some of the former are almost iron ores. The haematite schist and haematite-quarzites crop out in the ranges which form the boundary between the Dhalbhum and Manbhum. Another belt of ferruginous schist varying to banded haematite-quarzites and iron ore extends eastwards from the kharkai river at Samran in Seraikela. Though the region is not very rich in mangnese but its deposits occur both in the iron ore series and the Kolhan series. The largest deposits are of the lateritic type which may be also of high grade. At one place near Gua, an outlier of the basal Kolhan series conglomerate contains a little manganese.

The area is rich in copper. The copper belt extends from Duarparm to the north-east of Chakardharpur. Structurally there are two types of copper ores, a fine grained type and a coarse grained. The former type is much higher in Pyrrholite and Nickel minerals and is often full of fine unreplaced gangue minerals such as quartz, biotite, chlorite, etc. The latter type is usually richer in copper. Important mines are at Mosabani, Badia and Dhobani. The river bed of Subarnarekha contains gold alongwith sand. Asbestos is found in several places in Seraikela and Dhalbhum. The largest deposits are at Barabana and Rajnagar in Seraikela. A few atomic minerals are also found in this basin. Chromite deposits are also found in this region. Along the northern side of the copper belt kyanite occurs associated with kyanite-quartz, granulite and aluminous mica-schists along a belt about 120 km in length.

This region has recorded phenomenal increase in industrial development, three centres consisting of Jamshedpur, Ranchi and Ghatshila. In the upper Subarnarekha basin Ranchi and Muri have given rise to several metallic and non-metallic industries and forest based industries. Hatia has become the focus of Heavy Engineering Corporation. Ratu is noted for ball bearing and Muri is famous for aluminium industry. Among the forest based industries lac industry takes leading role. Jamshedpur has emerged as important industrial centre because of Tata Iron and Steel Industry and several other industries related to Tata Company. Ghatshila has unique position in terms of copper based industry.

Exploitation of minerals and development of industries have caused remarkable growth of transport. The area is linked with main metro cities of the country like Kolkata, Mumbai, Delhi, Chennai, etc. The main line connecting Kolkata and Mumbai passes through this region. Almost all iron ore and copper mines are linked with rail route which facilitate transportation of minerals and the movement of people. There are National and State Highways. The alignment of National Highway No. 33 covers Chaibasa, Chakardharpur, Ranchi road. Another National High passes through the south-eastern portion touching Baharagora. There are several state Highways which connect Ranchi-Jamshedpur-Rajnagar-Chaibasa, Jamshedpur-Seraikela etc. Rivers are not navigable. At places only ferry services are made available. Air strips are available at Jamshedpur and Ranchi.

As regards population characteristics this basin presents spatial variation. Areas around Ranchi and Jamshedpur have relatively dense population because of the impact of urban and industrial development. Areas lying between the eastern scarp of the Ranchi plateau and the Dalma range contain sparse population. In the south-eastern portion there is comparatively dense population based on homogeneous fertile land. The average density of Dhalbhum region ranges between 600-800 persons per sq, km, In the upper portion of the basin there is about 50% scheduled tribe population but in the lower portion it is about 45%. In Purbi Singhbhum, the share of rural and urban population is 44.44% and 55.56% respectively. In Ranchi district the share of rural and urban population is 56.86% and 43.14% respectively. In Purbi Singhbhum, the literacy is 75.49% whereas in Ranchi district the same is 76.06%. Important tribes of this region are Hos, Santhals, Bhuiyas, Bhumijis, Oraons, Mundas, etc. The concentration of Hos are marked in Kol and Seraikela. Bhuiyas are more in number in Seraikela and Kharsawan. Santhals are more concentrated in Ghatshila and Seraikela. Bhumij are well distributed in Ghatshila.

The area is marked with smaller nucleations in the lower basin and relatively dispersed settlements in the upper basin. The settlements are almost on the crest of gently undulating ground, but on very rare occasions they also build houses on the slopes or on the top of the hillocks. This is done for two reasons : the site remains dry during the rains while the lower fertile lands are more profitably utilized for agriculture. The settlement of the Kharias and Birhors are close to the jungles while the settlements of Hos, Santhals, Bhumij, Mundas who are in considerable number have settled close to agricultural land. In remote rural areas mud walled and thatched houses are in good number. Certainly these days tile roofed houses are becoming popular. Houses near the urban centres are mostly brick walled and tile roofed. The no. of houses owned by a family vary from one to four depending upon the economic condition and the size of the family. Important towns of this region are Ranchi, Jamshedpur, Ghatshila and Muri. Virtually Ranchi urban agglomerations has population of 10.73 lakhs while Jamshedpur urban agglomeration has the population of 6.77 lakhs.

References :

1. Ahmad, E. (1958) : Geomorphic outline of Chotanagpur, Geographical Outlook, vol. II, p. 17.

17

Geographical Background of Districts of Jharkhand

1. Bokaro District :

Bokaro district lies in the eastern portion of the Jharkhand state. It is bounded by the district of Giridih, Hazaribah and Dhanbad in north, West Bengal in South, Dhanbad in east and Ramgarh in west it comprises an area of 2883.00 sq.km and has 20,61,918 (2011) persons. This district came into existence on 1st. April 1991 with headquarter at Bokaro. This district possesses 2 Sub-divisions viz. Chas and Bermo, 9 C.D. Blocks and 2 statutory town and 25 census towns. There are 635 villages in the district.

This district forms the part of Lower Hazaribag Plateau, Panch Pargana and Damodar Basin. It consists of areas less than 300 metres in height. More or less the land surface exhibits little undulations. However a few smaller hills and riverlets waterfalls provide picturesque landscape. The river Damodar passes through the central parts of the district. Another important river is Jamunia which joins the river Damodar near Chandrapura. The river Ganga joins the river Damodar from the South. The general slope is from north-west to South-east.

This area provides climatic conditions of slightly different nature because of less elevation, less forest cover and development of industrial activities. As such the summer becomes hot and the winter season records low temperature. The highest temperature is recorded as 46°C. In winter season temperature comes down to 3 to 4°C when cold wind blows otherwise the temperature ranges between 10 to 15°C during winter months. The average rainfall is 157 cm but the highest rainfall ever recorded has been 195 cm.

The vegetation of this area has been considerably affected by the mining and industrial activities and growth of urban centres. However, isolated trees of Sal, Mahua, Gamhar, Semal, etc. dotted in the district. In addition to these isolated trees, there are patches of bushed which reveal merciless exploitation of vegetation in the area. The area has Domat Soil and sandy soil because of the sedimentation of Damodar river and its tributaries.

Although the area has less undulation suitable soils and favourable climatic conditions for agriculture but considerable growth in mining, industries and urban centres most areas have become waste land specially in mining areas. Industrial and urban areas have engaged substantial areas under non-agricultural uses of land. This district has engaged vast area in residential, industrial, commercial, development of transport network and other uses. However plantation has been developed in the vicinity of industrial areas to minimize the impact of industrial pollution. People residing in neighbouring villages have developed vegetable in the market. Virtually food crops are not given due importance

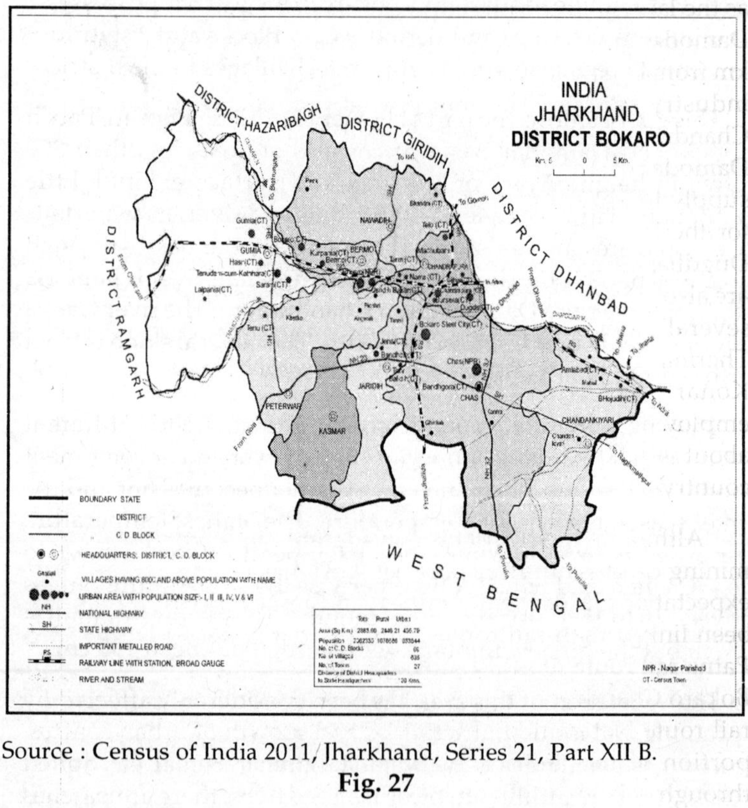

Source : Census of India 2011, Jharkhand, Series 21, Part XII B.

Fig. 27

because of urban demand of vegetables, fruits and dairy products. A good no. of people are also engaged in dairy activities.

The area is rich in mineral resources. Coal is the principal resource. Bermo region provides the best quality of coal. It has several important coal mines like Thori, Ghato, Kathara, etc. Bokaro-Kargali Region produce 24.8 lakh tons of coal. In addition to coal the area has also china clay and inferior quality of iron.

This district may be said to be the most industrialised in the state because the entire area provides industrial look having steel plant, coal washeries, thermal power stations and other industries. The site of Bokaro for steel plant was selected in the year 1952 by German specialists and in the year 1964 Bokaro Steel Ltd. was established. Russian government agreed to provide financial and technical assistance on 25th. January 1965 and this plant emerged as the largest steel plant in Asia. This plant is located on the river Damodar and state highway connecting Dhanbad and Ranchi 48 km from Dhanbad. Gumia of this district has the largest explosive industry of Asia. Thermal power stations have developed at Chandrapura, Bokaro and Tenughat. The earthen dam on the river Damodar at Tenughat is the largest in Asia where from water supply to Bokaro is made through canals. Bhandaridih is noted for the largest refractory in India. There are Coal washeries at Dugdha, Kargali, Swang and Kathara. Coal washeries of this area are also of very large size. In addition to these industries there are several other industries Konar dam provides water to Bokaro Thermal Plant. Bokaro Thermal Plant is located by the side of Konar river. It has five units. Bokaro Steel Plant has been employing over 60 thousand workers. Its annual production is about 40 lakhs metric tons. It produces 60 % steel sheet of the country.

Although, the district is considerably industrialised and has mining centres but transport net work has not developed as per expectations. Most of the mines and important industries have been linked with rail routes. This district is linked with Ranchi-Patna rail route. It is also linked with Kolkata. One rail route via Bokaro Chandrapura goes to Bankura-Mednipur. In addition to rail route National Highway no.2 passes through north-eastern portion of the district. National Highway No.32 also passes through this area. There are other roads connecting towns and service centres.

The area has dense population not because of agriculture but because of mining and industrial development. The average density is 715 (2011) persons per sq.km but there are development blocks where density is above 1200 persons per sq.km Bermo development block has the density of 1742 persons per sq.km. The Chas development block has the density of 1214 persons per sq.km. This district has the urban population of 9,83,644 persons. Important towns are Bokaro Steel City, Urban agglomeration consisting of Bokaro Steel City and Chas having population of 5,56,460 persons, Phusro urban agglomeration consisting of Bermo, Jaridih Bazar (C.T.), Phusro (N.P.), Chandrapura (C.T.), Bokaro (C.T.) and Kurpania (C.T.) has population of 1,85,555 persons. Gumla (C.T.) and TanuDom Cum Kathara (C.T.) has population of 48,141, 27,425 and 22,080 persons respectively. The areas has 14.51% Scheduled Caste and 12.40% Scheduled Tribe. Important tribes are Munda, Oraon and Santhals.

2. Chatra District :

The district of Chatra is located in the lower Hazaribag Plateau. It is bounded by the district of Gaya in north, Ranchi in south, Hazaribag in east and Palamu and Gaya in west.

It comprises an area of 3718 sq. km and population of 10,42,886 persons (2011). For the administrative convenience, the district has only one sub-division e.g. Chatra sub-division, which has 12 C.D. Blocks. In the entire district, there is only one statutory town (Chatra M) and one census town (Bachra). There are 96 Panchayats and 1471 villages in the district.

No doubt this district has recent origin being bifurcated from the district of Hazaribag but Chatra headquarter has old administrative headquarter in the form of Commissionary form 1771 to 1780. When Ramgarh district was formed in 1780 this centre remained as district headquarter and it continued upto 1834 when district of Hazaribag came into being and headquarter of the district was else shifted at Hazaribag. Afterwards this place remained important trade centre. In the year 1940 Chatra became subdivision of Hazaribag District. As a matter of fact Chatra town had got the status of Municipality in 1869.

Since this district consists of parts of Upper Hazaribag Plateau and Lower Hazaribag Plateau and northern scarps it presents diverse topographic features. It has an elevation of about 450 m. Kalua hilly area (520 m.) and Lahabar Hilly form higher parts of this district. The northern portions has scarp area. The land slopes from south to north. This region has given rise to a few rivers.

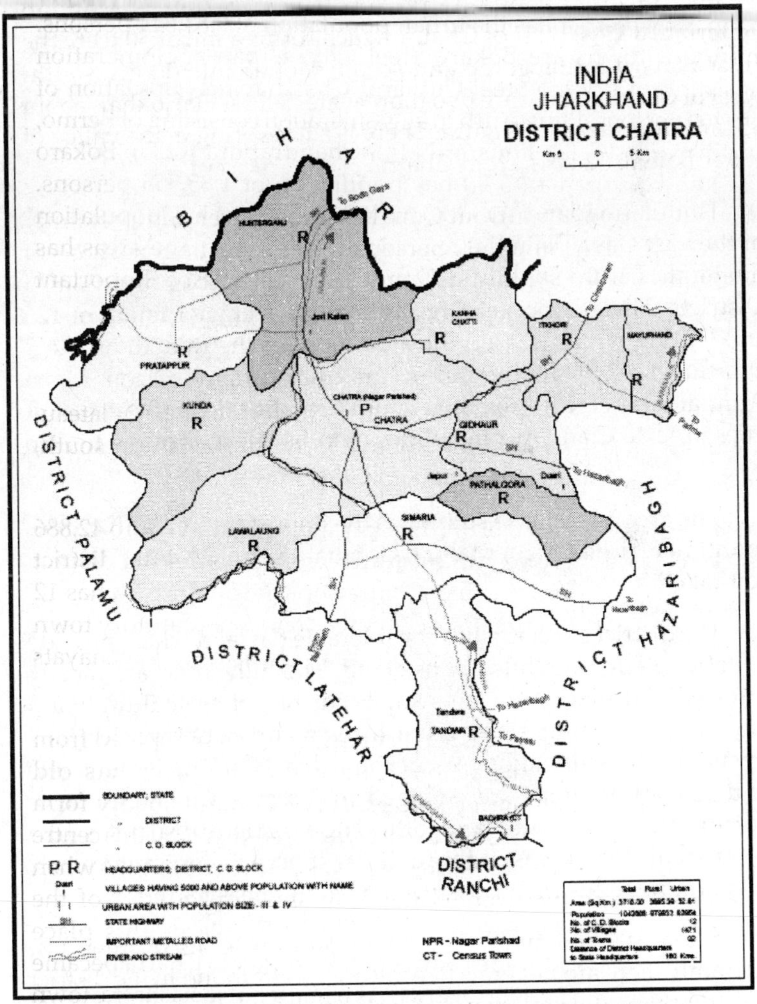

Source : Census of India 2011, Jharkhand, Series 21, Part XII B.

Fig. 28

Important rivers draining the district are Yamuna river, Barki river, Chako river at Garhi river etc. The river Yamuna flows in the north-east portion while the river Barki flows in south-east portion. The river Chako flowing in between Yamuna and Barki rivers joins the river Koel. The river Garhi joins the river Damodar near Tandwa. These rivers form deep valleys and waterfalls. The waterfalls of Keridah is located in between two hilly ranges, nearly 8 kms. north-west of Chatra. Another waterfall is Mahdah waterfall, 8 kms. west of Chatra which forms a fall of 50 feet. The river Banaru forms gorge and scenic spot about 10 kms. south-west of Chatra. There are two more waterfalls at Hariokhar, about 35 kms. north-east of Chatra. There is a hot spring in the valley of River Balbal, 32 kms. from Chatra.

Undulating land surface, partial forest cover and proximity of the vast Gaya plain the region provides climatic conditions of moderate nature slightly different from Hazaribag and Ranchi Districts. This district receives an annual average rainfall of 125 cm. During winter season the area receives the rainfall of 1 to 2 cm. But most of the rainfall occurs during rainy season. Mean Annual temperature remains about 25^{0}C but during summer it goes upto 46^{0}C and during winter months it comes down to 2^0 to 3^{0}C.

The area has laterite soil. Sal is the important tree. Other trees found here are Mango, Mahua, Gamhar, Kendu, Karanj, Jamun, Kathal etc.

The district of Chatra has considerable flat land which provide suitable site for agricultural use land. The hilly areas are mostly forested and scarp area have small patches of agricultural land and the remaining areas have forests and rocky projections unsuitable for cultivation. As such less than 50% land is cultivable and remaining areas are engaged in forest cover, waste land, culturable waste land and other non-agricultural uses like residential areas, areas engaged in transport and communication, playground, educational institutions, religious sites, etc.

The area is relatively less resourceful in terms of minerals. Certainly there are minerals like mica, coal etc. found in the district but commercial production has not developed. Even this district is devoid of major manufacturing industries. There are small scale industries related to bronze, lac, timber etc.

From the point of view of transport and communication this district may be considered backward. The area is devoid of rail routes. Even national highway does not pass through the area. There are a few state highways and other metalled and unmetalled which connect Chatra with other towns of the State. These roads include Chatra-Daltonganj, Chatra-Ranchi, Chatra-Chauparan, Chatra Hazaribag, Chatra-Hunterganj-Dobhi where it joins the Grand Trunk road. Chatra is also linked with Gaya and Lohardaga. There is no waterways and airways facilities.

The district of Chatra exhibits uneven distribution of population caused by heterogeneous land surface. Certainly flat plateaus have considerable habitation. The area is dominated by rural population having only one urban centre of Chatra (49,985 in 2011) persons. The district has only 62,954 urban population. Average density of the district is 281 persons per sq. km. There are 1377 inhabited villages and 94 uninhabited villages. Settlements are of scattered type. The area is inhabited by tribal and non-tribal population both. Tribals concentrate in hilly areas and non-tribals reside in flat plateau. Important tribes are Munda, Oraon, Kharia etc.

Important places of religious importance are Gharsimar having temple of Shiva, Vishnu, Ganesh and Durga nearly 6 kms. from Satgawa. There is an old temple of Kali on the peak of Kalua Pahar, 15 kms. west of Hunterganj. This site has several Jain statues also. There are caves of 17th and 18th century near Kunda village.

3. Deoghar District :

Deoghar district is located in the western portion of Santhal Parganas. It is bounded by Banka district in north, Dumka in east, Giridih in west and south by Jamtara district.

The district extends between 24°29' N. latitude to 24°48'N. latidude and 86°42' E longitude to 86°7' E longitude. It comprises an area of 2477 sq. km. It has a population 14,92,073 persons in 2011. It consists of 10 development blocks namely Deoghar, Madhupur, Mohanpur, Sarwan, Palolori, Sarath, Karon, Sona Rai Tharhi, Devipur and Margo Munda.

Deoghar has been named after the home of the Gods in a

modern name. In Sanskrit word other names like Baidyanatha or Hardapitha or Ravanavana are mentioned.

The district contains several clusters of rocky hills covered with jungles, and series of long ridges with intervening depressious. Most of there rolling uplands have been devided and are cultivated by highland crops. However, there are hills and

Source : Census of India 2011, Jharkhand, Series 21, Part XII 'B'.

Fig. 29

undulting tracts like Phuljori, Teor, etc. Their isolation makes them prominent and they stand up boldly, breaking the monotony of the land scape. Phuljori hilly range is about 750 m. high and Teor is 670 m. high. Degaria is another hill having 575 m. high, four km west of Jasidih. The average relief of the district is 247 m. above sea level. The general slope of the district is from north-west to south-east. Mostly crystalive rocks predominate the area. The undulating land surface with isolated peaks have given birth to a number of snake like rivers twisting over the entire district. Important rivers are the Ajay, the Paltro, etc. These two rivers merge together south of Sarath and the river Ajay becomes more wide. These rivers gather a large number of tributaries which form landscape full of Tarns and Dons.

As the region is not so high it has got climatic conditions not much different from the plains. Certainly rocky exposures become more hot during summer and relatively attain cold conditions during winter season. Winter months are very cool but the heat from the end of March to middle of June is severe and hot westerly winds are extremely disagreeable. During winter season average temperature remains 15°C and during summer it goes up to 35°C winter rain is negligible and average annual rainfall is 135 cm Most of the rain occurs during rainy days.

The area has recorded considerable decline in forest cover. Only rocky hills have vegetal cover but the gentle sloping land and the plain have been deforested.Important trees are Sal, Semal, Sisam, Mahua, etc. During recent years some hilly areas have received afforstation especially Teak, Sal, Sisam, etc. The hilly areas contain lateritic soils but the plains and valley have predominace of clay and mixed soils.

In case of land use considerable changes have taken place. Areas involved in forests have tremendously decreased especially on gentle sloping land and in the plains. Where people have developed agricultural land and other uses. In hilly areas there are still land under forest and remaining areas have waste land due to soil erosion and rocky projection, pasture, fallows, land under transport and residential uses are also available. In the plains more areas are given to cultivation, residential use and other non-agricultural uses. There are several religious sites which have also engaged some land.

The area is not significant in terms of mineral resources and industrial developments. However this district has got the presence of inferior quality of coal. Jayanti Coal field is 10 km south of Madhopur. Lead ores principally argentiferous galena exist at Behroki near Deoghar. This district has got a few industries based on forest resources and agricultural products. During recent years the development of Industrial Estate has encouraged a growth of a few industries.

Since Deoghar is a very old religious site it has got relatively better transport facilities. The rail route connecting Delhi-Howrah passes through almost central part of the district.Branch line links Jasidih with the district headquarter of Deoghar. Another branch line connects Madhupur and Giridih. The road transport is also well developed. Threre are roads connecting Deoghar-Dumka, Deoghar-Jamtara, Deoghar-Chakai, Deoghar-Godda, etc. Rivers are not navigable becuase of rolling land. This district also lacks air facility.

This district is moderately populated. The average density of population is 602 (2011) persons per sq. km. The development block of Devipur has the density of 401 persons per sq.kms. This district has 2354 inhabited villages and 308 uninhabited village. There are two towns - Deoghar (2,03,123), Madhupur (55,238) persons. The area has 12.74% scheduled caste and 12.13% scheduled tribe population. Important tribes of this area are Maler, Santhal, Malpaharies, etc.

This district has several religious sites of immense importance. Deoghar is noted for Shiv temple thereby Deoghar has commonly been known as Babadham. Virtually most of the activities performed at Deoghar concern religious funcions. Several lakh people from different parts of the country and abroad come here for worship. Other religious sites are Nandan Pahar, Trikutanchal Parabat, Tapoban, Navlakha Temple, Yog Ashram, Hindi Vidyapith, etc. Deoghar has also the centre of Dabour Medicine. On account of religious segnificance of Deoghar it has become the cluster of Pandas who participate and manage the religious affairs.

4. Dhanbad District :

The district of Dhanbad is located in the eastern portion of the state. It is bounded by Giridih & Jamtara in north, west Bengal

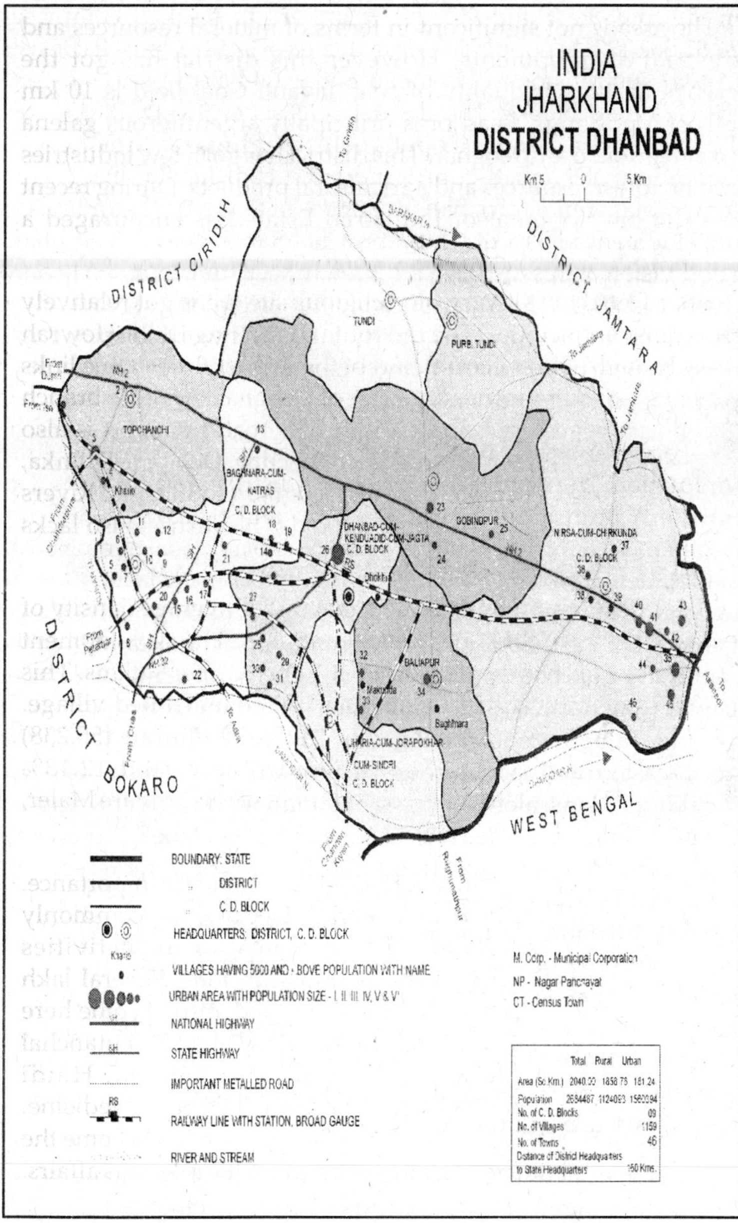

Source : Census of India 2011, Jharkhand, Series 21, Part XII 'B'.

Fig. 30

in South & East and Giridih & Bokaro in west. It comprises 26,82,662 (2011) population over an area of 2040 sq.kms. It consists of two subdivisions - Dhanbad Sadar and Baghnara and nine development blocks. The district consists of 9 C.D. Blocks viz, Tundi, Purbi Tundi, Topchanchi, Baghmara cum Katras, Gobindpur, Dhanbad cum Kenduadin cum Jagata, Jharia cum Jorapokhar cum Sindri, Baliapur and Nirsa cum Chir kunda. Two development blocks of Chas and Chandan Kayari of Baghmara Subdivision of this district have already been transferred to Bokaro district in 1991. This district came into origin on 1st.November. Its nomenclature concerns paddy land. The word Dhanbad has been coined from Dhan+Wayed meaning whereby paddy and paddy land of inferior order.

Topographically it is a lower plateau having relatively little undulation. The district consists of two district physical units. The northern undulating areas are hilly and forested while the Southern areas provides an appearance of plain. Coal mining areas exhibit depressions by the subsidence of the upper layer after the extraction of coal beneath. The highest peak of the area is Dhangi Pahar (1265') situated in northern hilly areas. Important rocks found are quartzite, gneiss, schists, etc. The general of slope of land is from north-west to south east. Important rivers of this area are the Damodar and the Barakar. The river Barakar forms the north-eastern and eastern boundary of the district. This river flows for about 76 kms. along the district boundary. The river Damodar formerly known as the river of sorrow has become boon for this region due to control over flood, construction of water reservoirs, afforestation along the river bank, development of hydel power, etc. Several dams have been constructed for water reservoirs. The water reservoir of Topchanchi is important one. There is another important dam near Panchet. In addition to these there are several tanks like Chapa, Bahudih, Bhara, Ainchas, Ashanlia, Pandara tanks in Nirsa etc. There are a few tributaries and springs. River Khudia joins the river Damodar near Chirkunda. The river Munia and Katari also join the river Damodar.

Being located in the eastern portion of the state and nearer to the Bay of Bengal and also having less elevation the area provides climatic conditions slightly different from the higher plateau areas of the state. During winter season, temperature ranges from 8.4^{o}C to 34°C and during summer season temperature ranges from 13.3°C to 45.5°C. During rainy season the temperature ranges between 15 and 36°C. Average annual rainfall is 127 cm.

This district has relatively less forest cover. Forest is confined in the northern hilly areas. Important trees are Sal, Sisam, Gamhar, Kend, Mahua, Bhelwa, Imaly, etc. Most of the trees in mining and industrial areas have been cut but during recent years afforestation has been done to minimize the pollution. The area has clayey and Domat soils in the lower areas and Sandy and lateritic soils in the higher areas.

No doubt the area is favourable for agriculture especially in the southern parts but because of excessive ruining most of the agricultural land have turned into waste land due to subsidence of coal mines and deposition of industrial waste materials. Even workers are more in non-agricultural activities. Certainly people have developed interest in growing more vegetables and seasonal fruits for supply in urban areas. Amongst food crops paddy, maize and wheat are important. There are minor irrigation schemes also. Other important uses of land are under mining, industrial, residential, transport, etc.

This district is noted for coal resources. There are over 200 coal mines in which more than one lakh people are engaged. Jharia Coal mine is the richest in good quality of coking coal. This region supply coal to the iron and steel industries of Rourkela, Bhilai, Durgapur, Bokaro, Jamshedpur, etc. There are many coal washeries also like Dugdha, Bhojudih, Gidhi, Pathardih, Kangali, etc. In addition to coal, chinaclay is also found in this area.

There is a good network of rails and roads. Grand Trunk road passes through the district. There are also other roads like Gobindpur-Chas-Jamshedpur (75 km), Gobindpur-Balia-Sindri (24km) Gobindpur-Tundi-Giridih (28km) Jharia-Baliapur (28km), Dhanbad-Pathardih-Sindri (26km) etc. This district has also good rail network. Important railroutes are Kumardhubi-Gomro, Pradhankhanta-Pathardih, Dhanbad-Phularitar, Dhanhad-Pathardih, Sudamadih-Gomo, etc. Most of the mines are linked with railways and roads. There is a small air strip at Susanileba for landing of the aeroplanes. There is ropeway for carrying coal.

Abundance of coal and industrial infrastructure has

encouraged industrial growth in the district. There are several coal washeries, chemical industries and other industries. There is a large size fertilizer industry at Sindri. Important products are Ammonia Sulphate, ammonia-nitrate,urea, etc. There is also a cement factory at Sindri. There is engineering industry at Kumardhubi. A silver industry located at Tundi. A phosphate industry lies at Chasnala. There are industry producing coaltar phenyl, pesticides etc. at Ludhiana.

The area is inhabited by both tribal and non-tribals. Out of 26,82,662 (2011) population of the district, 16.29% population is scheduled castes and 8.68% population is scheduled tribe. Kurmi caste shares 16.03% population of the district. Other communities like Bhumihar, Brahman, Kayastha, Rajputs are also in good number. They have come from other areas. Average density of population in this district is 1316 (2011) persons per sq. km. Important urban agglomeration are Dhanbad urban agglomeration which consists of Dhanbad (UA) - 11,96,214 persons, which consists of Dhanbad (M. Corp) (11,62,472) , Malkera (CT-8,232), Nagri Kalan (CT-9,410), Banakalan (CT-9,489) and Pondarkanali (CT-6,611).

Some religious sites are also located in the district. There is an old temple Kalayaneshwari at Maithon. There is a sacred place of Pindrajor on the river Guai, 59 Kms. from Dhanbad is supposed to be the abode of Pandayas who stayed for sometime while in exile.

5. Dumka District :

Dumka, the largest district of Santhal Pargana commissionary is located in north-eastern part of the state. It is bounded by Godda and Banka districts in north, Jamtara and West Bengal in south, Pakur and West Bengal in east, Deoghar in west. It extends between 23° 59' n. latitude to 24° 40' n latitude and 86°55'E longitude to 87°49'E longitude and comprises an area of 3761 sq.km. It has 13,21,442 (2011) population spread over one subdivisions- Dumka Sadar having ten development blocks and 5 towns. These development blocks include Saraihat, Jarmundi, Jama, Ramgarh, Gopi-Kandar, Kathikund, Dumka, Shikaripara, Ranishwar, Masalia.

The word Dumka has developed from Dumloh, Koh being a

common termination to names of village in the Santhal Partanas. It is also known as Naya Dumka except in official reports. This district was formerly subdivision of Santhal Pargana district. It was constituted sub-district or Subdivisions in the year 1872 and later on Dumka became headquarter of Santhal Pargana district.

Being located in the Rajmahal highlands, it consists of rolling open land surface with long ridges and intervening depressions. A large area is occupied by hills and forests in north where a part of the district is included in Damni-i-koh. There are two principal

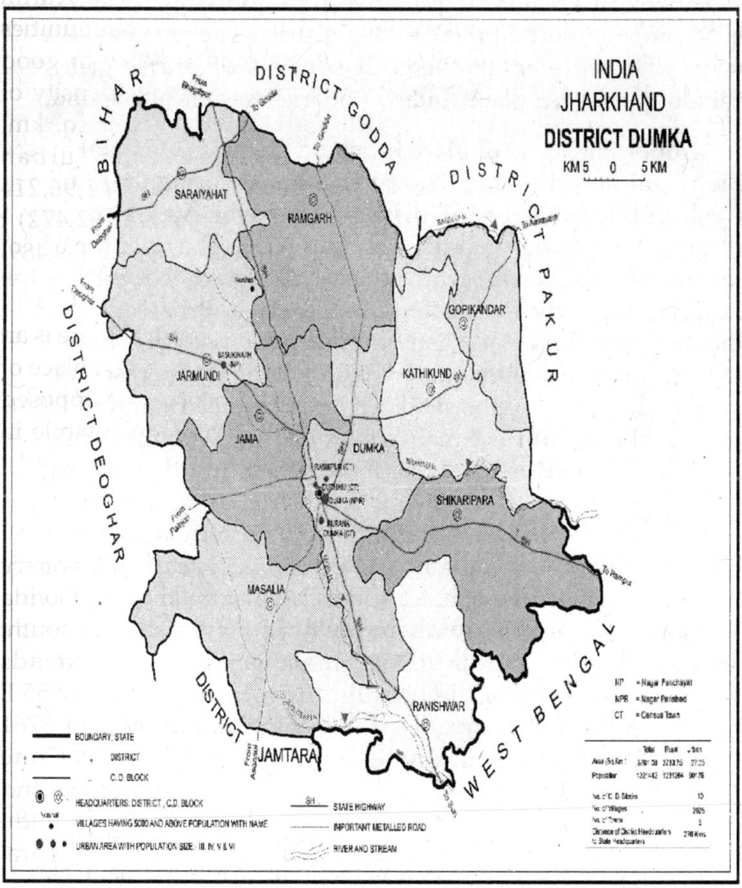

Source : Census of India 2011, Jharkhand, Series 21, Part XII 'B'.

Fig. 31

ranges running south ward from Pakur and Godda Districts and enter this district close to the Sahibganj Dak bunglow in the extreme north of Dumka Damin. They run parallel to each other and passing through the Damin terminate in the Nangal Bhanga Hill, over which the Rampur Hall road is carried. To the south-east is another extensive range known as the Ramgarh Hills, situated in Taluks Darin Manleshwar and Sulunga. The sceneries in these hills are very picturesque, hill after hill stretching away in the distance covered with dense jungle, while far down in the valleys are seen terraced rice fields green with paddy. The average elevation of the region ranges between 150 to 300 metres. Masanjor is the most important peak in this district. The area has basaltic traps and sedimentary beds. It consists of coarse grits and conglomerates. At places quartz and gneisses are also found.

Important rivers of this district are the Ajay, the Barakar, the Brahmani and the Mor. The river Barakar forms the southern western boundary of the district. The river Ajay also passes through the Southern portions of the district. The river Morkhi & Mor pass through the centre of the district. Its course lies the headquarter of the district Dumka. The river Brhamani flows in the north-eastern of the district. Brahamani rises in the Dudhua Hill in north of Dumka and flowing through Pherasemul and Sankara form the southern boundary of Dumka Damin and joins the river Bhagirathi The main tributaries are Gimro and Ero. The river Mor rises in the Tiur Hills at the extreme north east corner of Deoghar district and enter Dumka district at its north-western corner. After flowing in winding south eastern course close of Dumka it joins the Bhagirathi in West Bengal. The river Ajay rises in Kiul and enters the Jamtara subdivision at Kajara and flowing southwards forms the southern boundary of the district. There are six hot springs in this district. Jharia Pani near Gopi Kandar, Tatloi on the bank of the Bhurbhuri river near Palasi, Numbil near Kendghata, Susumpani, etc. The general slope of the area is from north-west to south-east.

Owing to its position near humid West Bengal and hilly landscape of the region climatic characteristics, to some extent, differ from the remaining parts of the state. In the undulating areas the winter months are very cold and the rains are not oppressive, but the heat from the end of March to middle of June is severe and the hot-westerly winds are externally disagreeable. During winter months temperature ranges between 60° and 70° F but during summer monthly mean temperature ranges between 70° and 88° F. The rainfall hardly exceeds 2 cms. between November and April but during rainy season the rainfall exceeds 150 cms. August and September are also rainy months with a fall of 30 and 25 cms. respectively.

Since the area consists of hilly landscape and receives more rainfall it has considerable vegetative cover. Certainly due to increasing habitation and ruthless exploitation, most of the forest areas have turned into bushes devoid of big trees. During recent years efforts have been made to develop plantation to restore ecological balance. Important trees include Sal, Semal, Kaha, Ashar, Mahua, etc. In certain areas inferior quality of Teak is also found. The impact of lava eruption, igneous intrusion and later sedimentary formations have influenced soil types of the area. Usually most of the areas have red soil having less fertility but the soils in the valleys and lowlying paddy fields are fertile.

Important sources of irrigation are well, tank and artificial reservoirs. Besides there is canal system on river Mayarakshi which also provide irrigation in some areas.

No doubt the district has considerable landscape having vegetative cover but because of population pressure most of the forest areas have been converted into culturable land. Apart from areas under forest cover and cultivation non-agricultural uses like areas under settlements, transport network, pasture, groves, religious sites, playground, waste land are also important. Imortant crops of this region are paddy, maize and wheat. Sometimes people grow Bhadai paddy and Aghani paddy also. During recent years the production of wheat in Rabi season and vegetables in Rabi and Garma seasons has considerably increased.

The area is devoid of National Highway but there are State Highway and other roads which connect district headquarter and other places. Dumka is the Nodal Centre of road network where roads from different directions coverage. Important roads radiating from Dumka are Dumka-Jamtara, Dumka-Rampur Hart, Dumka-Sahebganj, Dumka-Goda, Dumka-Amarapara, Dumka-Deoghar, etc. Jamtara subdivision of the district has rail connection.

This rail route connects Kolkata in east from Delhi in north-west. Rivers are not navigable because of undulating land surface. There is a small air strip.

Chief mineral products of economic value are coal, building stones, china clay, lead ores etc. Iron ores are found in considerable quantity in basaltic trap and trappean beds as well as in sandstones and are worked by iron smelters. This area has not encouraged any major industry. Certainly there are industries based on forest resources agricultural products and other daily needs in market centres and urban areas.

This district has moderate density having 351 (2011) persons per sq. km. The development blocks having more than 400 persons density include Saraihat, Jarmundi, Dumka, Jamtara and Narayanpur. The remaining blocks have less then 300 persons per sq. km. density. This district has 2688 inhabited villages and 237 uninhabited villages. There are 5 towns namely Dumka (M) (90,178 persons), Basukinath (NAC) 17,123 Rasikpur (CT) 8320, Dudhani (LT) 7,117 and Purana Dumka (CT) 10,034. This district comprises a population of 6.02% of scheduled caste and 43.22% scheduled tribe. Important tribes are Santhal, Paharia, Maler, etc. Santhal are predominent. Malpaharies and Sauria Paharias are also found in hilly areas. Settlements are linear dispersed in hilly areas and dispersed in plains.

Important places of this district are Basukinath as religious site, Masanjor as hydel power station and picnic spot, Jharia Pani hot spring near Gopi kandar, Tatloi hot spring near Palasi.

6. Garhwa District :

This district is located in north western part of the state. It is bounded by Bhabhua-Rohtas south Districts in north, M.P. in Uttar Pradesh in west and Palamu in east. It is physically demarcated by the Son river in north, Kanhar river in south and west and Koel river in east. It has an area of 4093 sq.km. and consists of 13,22,784 (2011) persons. It came into existence on 31st March 1991. It consists of 3 sub-divisions Garhwa, Ranka and Nagar-Utari. There are altogether 193 Panchayats and 888 villages. It has 19 development blocks namely Bhawanathpur, Majhaion, Nagar-Utari, Dhurki, Meral, Garhwa, Ranka, Bhandaria, Dandai, Chinia Kharoundhi, Ramna, Ramkanda, Bardiha, Bishunpura, Sagma,

Ketar, Dauda and Kandi.

The nomenclature of Garhwa is based on Garh. This district provides diverse physiographic features because northern part of

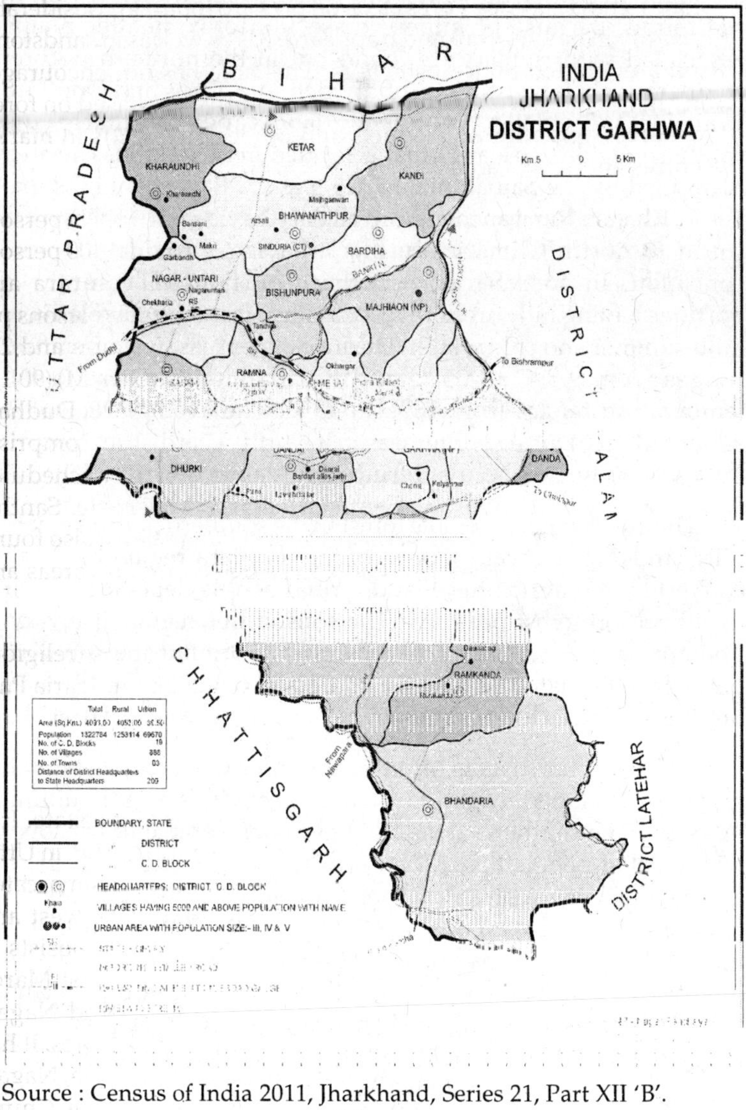

Source : Census of India 2011, Jharkhand, Series 21, Part XII 'B'.

Fig. 32

this district has relatively homogenous land formed by the river Son and the southern portion is hilly and undulating. There is a hilly range namely Banda Pahar having an elevation of about 400 metres above the sea level. It is 5 kms. north-west of Garhwa. In the southern portion the land is more dissected and narrow valleys are found. Because of intense dissection only smaller plains associated with rivulets are found but in the northern portion comparatively larger plains subdivided into Tarns and Dons are available and provide contrasting landscape than that of the southern region. Sarwar Parbat is located in Bhandaria. The area is drained by the Satbahinia, Banro, Panda, Bayochanki, Dhina, Tahle, Khajuri, Ranji and Sarswati rivers. The general slope is from south to north. Climatically this area also provides diverse conditions in northern and southern portions. The southern portions having hilly areas covered with forest provide relatively mild summer and cold winter. During winter season temperature ranges from 9.5°C to 35°C and during summer season the temperature ranges from 18°C to 45°C. The average rainfall is about 120 cms. During winter season it hardly records 1 cm. rainfall but most of the rain occurs during rainy season.

The area has considerable forest in the southern portion where hilly landscape prevails. Forest of the northern region has been converted into agricultural land. Tribal people depend for their livelihood on forest products. The soils of this region also vary. The northern areas have Domat and clay soil but the southern parts have red lateritic soils. Tarn areas provide less fertile soils but the don areas have fertile soils.

The impact of diverse physical conditions inadequate rainfall and less developed irrigation facilities is obvious on agricultural land use. The southern areas still have considerable forest cover where tribal people depend on forest products but in northern areas people have converted forest covered areas to agricultural land and grow rice, maize, Ragi, Millets etc. Plain areas of the north have rice areas having irrigation facilities are also engaged in growing wheat and vegetables. In addition to culturable land use, there are areas engaged in pasture, dwelling purposes, playgrounds, orchards, fellows, etc. limited land is involved in industrial and transport network.

This district has the facility of rail route which connects all important places of the country. Garhwa headquarter is linked with Daltonganj, Ranchi, Husainabad, Netarhat, etc. Rivers are unnavigable and there is no air strip.

The area is devoid of important minerals. Certainly there is limestone, dolomite, graphite, felspar, etc. in limited amount. Bhawanathpur mine is noted for limestone production from where limestone is supplied to Bokaro steel plant. As such the area is poor in mineral resources. Industries have not developed. Certainly industries related to forest and agricultural resources have developed. Garhwa has several timber industries. There are industries concerning carpet, Leaves Plate, Leather, etc.

This district has an average of 323 (2011) persons per sq.km density but in rural areas the density is slightly less (250). Development blocks having higher density than the regional average include Utari (519), Garhwa (596) because of urban growth. Development blocks having less density include Bhandaria (100) because of hilly land scape, Ranka (212), Dhurki (236) etc. This district has 844 inhabited villages and 44 uninhabited villages. It has only two town i.e. Garhwa having population of 69,670 and Sinduria (CT) having a population of 5,262 persons. Important tribes residing in this district are Oraon, Munda, Gonda, Korba, Parahia, Asur, Birhor, etc.

Important places of tourist interest are Nagar Utari which is a small service centre having the temple of lord Krishna. The Banda Pahar has a waterfall noted for curing skin diseases, The peak of this hills has a Shiva temple known as Bandeshwarnath.

7. Giridih District :

The district of Giridih is located in the northern part of the state. It is bounded by Nawada & Jamui in north, Bokaro in south, Deoghar, Jamui, Jamtara & Dhanbad in east and Hazaribag and Koderma in west. This district's delimited by 23°28' N. latitude to 24°47' N. latitude and 85°35'E. longitude to 86°35' E. longitude and comprises an area of 4962 sq. kms. This district has recorded 24,45,203 persons in 2011. It has one subdivision - Giridih Sadar and 13 development blocks.

This district came into being on 6th Dec. 1972 being separated from Hazaribag. Its headquarter is at Giridih. This district has also three types of topographic areas - central plateau having moderate elevation, lower plateau having lower elevation and the trough basins of Damodar. The lower plateau areas have relatively rough terrain having an elevation 390 metre. In the north and north-west there is a table land having an elevation of 250 metre where steep scarp is found. Areas of Damodar trough basin lies

Source : Census of India 2011, Jharkhand, Series 21, Part XII 'B'.

Fig. 33

in the south. The highest peak of Parasnath of this state is located in this district. The area is drained by the rivers in Damodar, Barakar, Sakri, Usri etc.

Because of moderate elevation and considerable deforestation and mining and industrial activities climatic conditions exhibit some differences from the higher plateau areas of the state. The highest temperature goes upto 40°C but the average temperature during winter season remains 10°C. Average annual rainfall is 135 cm. Certainly areas around Parasnath hills receive more than 150 cm. rainfall. Average annual relative humidity remains 60%.

Soils of the area contain good proportion of sand. Certainly areas in the river basin has alluvial soil but higher areas have lateritic soil. The impact of relief is also quite obvious on natural vegetation. Higher areas having more undulations have more forest cover. Important trees are Sal, Khair, Semal, Mahua, Palas, Kusum, Kend, Ashar, etc.

The area is also suitable for agricultural purposes especially in the lower plateaus and the trough basins. Certainly mining areas are devoid of dominance of culturable land. Because of rough terrain irrigation facilities are not developed. Irrigation by tanks and reservoirs enable cultivators to irrigate crops and vegetables. Rice is the principal crop. However, maize, wheat, pulses are also grown. During recent years cultivation of vegetables and seasonal fruits has considerably increased. The other uses of land include residential areas, pasture lands, fallows, orchards, playground and other non-agricultural uses.

This district has also considerable waste land in mica mining areas because of intensive mining abandoning the land after the exhaustion of mica.

Transport network of this district is relatively better developed. The headquarter is linked with Eastern Railway by Madhupur-Giridih Branch line. There is small air strip for the landing of planes. The area has better road linkage. Two National Highways - Grand Trunk Road and National Highway No. 23 pass through the district. There are other roads connecting important places like Giridih-Madhupur (39 kms.), Giridih-Dumri (45 kms.), Giridih-Gobindpur (23 kms.), Giridih-Jamua (26 kms.), Giridih-Jamua-Sarawan (49 kms.), Dhanwar-Sarai (46 kms.),

Pataldih-Satgama (64 kms.). Rivers are not navigable.

The area having coal and mica resources have given rice to mica processing industries and coal mining activities. This region has good quality of coal. In addition to mica and coal based industries there are leather industries, agro-based industries also.

The area has the density of 493 persons per sq. km. (2011). This district has one statutory town and eleven census towns. This district has 2558 inhabited and 191 unihabited villages. Scheduled Caste and Scheduled Tribe population share stands at 13.31% and 9.74% respectively. Important tribes of this district are Munda, Oraon and Santhal. Settlements are relatively dispersed.

Important sites are Parasnath hill having temple of Jain, Usri waterfall on the river Usri, 13 kms. east of Giridih; Madhupur is a sacred place having Jain Temple of 2 thousands years old. There is a picnic spot near Khandoli Pahar, 8 kms. from Giridih where a big tank and a picnic spot attract the tourists.

8. Godda District :

Godda district is located in north-western portion of the Rajmahal highlands of the state. It is bounded by Bhagalpur (Bihar) district in north, Dumka district in south, Sahibganj and Pakur districts in east and Bhagalpur (Bihar) in west. This district lies 24°30' N.lat. to 25°14' N.lat. and 87°3' E.long. to 87°36' E.longitude. Its length from north to south in over 75 km and its breadth varies from 18 to 36 km. It comprises 13,13,551 (2011) persons over an area of 2266 sq. kms. Out of total population 8.80% population is Scheduled Caste and 21.26% population is Scheduled Tribe. It consists of nine administrative blocks namely Godda, Poryahat, Pathargama, Mahagama, Meherma, Boarijor and Sundarpahari Thakur Gangati and Basant Rai. The eastern portion of the district along the skirts of Rajmahal hills is included in Damini Koh and consist of hilly country, with forests in the southern half. The western portions of the area merely comprising the Jamindari tract is different in Character Tappa Manihari to the north is an alluvial plain very similar to the adjacent level land in the district of Bhagalpur. Next to it comes Tappa Patsuna, and upland tract sloping downwards towards the west. Towards the south in Parganas Pasai, Gudda and Barkop, the land is undulating. The Chief hills of the district are in the Damini-Koh and belong to the

Rajmahal Range, of which there are a few outliers. In the west of Pasai there is a small range which continues westwards into Bhagalpur district. In Barkup there is a cluster of small hills of an isolated hill at Bisaha, while other detected hills of black rock are found in Patsunda and Manihari. The rivers of the district which rise, for the most parts in the uplands to the east, receive the drainage of the western slope of the hills and flow generally in a

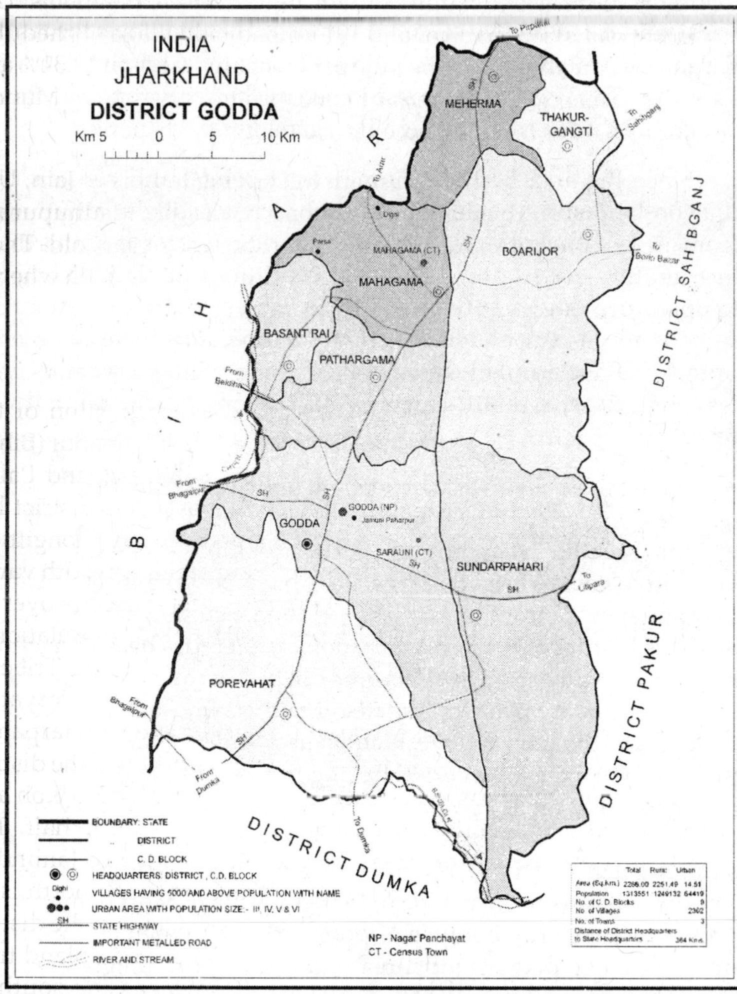

Source : Census of India 2011, Jharkhand, Series 21, Part XII 'B'.

Fig. 34

westerly direction into the Girua which flows northwards to the Ganges. In the manihari plain there are two streams - the Koa and Bhorai which have scoured out deep channels in the alluvial soils. Tappa Patsunda is separated from Paragana Barkup on the south by the river sundar, which rises in the hills of Telo region. South of Sundar there are three streams, the Sapani, Bheriya and Harna, the two later flowing pargana Godda. The largest river of the district is the Kajia, which rises in the highland of Belbathan and Pasai, and is fed by five separate streams. The Lilji flows through Parganas, Amlamatia and Godda and joins the Chir river above Panjwara. The later stream is also joined by the Pareya and Sugthan which flow westerned through Paragana Pasai.

Since the area lies near humid west Bengal and consists of uneven land surface climatic conditions reveal slight variations from neighbouring areas. During. During winter months the temperature comes down and cold conditions prevail. Average temperature ranges between 15°C to 25°C but during summer seasons hot conditions develop and the temperature increases even upto 43°C. The rainfall is negligible during winter season, even less than 2 cm rainfall. During rain season 150 cm rainfall is recorded.

Vegetative cover is also dependent upon the existence of rough terrain. Plain areas have become almost devoid of vegetation except isolated trees but the hilly areas have considerable vegetative cover certainly large trees have become insignificant in number. Important trees are Sal, Mahua, Semal, Sisam, Kathal, etc. This district also reveals variation in soil types based on land surface. The hilly areas have leteritic soil because of lava formation and the plain areas have domat soil and clayey soil. Tarn areas have less fertile soil where residential uses and growth of orchards prevail. While the lower areas having fertile soils have culturable use. The holly areas have considerable land under forest cover and waste land having rocky projects and pastures but the gentle slopping land and the plain areas have more culturable use. Important crops are rice, wheat, Maize and Ragi. Most of the area grow only one crop because of poor irrigation facilities. Besides other uses like playground, area under transport, religious sites etc. are also found.

The area is devoid of important minerals and major industries. However, there are industries related to stone crushing because of availability of basalt rock- China clay has also encouraged pottery. In addition to these there are small industries based on forest and agricultural products.

The area is poor in terms of transport. Rail route is non-existent even road network is not well developed. However the headquarter of Godda is linked by metalled with neighbouring headquarters of Deoghar, Dumka, Pakur, Sahibganj, Bhagalpur, etc. There is no navigable water ways and presence of air strip.

This district has moderate concentration of population. The average density is 580 (2011) persons per sq.km. The development blocks having higher density more than the district average, include Thakur Gangati (1176), Meherma (1111), Mahagama (1206). The remaining blocks have density less than the district average. This district has 1688 inhabited villages and 614 uninhabited villages. There are 3 towns namely Godda (48,480), Mahagama (10,969) and Sarauni (4,970) persons. Important tribes are Santhal and Paharia. Settlements are linear dispersed in hilly areas and relatively dispersed in the plain.

9. Gumla District :

This district is located in South-western portion of the state. It has got the status of a district on 18th May 1983. It is bounded by state boundary on west and south. Chattisgarh lies in the West and Simdega in the South. In north of this district there lie the districts of Latehar and Lohardaga and in east there lie the districts of Ranchi and Khunti. It extends between 22°35' N. to 23°33' N. latitude and 84°40'E. to 85°10'. It comprises an area of 5360 sq. km and consists of 12 development blocks. It consists of 10,25,213 persons out of which 68.94% people are scheduled tribe mainly Oraon, Kharia and Munda.

As regards the nomenclature of Gumla there has been differential opinions. Some people hold the view that the word Gumla has developed from the word Gomela (Go + Mela) meaning thereby cattlefair. Others adhere the view that it has been coined from the Mundari language Gumla meaning thereby separating the paddy husk from rice. In support of the view people opine that the area has remained notable for good variety of rice. This place has developed on exposed rock surface. The township of Gumla has developed on the land having 630 metres elevation above the sea level.

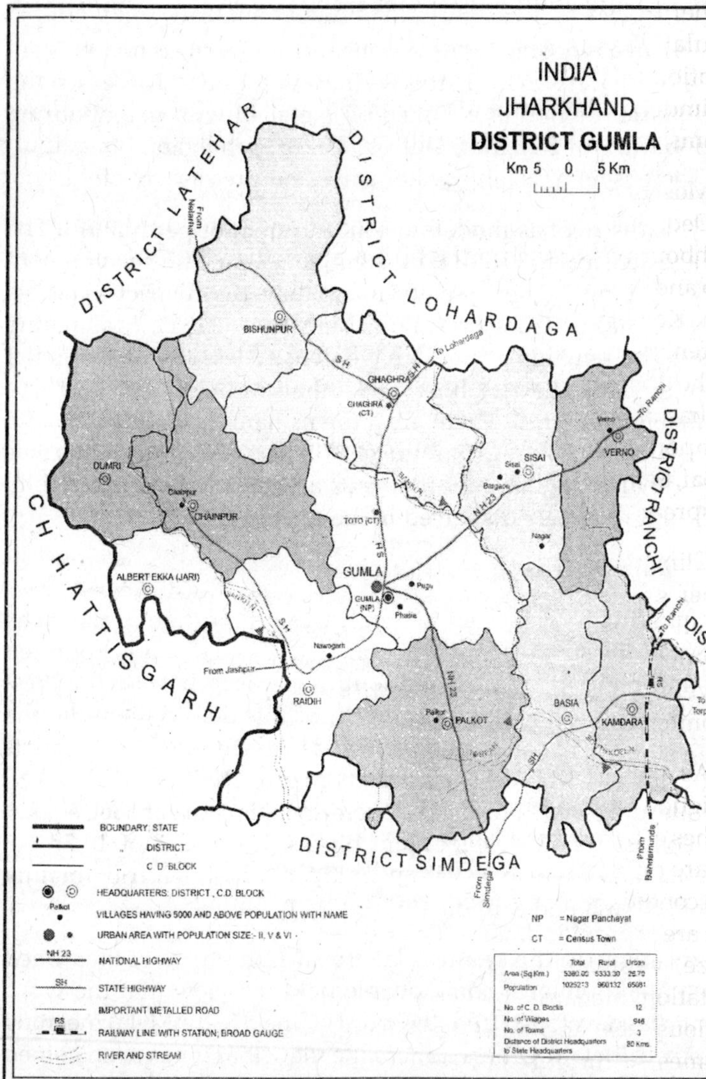

Source : Census of India 2011, Jharkhand, Series 21, Part XII 'B'.

Fig. 35

The district consists of partly higher Ranchi Plateau in the north, rugged Simdega Plateau in the south and two river basins of Sankh and South koel. As such the topographic diversities reveal more impact on human habitation, economy and culture. The northern parts of the district has higher plateau areas where undulations are more marked and the land has an average elevation of 700 metre. The southern parts of the district is known as Simdega Plateau is more rugged and hilly with turbulent streams, steep slopes, high cliffs and narrow valleys.

Most of the areas lying between two rivers the South koel and the Deo Nandi have created a rugged landscape different from neighbouring areas. The chief tributary of the South Koel are the Karo and the Chata. The Sankh river has the tributaries of Chirra, Lava, Kok etc. Other rivers are Paju, Chetar, Nala etc. There is Nagfeni waterfall nearly 14 km away from Gumla. The district has the dominance of laterite soil. Since the area is more dissected and has rough terrain there is considerable forest cover. Saru is an important hilly range having dense forest clad. Important trees are Sal, Semal, Mahua, Kathal, Kendu leaf, etc. Gumla forest area has spread over 1194.5 sq. kms.

Climatically this district is suitable for comfortable living and proper growth of vegetation and agriculture. Average annual rainfall is about 90 cm but in areas having higher elevation, the amount of precipitation increases. The average temperature ranges between 16^0C and 29^0C. Summer season hardly becomes uncomfortable but winter season becomes a little cold.

As the area is highly dissected and has rough terrain, contiguous agricultural lands are found in limited areas. Certainly patches of agricultural land alongwith rivers and rivulets of varied size are diffused in the district. The low lying Don areas provide ideal conditions for the production of rice but comparatively higher Tarn areas provide suitable conditions for the production of Ragi, Maize, vegetables and pulses. Other uses of land concern habitation, orchards, playground, threshing ground, graveyard, religious sites and other uses. This area has smaller irrigation schemes in the form of reservoirs. Important reservoirs are Chhinda in Simdega, Paras in Bharno, Ramrekha, Kansjor, Katri etc. The koel-karo project has not taken shape uptil now.

This district has poor network of transport system because of rough and dissected terrain. But the headquarter of Gumla is linked with Ranchi by State highway of 96 km, Patna with 351 km. There are several religious and historical sites which are connected by roads.

Since the area lacks mineral resources and good transport network only small scale industries based on agricultural and forest products have developed forest based industries are saw mills, lac industry, biri industry etc. Agro based industries are small rice mills, wheat flour mills, oil pressing mills, etc.

This district is sparsely populated having substantial percentage (69%) of tribal population. Important tribes are Oraon, Kharia, Munda etc. Hilly areas of the district have Asurs, Birhor, Korba etc. There are people of other regions like Haryana, Rajasthan, Bengal, Orissa having people of varied religious.

Several tourists sites have also developed. Nagfeni waterfall is only 14 km away from Gumla. Other picnic spots are Rajadera, Hiraeah, Keraghag, Sadani, etc. Ramrekhadham is only 13 km west of Simdega. It is supposed that Sri Ram visited this place while in exile. There is a cave temple of Hanuman at Angani which has become important religious site. It is also considered as the birth place of Hanuman. Nearly 3 km away from this site is the hilly areas of Damdamia where Shiva temple is important. Palkot nearly 25 km away from gumla being surrounded by forest covered hills had been the capital of Nagbanis. There are caves like Sitalpur, Malmalpur, etc. which remain cool during summer. These sites have hot and cold springs and provide ideal picnic sites. Another important site is waterfall of Baghmunda in Basia development block where there is temple of Sagarnath nearly 400 years old.

10. Hazaribag District :

Hazaribag district comprising considerable areas of the Hazaribag Plateau is bounded by the districts Gaya and Koderma in north, Ramgarh in the south, Giridih and Bokaro in the east and Chatra in the west. It extends between 23°25' N. lat. to 24°48' N. lat. and 84°29' E. longitude to 86°38' E. longitude and contains an area of 3555 sq. kms. and has got 17,34,005 persons (2011) after bifurcation of Chatra district. The nomenclature of this district as

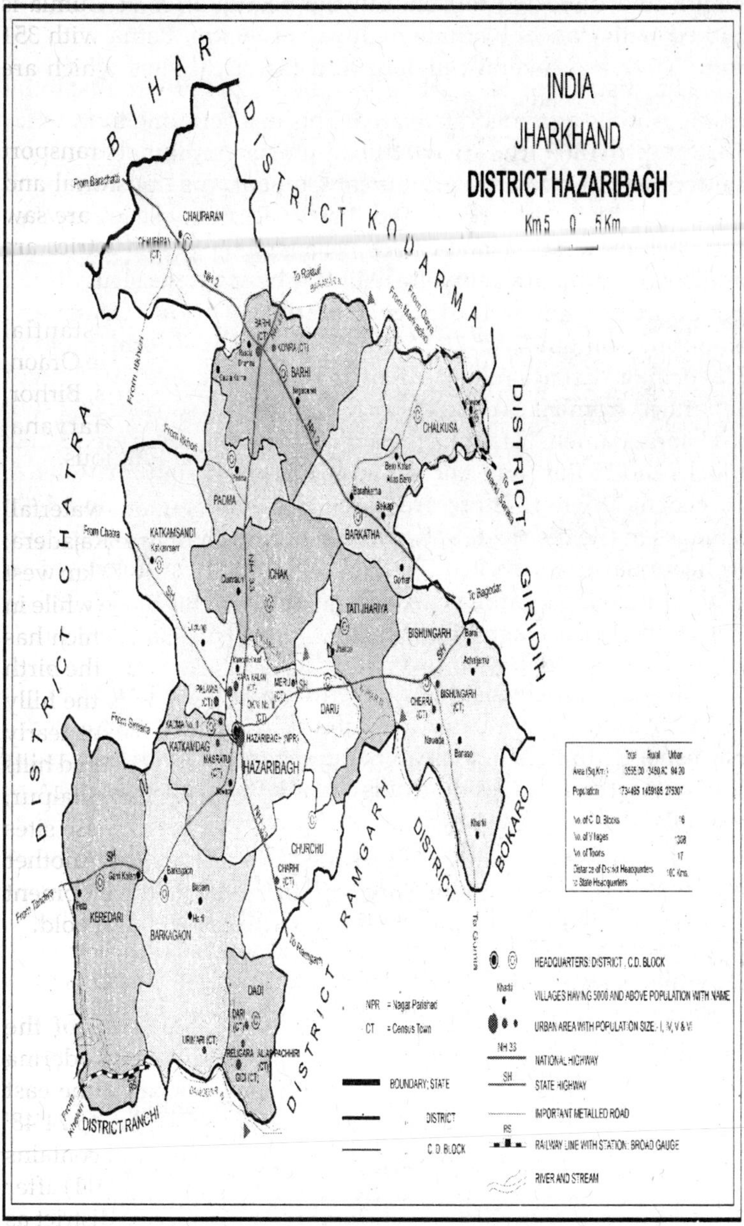

Source : Census of India 2011, Jharkhand, Series 21, Part XII 'B'.

Fig. 36

well as its headquarter is based on the name of Hazari village having mango orchards.

The district of Hazaribag has also registered substantial changes in its administrative area. This district came in the year 1833. Upto 1869, the district headquarter remained military cantonment and afterwards this cantonment was transformed to Ramgarh. Then the area of the district further increased after inclusion of Ramgarh and Kharaydiha and Hazaribag continued as district headquarter as well as Military headquarter. This district got bifurcated into four districts Hazaribag, Giridih, Chatra and Koderma. On 6th Dec., 1972 Giridih district was formed out of Hazaribag district. On 9th April, 1994 Koderma district got separated from Hazaribag district. This district consists of 2 sub-divisions, namely Hazaribagh Sadar and Barhi. These are 16 C.D. Blocks and 15 Revenue Anchals and 17 towns (1 statutory town & 16 census towns). There are 257 Gram Panchayats and 1308 Villages in the district (2011). This district derives its name from its headquarter town Hazaribagh which literally means the place with 'a thousand gardens'. A village named 'Hazari' is merged within the Hazaribagh Municipality.

This district consists of uneven land surface having areas characterised by plateaus and hills. Virtually its topography corresponds to Ranchi Plateau being separated by the Damodar Valley. As it has also residual hills and inter plateau valleys. There are hills even near the township of Hazaribag namely Chendbar (2817) and Kenari hills. The highest hill of the district is Marangburu (34551) south of Damodar. Hills have steep scarps. The general slope of the district is from north west to south east. Important river is Damodar but there are a few tributaries namely Jamunia, Baranki, etc. The tributary Bhera joins the river Damodar at Rajrappa where there exists the temple of Chhindmastika and Rajrappa waterfall.

The climatic conditions of the district are similar to Ranchi District. During summer season the temperature is in extreme cases increases to 44^{0}C and during winter season the same comes down to 2 to 3^{0}C. It receives an average annual rainfall of 135 cms.

Soils of the district is of lateritic types but valleys are occupied by fertile soils. The district has considerable forest cover. The forest is highly distributed. Sal is the main tree of the forest other trees are Palas, Kusum, Semal, Gamhar, Karam, Pasar, Asan, Kendu etc. Bamboo thickets are also well distributed.

The region has considerable agricultural land. The don areas are used for growing paddy but Tanr areas are used for growing Ragi, Maize and other Bhadai crops. Homogenoous areas are also used for growing wheat and other Rabi crops. The area as recorded phenomenal increase in the production of vegetables and seasonal fruits. Farmers grow early variety of Potato for getting more money. People of the area also use higher tanr lands for constructing, dwellings, orchards, etc. Non-agricultural uses of land are also available in mining, industrial and urban areas.

This district has several coal mines and associated coal washeries. In addition to coal mica, china clay etc. are also found. This district has given rise to many industries - Ceramic industry near Hazaribag Road Station, Bhukunda Glass Factory, Lac industry in Gola, Brass industries at Bishnugarh, Ichak. Saw mills, flour mills and other smaller industries have developed in almost smaller towns and service centres.

The district has poor railway facilities but there are good roads. Two national highways - No. 2 and 33 pass through the district. Important roads are Hazaribag-Ranchi, Hazaribag via Barhi to Koderma and Patna. It is also linked with Barkakana, Patratu, Bagodar etc. Hazaribag town has the facility of landing of small planes. Rivers are non-navigable.

The area is moderately populated, certainly the areas having rough terrain have sparse population but the homogeneous areas and mineralised pockets have attracted large size of population. The area is inhabited by tribals and non-tribals. Among tribals Munda, Oraon are dominant. Among non-tribals trade communities are in dominance. Average density is 488 (2011) persons per sq. km. Hazaribag has 17 towns where 15.87% (2011) urban population people live. This area has encouraged the growth of Hazaribag town. Hazaribag town is located on the nodal point of three important roads connecting Ranchi, Bagodar and Barhi. Hazaribag town is also divisional headquarter (North Chotanagpur commissionery). This town has several important temples, tanks and scenic spots. There are many institutions of higher learning. St. Columba's College is a famous college. It was established in 1895. It has also the university headquarter of Vinoba Bhave University.

11. Jamtara District:

Jamtara is a new district which was carved out from the parent district Dumka on 26th April, 2001. Earier it was a sub-division in the Dumka district. It is bounded by Deoghar district in the north, Dumka district & W. Bengal state in the east, Dhanbad district & W. Bengal in the south and Giridih in the east. It has a population of 7,91,042 persons and area 1811 sq. km. as per census 2011.

The history of Jamtara is inseperable from the parent district of Dumka and has witnessed many administrative changes. Earlier the Dumka district was a sub-division of the Santhal Pargana district and when Dumka became a district, Jamtara was constituted as a sub-division. Presently, the district of Jamtara possesses one sub-division - Jamtara, 6 C.D. blocks, 118 Gram Panchayats and 1161 revenue villages. The name of 6 C.D. blocks are Narayanpur, Karma Tanr Vidyasagar, Jamtara, Nala, Fatehpur and Kundhit. There are 02 statutory towns namely Jamtara and Mihijam and one census town namely Karma Tanr.

This district lies in the Rajmahal highlands and hence it consists of rolling open land surface with long ridges and intervening depressions. The average elevation of the region ranges between 150 to 300 metres. The area has basaltic traps and sedimentary beds, which consists of coarse grits and conglomerates. At places, gneisses and quartz are also found. The important rivers of the district are Barakar & Ajay. The river Barakar forms the southern western portion of the district. The river Ajay flows through amost in the central part.

Climatically, it differs from the rest of state to some extent, as its proximity to W. Bengal & hilly landscape causes more rainfall. The total rainfall exceeds 1300 mm/year. The average temperature reaches above 25.5°C while the maximum and minimum temperature reaches 44.5°C and 4.9°C respectively.

Earlier, the district had considerable area under forest cover, but most of these have been converted into culturable land. The economy of the district is mainly dependant on agriculture and allied work. More than 64% of the total workers are engaged in Primary sector. The important crops of the district are paddy, maize & wheat which are cultivated during 3 seasons. It is included in the list of most backward district consisting of 250 districts of the country, as announced by government of India and hence

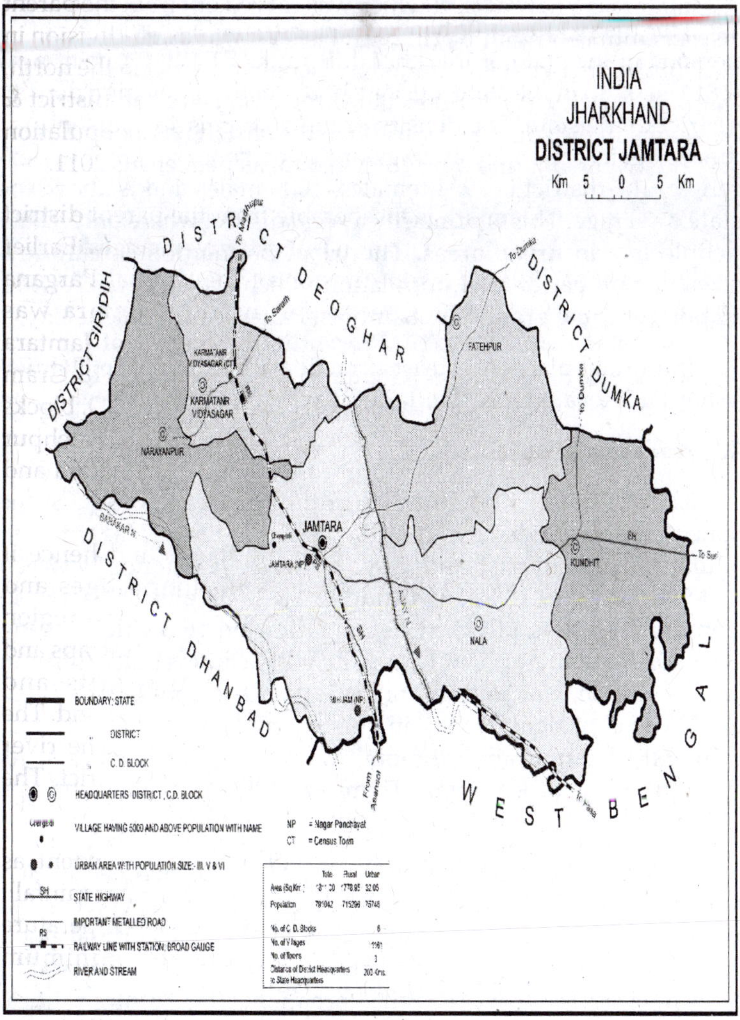

Source : Census of India 2011, Jharkhand, Series 21, Part XII 'B'.

Fig. 37

receives founds from the Backward Regions Grant Fund Program.

This district is devoid of good communication connectivity especially in terms of railway & NH services. Only one branch of railway broadgauge connects the district headquarter Jamtara to Asansol - Madhupur. There is no NH at present and it is only served by SH & district roads.

This district ranks 19th in terms of population having 7,91,042 persons in the state of Jharkhand. It ranks 21st in terms of area (1811 sq. km.) in the state. It is one of the moderatley high density district in the state. The density of the district is 437 persons/sq. km. as against the state average of 414 persons/sq. km. The sex-ratio of the district i.e. 954 females/1000 males and is above the state's average. This is primarily a rural district where only 9.58% people live in urban areas. The tribal population constitute a sizeable number of total population which accounts for 30.40%. Important tribal groups are Santhals, Paharia and Maler etc.

Important places of tourist attraction are Ladana dam, Parwat Vihar Park located near Jamtara town.

12. Koderma District :

The district of Koderma lies in the northern portion of the state. It is surrounded by Nawada district in north, Hazaribag in south, Giridih in east and Gaya in west. It extends between $24^0 15'46''$& $24^0 49'18''$ N Latitude & $85^0 26'0''$ and $85^0 54'16''$ E. Longitude. It has physical demarcation on two sides having Nawada Plain in north and Hazaribag Plateau in south. It has an area of 1655 sq. kms. and comprises 7,16,259 persons (2011). There are 6 development blocks in this district, namely Koderma, Jainagar, Chauparan, Satgama, Markacho and Barkatha. Its headquarter is at Koderma. There are 109 Panchayats and 699 villages.

This district originated on 9th April, 1994, being separated from the district of Hazaribag. It has been an important place from point of view of historical events, religious sites and cultural glory. It has been sacred site of Jains where Jainis travellers arrived in large number. More over it is an international trade centre of Mica.

Since this district occupies most areas of Koderma Plateau, it has topographic features slightly different from the Hazaribag Plateau. This area has more slope than the Hazaribag Plateau which lies in the south. The northern scarp has steep slope. This plateau has Dharwanian rocks like quarzite and schists containing Pegmatites veins in which good quality of mica is found. Areas having quarzite rocks are less subjected to erosion due to resistant rocks and forms peaks. The area has several rocky exposures of

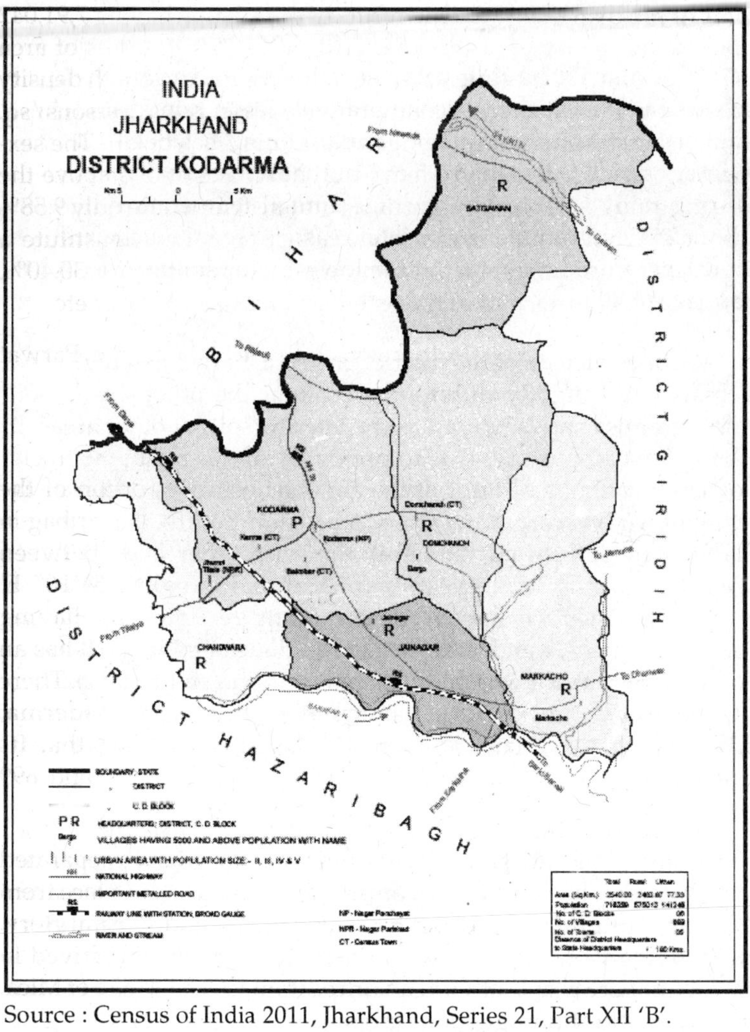

Source : Census of India 2011, Jharkhand, Series 21, Part XII 'B'.

Fig. 38

hard rocks but areas having schists are more eroded. There are isolated residual hills. The Mahabir Range (2210') lies south of Satgama, Nero Hill (1737') near Koderma, Durbasa Hill (2209') and Mormoko (2050') in Koderma forest. Average height of the area ranges 300 to 500 metres. The general slope of Koderma district if from south to north. Important rivers of the district are Sakri, Dhadhar, etc. The river Barsoni passes through the central part of the district.

The district having relatively lower elevation than the Hazaribag Plateau receive comparatively lesser rainfall and show higher temperature. Average annual rainfall has been 120 cms. winter rainfall is less than 1.5 cm. but most of rainfall takes place during rainy season. The average annual temperature remains about 25^{0}C but summer and winter seasons record much variation in temperature ranging from the lowest temperature 2 to 3^{0}C in January to 40 to 45^{0}C in May.

The flat plateau of the district having Dharwanian rocks and schists have slightly different soils than the other areas. Soil contains mica and is not red as the lateritic soils of other areas. It has red micaceous soil. Due to proximity of this plain, relatively lower elevation, the forest of the area has got more destruction. Only hilly and scarp areas have forest land but the flat plateaus have been deforested for habitation and other uses of land. Certainly during recent years afforestation has been made and vegetative cover has improved. Important trees are Sal, Mahua, Gamhar, Karanj, Kend, etc. The Sal has developed in humid and lowlying situations where thick soil is found but areas having steeper slope or rocky surface areas having thin soil have grasses like Sabai grasses, Harra and Dhak. Palas is found in waste land areas.

The land of this district is used according to the nature of land and transport facilities. Hilly areas and scarp areas have more land under forest while the flat plateau areas have more culturable land. Non-agricultural uses of land have also increased to due mining development, transport network, increasing population and residential areas and other uses. Important crops of the area are paddy in the homogeneous areas or the Don Areas. Maize vegetables are grown mostly in Tanr areas.

The area is noted for mica mines. There is a belt having huge reserve of good quality mica. Rubi Mica of this area is famous in the world. Virtually intensive mica mining has converted most of the fertile land into barren land. Important centre are Jhumri Tilaiya, Domchanch, Masanodih etc. Mica industries have developed in the form of cottage industries where mica splitting is done by Santhali Females. There is Mica-Mite Industry at Jhumri Tilaiya. It produces Micamite, Micalinen, Micasilk, Micapoper etc. In addition to mica industries there are several timber industries and agro-based small scale industries.

No doubt this area is famous in the world in terms of mica. Mica mineral but the transport network has not developed as per expectation. Certainly Koderma is linked with railroute connecting Patna and Ranchi and has also linkage with major cities of the country through Grand Cord Rail Route. Interior areas are devoid of rail routes. Even road transport is not well developed. However, Koderma is situated on the National Highway-33 which connects Ranchi, Patna, etc. Other roads are unmetalled. Roads connecting mica mines are most unmetalled.

The area is moderately populated having the density of 282 persons per sq. km. Development blocks having high density are Jainagar, Koderma and Markucho where per sq. km. density is more than 350 persons. The remaining development blocks of Satgama, Chauparan have density less than 250 persons per sq. km. This district has two important urban centres - Jhumri Tilaiya and Koderma having population of 45,904 and 12,941 persons respectively. Other important census towns are Bekobar, Karma and Domchanch. The areas is inhabited by both tribals and non-tribals. Among tribals Santhal, Munda and Oraon are found in good number. Settlements are of dispersed type.

Important visiting sites are Dhawajadhari Pahar having a temple, Burhwa Mahadeo Pahar, near Koderma, Tilaiya Dam, Tilaiya Sainik School etc.

13. Khunti District:

The newly created Khunti district, earlier a sub-division of Ranchi district, was carved out on 12th Sept, 2007. It is situated in South Chotanagpur division of Jharkhand. It is surrounded by Ranchi & Gumla districts in the north, Chaibasa & Seraikela-

Kharsawan districts in the south, Ranchi & Seraikela-Kharsawan in the east & Gumla & Simdega districts in the west. It ranks 23rd in terms of population (5,31,885 Person) and 16th in terms of area (2535 sq. km.) in the state of Jharkhand as per the 2011 census.

The history of Khunti district is inseperable from the history of parent district of Ranchi and has registered many administrative

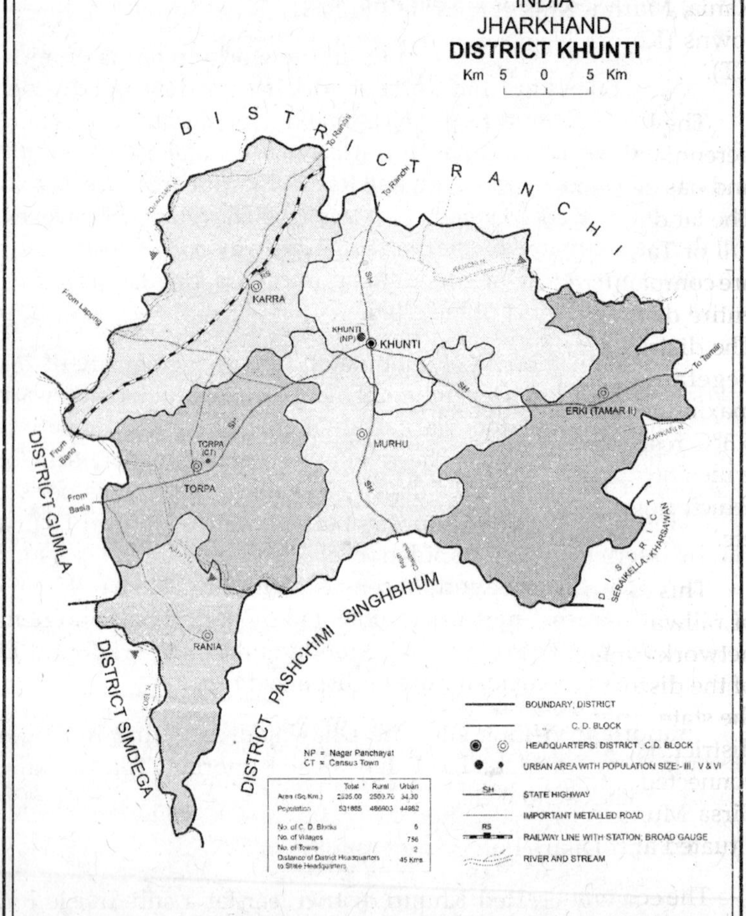

Source : Census of India 2011, Jharkhand, Series 21, Part XII 'B'.

Fig. 27

changes. Earlier the name of the erstwhile district Ranchi was Lohardaga as seat of the district was at Lohardaga. The name of the district was changed from Lohardaga to Ranchi in 1899. It became a sub-division of the Ranchi district and later on carved out of Ranchi district as 23rd district of Jharkhand on 12.9.2007. A village of Khunti district called 'Ulihatu' is the birth place of Bhagwan Birsa Munda. At present, it possesses one Khunti sub-division and the sub-division is further divided into 6 C.D. blocks, 86 Panchayats and 756 villages. The 6 C.D. blocks are Karra, Torpa, Rania, Murhu, Khunti & Erki (Tamar II). It has also 01 stautory towns (Khunti Nagar panchayat) and 01 census towns (Torpa-CT).

The lanscape of the district is uneven and mountainous with perennial rivers flowing at seasons. It consists of tabular landmass and has even flat surface with isolated hillocks known as Tongri. The land is formed of gneiss, granite & schists. The entire area is full of Tanrs and Dons on account of rolling topography. Tanrs are comparatively higher lands and Dons are the lower lands. The entire district has red soils except some patches of river valleys. The district is endowed with healthy climate and there is dense vegetation. The mean average temperature is 25.8°C. The maximum & minimum temperature has been recorded 43.6°C & 5.3°C respectively. The average total rainfall is 1285 mm out of which 86% falls during June-Sept months. The important rivers which flow in the district are Karo, Lohagara, Kanchi, Karkari etc.

This district is devoid of good communication links in terms of railway & NH connectivity. Only one branch of Broadgauge network conncting Ranchi to Bano passes through western part of the district linking Karra C.D. headquarter with other parts of the state. There is no network of National highway (N.H.) in the district. However, the district headquarters i.e. Khunti is well connected to Ranchi, Chaibasa, Tamar, Basia with state highways. Birsa Munda Airport in Ranchi is the narest airport which is situated at a distance of 33 kms from Khunti.

The economy of the district is primarily agricultural or related work. More than 83% of the total workers are engaged in the primary secor. It ranks 23rd in terms of population (5,31,885 persons) in the state of Jharkhand. It is 20th densily populated district in the state with 210 persons/sq.km. as against the state average of 414 persons/sq.km. The concentration of tribal population is very high (73.25%). The major inhabitants of tribals are Munda tribe. The condition of the tribal population is not very good and they are still struggling for basic facilities.

This districts has got natural beautiful landscape, which is well supported by cultural & religious centres. The Banai rivers branches itself into 5 different streams giving birth to "Panchgagh Waterfall'. The natural beauty around the waterfall makes it very attractive. There are also religious centres like Anganbari Shiv Temple, situated on Khunti-Torpa road. The picturesque hills of Dombari Buru is also famous situated close to 'Ulihatu' village-the birth place of Bhagwan Birsa Munda.

14. Latehar District:

The newly created Latehar district earlier a sub-division of Palamu district, was carved out on 04th April 2001. It is situated in the north western part of the state of Jharkhand in the Palamu commissionary. It is surrounded by Palamu in the north, Chatra & Ranchi in the east, Lohardaga & Gumla in the south and Garhwa & Chattisgarh state in the west. It has a population of 7,26,978 and area of 4291 sq. km. It is famous for its natural beauty, forest, forest products & mineral deposits.

The history of Latehar is inseparable from the parent district of Palamu and has witnessed many administrative changes since 1924. It was a sub-division of the Palamu district which was established as a district in the year 1892. Latehar has been named after the village of the same name situated on Ranchi-Daltonganj road. At present, it possesses one sub-divison namely Latehar, 9 C.D. blocks and 769 villages. The 9 C.D. blocks are Latehar, Chandwa, Balumath, Bariyatu, Herhanj, Manika, Barwadih, Garu and Mahuadar. There are 01 statutory town Latehar and 03 census towns namely Chandwa, Alaudia and Barwadih.

Latehar is highly hilly and characterised by rugged topography, steep slopes, high cliffs with turbulent streams. It is located in the Pat regions of Chotanagpur Plateau. The major part of the landscape is formed of hills & plateau. The average relief is more than 400 metres. Some of the hills are of metamorphic nature. The other hills are composed of snadstones, conglomerates and lava cappings with thick Laterite soils.

As this area consists of higher lands, so climatically it is more suitable for comfortable living and adequate growth of vegetation. The average total rainfall is 1285 mm and most of this occur during June to September months. The average temperature is 25.8°C which rises to 43.6° during summer months. The minimum temperture is 5.3°C during winter months.

The economy of this region revolves around the forest,

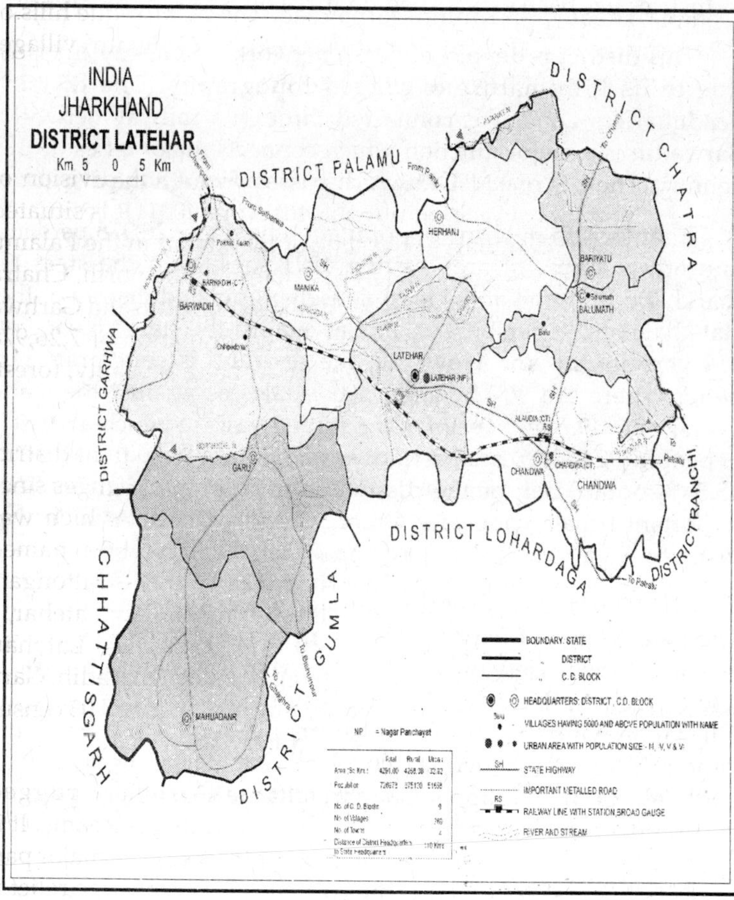

Source : Census of India 2011, Jharkhand, Series 21, Part XII 'B'.

Fig. 40

agricu!ture & minerals. A large number of peope are engaged in agricultural activities. The cultivation of paddy, maize, cereals, wheat, oil seeds are very common. The Kharif & Rabi are the main agricultural seasons. More than 78% of the total workers are engaged in primary sector. The economy of the people mostly tribals are also supported by forest products which include Kendu leaves, Mahua fruits & leaves, lac etc. Geologically, this region is very rich in various mineral deposits like Coal, Bauxite, Dolomite, Graphite, Granite, Quartz, Fireclay etc. However, these minerals are being used in a limited scale as no large & medium size industries has been established on these minerals in the district.

This district is devoid of dense network of railway & roads due to its hilly nature & rugged topography. The district headquarters Latehar is connected through a railway network. Barwadih is a major junction which connects major cities of the country. There is one NH-75 which passes through the district.

It ranks 20th in terms of population having 7,26,978 persons and 6th in terms of area having 4291 sq. km. in the state of Jharkhand. It is one of the least densely populated district of the state having 169 persons/sq. km. as against the state average of 414 persons/sq. km. However, the sex-ratio is comparatively better. There are 967 females/1000 males as against the state average of 949. This is primarily a rural district in which majority of people (92.87%) reside in rural area. The tribal constitute about 45.54% population of the district as per the 2011census. The important tribal groups are Oraon, Munda, Gond, Asur in the district.

The district has several important places of tourist interest. The Betla National Park is just 65 km north-west of Latehar. It boosts a wide variety of wild life. Neterhat is another important town which is referred as 'Queen of Chotanagpur'. It is a popular hill station which is famous for 'sunrise' & sunset as well as the famous Neterhat Residential school is situated here. Besides this, Lodh fall, Sadni fall, upper Ghaghri falls are also other places of tourist attraction.

15. Lohardaga District :

The district of Lohardaga is located in western part of the Ranchi Plateau. It is bounded by the districts of Latehar in the north, Gumla in the west and south and Ranchi in the east. This district has also clear cut demarcation in the west by Pat Region. This district extends between 23⁰30' N. latitude to 23⁰40' N. latitude and 84⁰50' E. longitude to 84⁰50' E. longitude and comprises an area of 1502 sq. kms. and population of 7,26,978 persons (2011).

This district has old history and has registered many

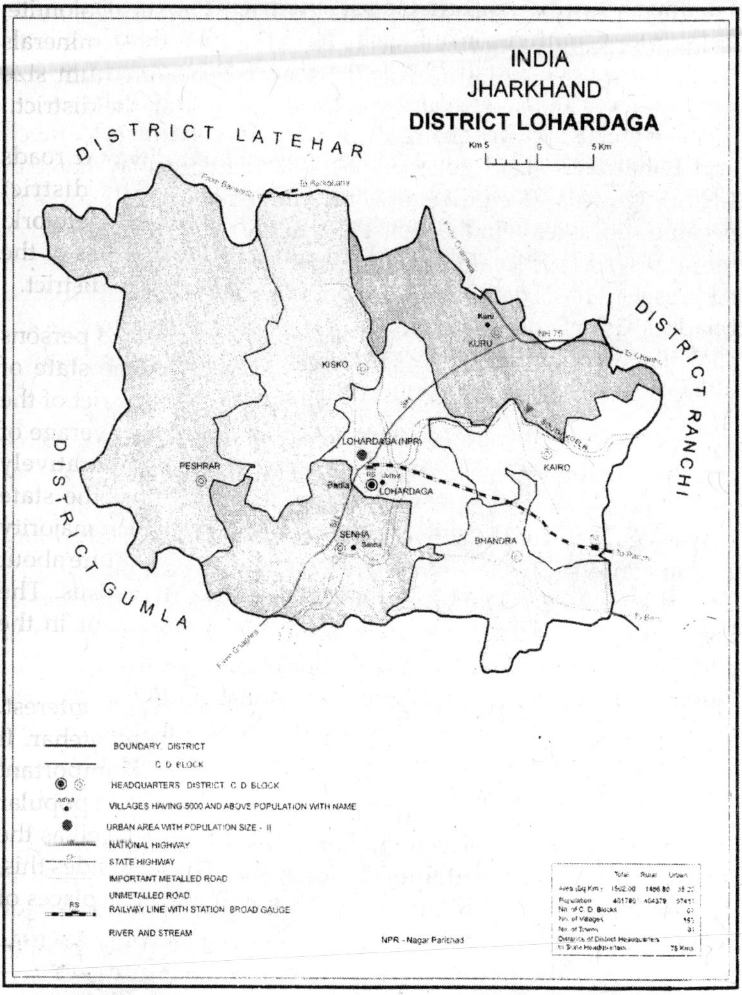

Source : Census of India 2011, Jharkhand, Series 21, Part XII 'B'.

Fig. 41

administrative changes. When there was South-West Frontier Agency consisting of Parganas of Palamu, Dhalbhum, Hazaribag etc. Lohardaga was the administrative headquarter. On the abolition of the South-West Frontier Agency in 1854, Chotanagpur became a non-regulation province under a commissioner and the district which was called by the name of the former headquarters of Lohardaga consisted of two subdivisions- namely Pargana, Palamu as the Mufassil and the present district with Pargana Tori as the Sadar subdivision of Ranchi. In 1892 Palamu and Tori, with three Parganas which had been transferred from Gaya to Lohardaga in 1871, were formed into the district of Palamu. The name of the district was changed in 1899 from Lohardaga to Ranchi and Lohardaga again got the status of a subdivision. Later on Lohardaga was given the status of a Revenue Thana and again became subdivision in the year 1972. In the year 1983 it achieved the status of a district after getting separated from Ranchi District. At present it has 5 development blocks namely Lohardaga, Sadar, Bhadra, Kiska Senha and Kudu and consist of 67 Panchayats and 354 villages. Ninety percent people reside in villages and only 72 villages have got electric connection.

As the district is located in the higher Ranchi Plateau it has more undulations. The western part of the district is hilly known as Pat and the eastern parts comparatively flat plateau, known as Lohardaga Plateau. The above two topographic areas get separated by the contour of 400 metres. This region has several hills and Tongri known as Bagaru Pat, Dhulia Pat, Lalmatia Pat, Rudani Pat, Pakhar Pat, etc. In addition to these Pats there are several residual hilly ranges and isolated hillocks. Some of them are Rani Pahar, Chilam Pahar, Chongri Pahar, Mahabani Pahar, etc. The land slopes from west to east and consists of igneous and metamorphic rocks. Important rivers are the Auranga, the North Koel, the South Koel, etc.

Since the area consists of higher plateaus and considerable forest cover it has suitable climatic conditions for natural vegetation, human habitation, wild life, etc. Summer season remains comfortable and winter season remains cold. Average temperature of the area is 23^{0}C. The highest temperature goes upto 36^{0}C and the lowest temperature comes down to 1^{0}C. The rainfall is about 100 cms. in the lower areas and 200 cms. in the higher areas. It has dominance of laterite soil. Nearly one fourth area of this region has forest cover where trees of Sal, Mahua, Jamun, Neem, etc. are mostly found. This region is rich in resources. Bauxite mineral is the most important and is famous for its production. There are 18 bauxite mines mostly located in Bagru Pat. Nearly 200 loaded trucks of bauxite and a few loaded railway wagons are transported from here every day. There are also mines of stones, limestone, chinaclay, etc.

The impact of hilly terrain, considerable forest cover and rocky exposures have obvious impact on land utilisation. Nearly 1/4th area is covered with forest and vegetation. Less than half area is under cultivation. The remaining area is under cultivation. The remaining area is under other categories of land use like habitation, pasture, orchards, playground, culturable waste, etc. Amongst these the pasture land is important and has encouraged dairy activities. There are 78 dairy co-operative societies and nearly 4000 litres milk is produced and sold every day. Out of total culturable land only 10% is irrigated which reveals poor development of irrigation of the cultivated area 56.5% land remains engaged under Aghani crops, 30.93% land under Bhadai crops and 8.84% land under Rabi crops and 1.40% land under Garma crops. Paddy, Ragi and Maize are important crops in the district. Certainly during recent years people have given due importance to the production of vegetables and fruits. There are a few minor irrigation schemes like Nandi Irrigation Project in Bhandra, Dahar, Bati Irrigation Project in Kisko and Phulyar Nala Irrigation Project in Senha.

The area is linked with narrow rail route which mostly carries goods and minerals. Important railway route stands between Ranchi and Lohardaga. The only state highway No. 48 connects this district but there are unmetalled roads linking the block headquarters, service centres, mining centres, etc. Rivers are rainfed and non-navigable on account of presence of rolling topography.

Although this district has huge reserve of bauxite it has not given rise to aluminium industry in the district. Certainly there are, small scale cottage industries, engaged in copper, bronze, utensil production. There are forest based industries and dairy industries.

Nearly 90% population get their livelihood from agriculture. Nearly 76% workers are cultivators and 16% workers are agricultural labourers. Only 1% workers are engaged in industries. 10% workers are migrate for works elsewhere. The area has uneven distribution of population as well as settlements. A good no. of settlements concentrate along the road connecting the townships of Ranchi and Gumla. The maximum concentration of settlements is in the Kundu development block.

Important tribes of this district are Asurs, Kharia, Lohra, Mundas, Oraons etc. Important dialect and languages are Asuri, Kharia, Kuruk, Mundari etc.

Lohardaga is only town in Lohardaga district where trade, administration, industries, educational institutions, etc. are concentrated. There are a few temples like Chutia Temple, Boraia Mandir, Jagannath Mandir. Korambe 16 km south from Senha near the river South koel, is known as Haldiaghati where battles between Nagbansi and Raxel rulers took place. This place has also a temple of Basudev Rai containing statues of several God and Goddesses. Pakhar Pat nearly 15 to 20 km is noted for natural scenaries having a few waterfalls.

16. Pakur District :

The district of Pakur lies in the eastern portion of Rajmahal hills. It is bounded by Sahibganj in north, Dumka in south, West Bengal in east and Godda in the west. In the north this district is separated from Sahibganj by a long chain of hills extending from river Gumani. In the east there lies a plain. It extends between 23°40' N. latitude to 25°18' N. latitude and 86° 28' E longitude to 87°57' E. longitude and comprises an area of 1811 sq.kms. It consists of 9,00,422 (2011) persons. It has six development blocks namely Litipara, Amarpara, Hiranpur, Pakur, Mahaeshpur and Pakuria.

This district contains three distinct tracts viz. 1. A portion of the Domini Koh 2. The Paragana Ambar and 3. The Pargana Sultanabad. The Damini joh portion of the district occupies the north-western corner of the district and covers 31% area of the district. The Pargana Ambar tract lies in the east and Pargana Sultanabad occurs in the south and these two contain the remaining areas. The greater part of the Damini Koh is a hilly tract populated by the aboriginals. The Pargana Ambar is a level rice plain studded with hills higher and thither in which the conditions are the same as in Sultanabad except for a strip of about 6 km adjoining the Dumka Damini where the country is rocky.

The principal rivers are the Bansloi, Brahamani, Pagla, Torai and Gimani. The Bansloi and Brahamani which are used extensively for irrigation are the largest rivers in the district, each

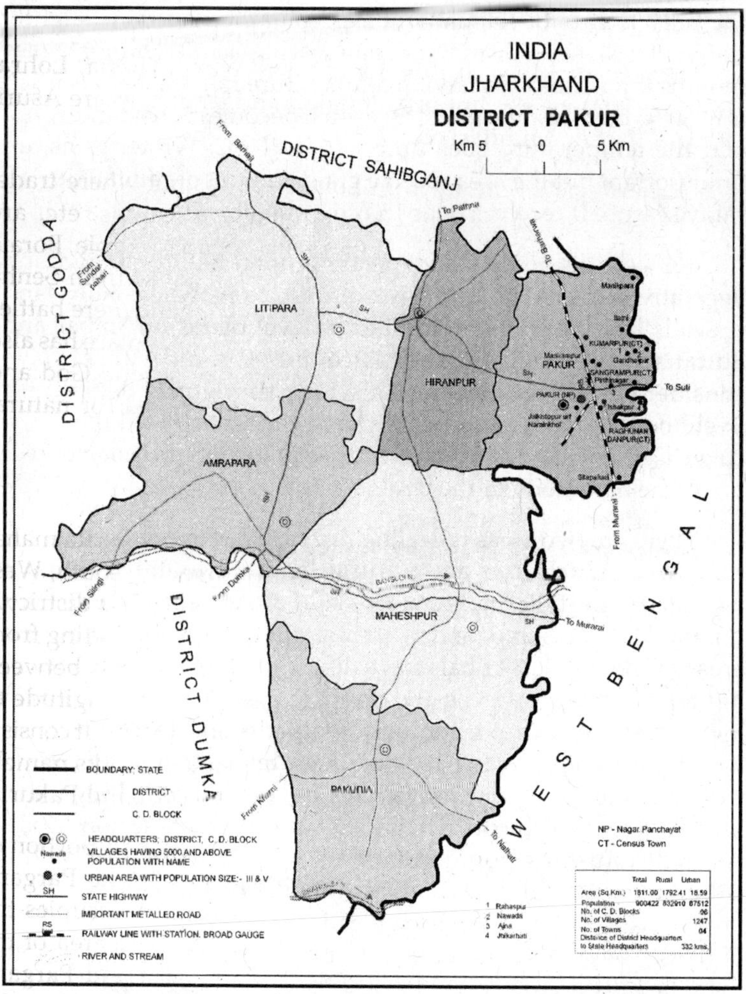

Source : Census of India 2011, Jharkhand, Series 21, Part XII 'B'.

Fig. 42

having an average breadth of about 65 metres. Both flow through the Sultanabad Pargana and so does the Pagala or Mad river a stream notorious for its violence. The Torai is a narrow river in Pargana Amber of some importance on account of its fisheries and the Gumani is a hilly stream in the Damini Koh running along the boundary between this district and Godda district.

As the area lies in the vicinity of humid Bengal plain and near the hilly ranges of Rajmahal it has climatic conditions slightly different from other districts of Rajmahal highlands. During winter months it remains cold having temperature ranging between 15° C to 30° C but during summer season it becomes considerably hot and the temperature goes upto above 40° C. Winter rains are unimportant but the area receive good amount of rainfall during rainy season. It receives over 150 cm rainfall.

Soils of this area vary depending upon nature of land and vegetative cover. The north-western portions where hilly land prevails lateritic soil is found but in level plains of Ambar and Sultanabad Parganas clayey is found. The hilly areas have considerable forest cover upon which aboriginals depend for livelihood. In plain areas people have almost depleted the forest cover. Certainly orchards attached to settlements and isolated trees break the monotony of the land.

There are two areas revealing distinct land use pattern. In the hilly areas patches of agricultural land are found which are available in the hills. In these areas land under vegetative cover is dominant. Culturable land is comparatively less than the plain areas of the district. Tribals have developed their settlements on hill tops and higher areas and they are practising agriculture in lower areas. In this portion only footpaths and cart-track have developed and involve less land under transport network. Certainly there are pastures, fallows and other uses in hilly areas. The plain areas of the district having settlements of the non-tribals specially Muslims population have used most of land under culturable use, settlemtns and other non-agricultural uses. In this area transport network has engaged considerable land. Because of dense population non-agricultural use of land is much more than the hilly areas.

There is a absence of mineral resources and any formidable

industry. There are, howerver, several stone crushers and agro, and forest based industries.

This district has the facilities of both railways and roadways. There is a rail route connecting Sahibganj, Rampur Hat and Howrah. The headquarter of Pakur also has the facility of railway. There is a good network of roads in the area. Roads connect Pakur with Gudda, Rajmahal, Dumka, etc. The plain areas have comparatively better tranoport system than the hilly areas.

The hilly areas of this district is populated by aboriginals and the plain areas are mostly inhabited by Muslims and Hindus. The average density of population in this district is 497 (2011) persons per sq.km. but developments blocks having very high density are Pakur (1270) and Hiranpur (493). The remaining development blocks have density of population less than the regional average. This district has 1141 inhited and 106 uninhabited villages. There are four towns Pakur with a population of 45,840, Raghunandanpur (8335), Kumarpur (7153), Sagarmpur (6184). This area has 42.10% Scheduled Tribe and 3.16 % Scheduled Caste. Important tribes of this area are Santhal and Paharia. They are more concentrated in Amarpara and Litipara. The district consists of a good no. of traders. Some of them are involved in smuggling activities from Bangladesh because of its proximity.

17. Palamu District :

Palamu district is located in the north-western part of the state. It is surrounded by the districts of Rohtas and Aurangabad in north, districts of Latehar and Lohardaga in south, Chatra in east and Chattisgarh state and Garhwa in west. It extends between 23°20' N.lat. to 24°39' N. lat. and 85°58'E longitude 86°20'E longitude. It comprises 19,36,319 (2011) persons over an area of 4393.00 sq. kms. It contains 3 sub-divisions viz. Daltonganj, Chattarpur and Hussainabad : The district consists of 20 CD Blocks. There are altogether 1870 villages in the district. After 2001 census on 17.4.2002, the erstwhile Latehar sub-division of the district was notified as a separate district of the state of Jharkhand.

This district came into being in the year 1892 but before it was a subdivision of Lohardaga district. Its headquarter at Daltonganj was established in the year 1861 by Colonel Delton. As regards the nomenclature of Palamu there has been different opinions.

Some people hold the view that Palamu has been named after house of frost which used to take heavy toll. Some people considered its name based on the abundance of Palas, Lane and Mahua. Some people considered its nomenclature on account of Goddess Palamu on Auranga river which is due course of time the word Palama was being called Palamu. The reality of this fact is attested by the rock Sculpture available in ruins of Palamu fort. Palamu district portrays a formidable historical background. It has witnessed many ups and downs. King Medani Roy's works

Source : Census of India 2011, Jharkhand, Series 21, Part XII 'B'.

Fig. 43

had put to this district in utmost top position. Palamu fort was constructed by him on Auranga river. The battle fought by Nilambar and Pitambar brothers in 1857 may not be forgotten. The people of Palamu have highly participated in Sardar battle and Quit India Movement. During the British rule it was used to be the summer capital of Bihar. A number of Sanskrit Schools had been established in Palamu district which produced many scholars of repute.

Palamu district is highly hilly and forested. The average relief is 400 metre. The highest spot is Netarhat hills found in Palamu area of three types. Some of these hills are made of metamorphic rocks. The remaining hills are composed of sandstones, conglomcerates and lava cappings having thick mantle of laterite. This district is drained by the river North Koel and its tributaries - Auranga and Amanat. There are other smaller rivers like Burhi, Banki, Panda, Danro, Tahle, Malay, Anaraj, etc. The river Koel is about 225 km in length. The river Auranga originals near Sohida and joins the river Koel after flowing 80 km. The river Amanat originating from Hazaribag plateau joins the river Koel. Other rivers are Kanhar and Sadama. There are several waterfalls also. Lodh waterfall is the biggest in the state. It forms a fall of 468m from Porsapat.

As this area consists of higher lands in south and almost plain areas in the north, it presents diversities in climatic conditions. The southern portions provide bitting cold and mild summer but in the plains it records cold winter and hot summer. During winter season the area records 16° to 18°C at Daltonganj and during summer season the temperature increases upto 41°C. The higher areas of the south even receives 200 cm rainfall but the northern areas remain in rain shadow and receive rainfall less than 120 cm.

The area is well forested in the southern parts but almost devoid of forest in the northern portions. Dense forest has been in the southern Pat region where steep slope and negligible human interference has been marked. There are reserve forest also where wild life is taken care. Important trees are Sal, Ashan, Karam, Sisam, etc. In addition these trees there are Mango Amla, Mahua, Kend, Jamun, Bel, Fruit trees etc. Bamboo thickets are also found in this area. Here, Betla park is noted for wild life. Soils of this region differ from south to north. Southern areas have lateritic soil because of the decomposition of laterite. In the northern portions, plain areas provide Domat and Clayey Soils.

Since the area has partly rough terrain and homogenous plain, land use also varies. In northern areas more land devoted to agriculture, residential use, transport and other non-agricultural uses but in the southern portion more areas are engaged in forest cover. Patches of agricultural lands are also found. The areas having rocky projection are waste land. There are very inaccessible areas where isolated dwellings are found and very limited lands are under cultivation.

The area contains a poor transport network. There is no any national highway. Certainly, there are a few state highways which connect Daltonganj and other important places from other areas. Important roads are Ranchi-Daltonganj (155 kms.), Daltonganj-Aurangabad (70kms.), Chandwa-Balunath-Chatra (28kms.), Rahla-Godarman (57kms.). The area has railway connection to Paha via Gaya and Baraka kana. There is a small air strip at Chianki near Daltonganj.

This area has also mineral resources such as coal, dolomite, graphite, quartz, felspar. There are coal mines at Rajhara and Pandua. Netarhat Plateau provides enormous quantity of bauxite. Important bauxite mines are Jora, Dumar, Gari tahr Pahar, Jarmira Pat, etc. Iron ore is found at Datur and Sua. Dolomite and limestone are found at Kothia Balumath, Pandwa, Majholia Graphite, & China clay are also found in this area.

The area is not developed in terms of industries. It is however, a few industries which have developed based on local resources. Cement factory has developed at Japla. Lumbering industries are found all over the district. Stone crushing industries have also been developed.

As such the physical condition of the district is not uniform the distribution pattern of population is also not even. The average density of population is 442 (2011) persons per sq.km. Relatively higher density is found in plain areas like Hussainabad 4859 (2011) persons. This district has 1711 inhabited villages and 159 uninhabited villages. There are 3 statutory towns i.e. Hussainabad, Bishrampur and Medininagar (Daltonganj) and 6 census towns. The scheduled caste population is 27.65 % and scheduled tribe is

9.34%. Oraon, Munda Chik Baraika Gond, Korba, Parahia, Asur, Birhar are important tribes in the district.

There is one site of importance is Burhaghagh also know as Lodha waterfall on the river Burha and water fall from a height of 450 feet. There is a hot spring at Tatha, 51 km away from Daltonganj. There is Sukhdhari waterfall on the river Kanhar, 35km away from Nagar Utari. There is Saptbahini waterfall at Demu where several temples have been constructed. There is Palamu Fort 5 km from Betala. There is ruin site of Chero Fort at Nawgarh.

18. W. Singhbhum District :

This district is located in the Southern portion of the state. It is bounded by Saraikela-Kharsawan and Khunti districts in north Orissa in South, Saraikela-Kharsawan and Orissa in east and the district of Khunti, Simdega and Orissa in west. It extends between 21°58' N. Lat to 22°55' N.Lat. and 85°0' E. Long. to 86°2' E.Long. It comprises an area of 7224 sq.km and has 15,02,338 persons in 2011. This district has 2 subdivision and 18 C.D. Blocks - Chakardharpur subdivision having Bandgaon, Khuntipani, Anandpur, Chakradharpur, Sonua, Goilkera, Gudri and Manoharpur, Chaibasa Sadar subdivision having Noamundi, Jagannathpur, Majhgoan, Kumardungi, Manjhari, Tantnagar, Hat Gamharia, Jhinkpani, Tonto and Chaibasa. This district came into existence in 1990 after the separation of old Singhbhum district. Chaibasa is the district headquarter.

This area consists of five distinct units. Hilly areas stretching from Deo river in the west to Dhunadih in the east and Sanjay Karo water divide into the south to southern scarp of Ranchi Plateau in the north. This area has hilly ranges, valleys, plateaus, etc. Hilly and steep sloping areas provide dense forest cover. This area includes the Porahat region. In the east of Porhat hilly region there lies the Chaibasa Plain which is made of Sanjay and Kharkai rivers. This is a flat plain area. North to the Chaibasa Plain there lies Kharsawan and Saranda hilly areas which provide dissected topography. In the southern portion of the district there lie Kolhan upland and the east lie Dalama Range and Dhalbhum highland. The plain areas have the elevation about 150 to 300 metre but the hilly areas have about 300 to 500 m. elevation. There are a few hills having elevation of 650 m. The Saranda hilly area consists of 700 hills. Dalma range in the northern parts of the district has stretches from Chandil to Ghatshila. It has a crescent shape extending from west to east. There are several ridges. Important ridges are Desbar, Dalma, Chandri Pahar, Raisindri, etc. The river Subarnarekha cuts this range and separates it into two parts. The eastern portion have high peaks. Dalamu peak is 1130 m. high

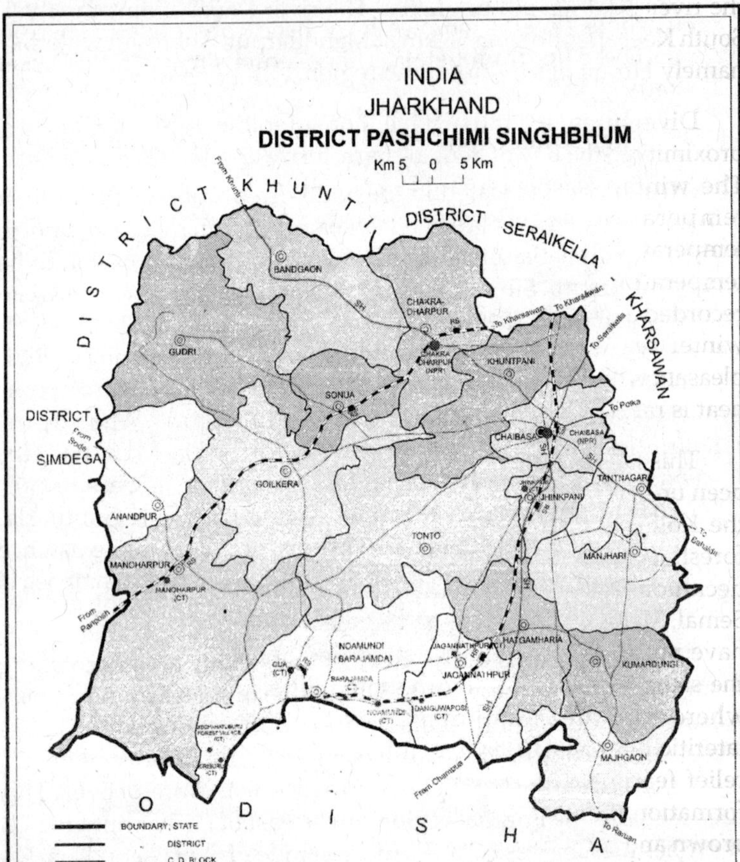

	Total	Rural	Urban
Area (Sq Km)	7224.00	7110.06	106.04
Population	1502338	1284304	218034
No. of C. D. Blocks			18
No. of Villages			1673
No. of Towns			11
Distance of District Headquarters to State Headquarters			143 Kms.

Source : Census of India 2011, Jharkhand, Series 21, Part XII 'B'.

Fig. 44

which is about 830 m. above the bed of Subarnarekha. The hilly ranges have steep slope. Dalamu trap range has a few peaks like Bicha (894 m.), Tatkora (938 m.) and Narji (802 m.). The Quarzite range of this region has peaks like Dharagiri (460 m.) and Laxni (526 m.). This region has few important rivers like the South Koel, Sanjay Baitarni, Roso, Brahamini, Deo, Koyana, etc. The river Sanjay rises in Porhat hilly areas and covering 48 km distance joins the river Kharkai at Lungta Sai. The river Koyana joins the river South Koel after flowing 56 km at Manotiarpur. There is a waterfall namely Hirani lies in limestone region.

Diversities in landsurface, considerable forest cover and proximity to the Bay of Bengal characterize the climatic conditions. The winter season remains reasonably cold when minimum temperature is generally recorded 3 to 4°C and average temperature remains 16°C. But during summer season the higher temperature even goes upto 43°C but the average temperature is recorded 33°C. This area receives annually 142 cms. rainfall. The winter season receives negligible rains. The weather remains pleasant with clear sky. Rainy season records heavy rain. Scorching heat is rarely witnessed.

This area has considerable forest cover. Nearly 20 % area has been under forest land. But most of the forest covered areas lie in the Kolhan upland, Dalma range and Saranda hilly areas. The forests contain thorny bushes semi-deciduous and humid belt deciduous forests. Important trees are Sal, Ashan, Jamun, Karam, Semal, Mahua, Kendu, Arjun, Gamhar, etc. Areas near settlements have mostly Mahua and Kusum trees and bamboo thickets but the steep sloped areas and remote hilly areas have dense forest where wild life is important. Generally soils of this region is lateritic but soils differ depending upon formation of rocks and relief features.The plain areas have black soil becuase of lava formation. Laterite soil is found in most hilly areas. Yellow to brown and brown soils are found in gentle sloping areas specially in Seraikela area.

The area reveals variation in land use pattern. Nearly one-fifth area is forested and three-sixth is culturable land but in Kolhan upland, Dalma Range, Saranda hilly regions there is dominance of forest cover and only patches of agricultural lands are seen. In the Saraikela and Chaibasa Plain most areas are culturable certainly non agricultural uses of land like habitation, transport, fallows, wasteland etc.are observed. Rice is the principal crop. Three-sixth area is engaged under paddy cultivation. Limited land is devoted to wheat cultivation. Other than paddy and wheat oilseeds and vegetables are also grown. These crops are irrigated by minor irrigation schemes like natural reservoirs, springs, waterfalls, tanks, etc.

The area is not so developed in terms of transport network. Certainly there are a few important roads traversing this region. National highway No.33 passes through this region which is joined by Chakradharpur-Chaibasa-Ranchi Road and Chaibasa-Haldipokhar road. Even National Highway No.32 passes through Chandil for a short distance. There are other roads like Chaibasa-Ranchi, Chaibasa-Jayantgarh, Chaibasa-Haldipokhar, Chaibasa-Saraikela-Chandil, Seraikela-Kharsawan- and Kharsawan Amda. In addition to these roads there are roads connecting mines and forests. This area has also rail network. Important rail routes are Sini-Chakradharpur-Manikui, Chakulia Soni, Chakradharpur-Goilkera, Goilkera-Saraikela, Bandra-Gamharia-Rajkharsawan, Dongaposi, etc. In addition these rail routes there are rail routes connecting mines. Rivers are not navigable. There are air strips at Chaibasa, Noamundi and Chakulia.

From the point of view of minerals this district is more important. It is noted for iron ore, manganese, copper, chronite, vanadium, tungstun, limestone, kyanite, dolomite, etc. Important iron ore mines are at Gua, Jameda, Noamundi, Gurumahisani, Badampahar, Kiriburu. Virtually it forms an iron ore series which provide two -third iron ore of the country. It is located in Kolhan region. There is a belt of copper mine near Bimani river which stretches from Dwarparam Kharsawan. Manganese is found it southern portion at Dua, Livuti, Jamda chronite is found at Jojohatu and Keraikela. Limestone is found at Jagarnathpur and Rajanlia.

Although this district is rich in mineral resources there is less industrial development. There is a cement industry at Jhinkpani. There is a railway workshop at Soni. Bihar Spounage Iron Ltd. at Chandil. There are other agro-based and forest based industries.Chakradharpur and Manoharpur are noted for Biri production.

The district is sparsely populated because of considerable areas having dissected terrain and forest cover. The average density is 208 persons per sq. km. Development blocks having relatively higher density include Chaibasa (428) and Chakradharpur (374) becuase of urban and industrial development and other development blocks have lower density. This district has 1642 inhabited and 31 uninhabited village. There are 15 towns. Important urban centres are Chakradharpur (56,531), Chaibasa (69,565), Adityapur (1,74,355), Seraikela (14,252) etc. This district comprises 3.79% scheduled caste and 67.31% scheduled tribe. Important tribes are Mundas, Bhuiyas, Hos, etc.

There are waterfalls at Lupungbutur, nearly 2 km away from Chaibasa and Hirni waterfall. There is a Shiva Temple at Chainpur near Chakardharpur. There are old temple of Lord Shiva at Harina and Jagannathpur.

19. E. Singhbhum :

Purbi Singhbhum is situated in the south-eastern portion of the state. It is bounded by West Bengal in north, Orissa in south, West Bengal in east and Saraikela-Kharsawan in west. It extends between 22°11' N to 23°01' N. Latitude and 86°04′ to 84°54′ E. Longitude. It has an area of 3533 sq.kms. and 22,93,919 (2011) persons. This district consists of two subdivisions - Dhalbhum having Patamda, Golmuri cum Jugsaloi, Boram and Potka development blocks and Ghatshila subdivision having Dumaria, Musabani, Ghatshila, Dhalbhumgarh, Chakulia, Gurabandha and Bahragora development blocks.

As regards the nomenclature of this district there is controversial opinions. Some people hold the view that the word Singhbhum is based on the word Singhbonga, the chief God of Ho tribe. Some people attatch this word with that of ruler of Poshat. Even some people link this word with the abundance of lions in early days. This district came into being on 16th November 1990 by getting bifurcated from the main district of Singhbhum.

This district is noted for varied land surfaces. There are high hills ranges eroded valleys and undulting land diferential forest cover. This district is surrounded by the Dalma Range and

Dhalhun hills. The ancient Dalma Lava flows have produced dominant linear range with epidiorite backbone. The Dhanjuri Lava flow is conspicious in in its layout in South Dhalbhum. Newer Dolerite Dykes stand out as lines of Dunjris or low hills in the region. The picturesque scenery of the Dalma-Dhanjori highlands stand a great contrast to the Mica-Schist Plain. The analysis of tis relief has solved various complex geomorphological problems. The pre-cambrian fold mountains which have been denuded to a peneplain are composed of mainly mica-schist and phyllites of the iron ore series. Phyllites and Schist hills bordring the Dalma mountains are in its remains. The relief of the area varies between

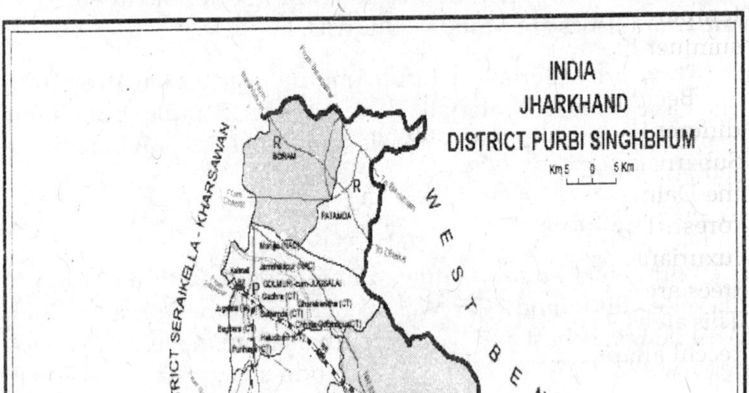

Source : Census of India 2011, Jharkhand, Series 21, Part XII 'B'.

Fig. 45

150 m. to 300 m. The relief is however, higher in southern portion where Dhanjori Range is located and the north-western portion where Dalma range exists. The lower areas lie between the above two ranges known as Dhalbhum plains mainly created by the river Subarnarekha and its tributaries.

This district has its location in south-eastern portion having nearness to the Bay of Bengal which facilititate this area with comparitively better amount of rainfall. This area receives about 150 cm rainfall annually and record main annual temperature above 26° C. During winter this area receives negligible rainfall but heavy downpour takes place during rainy season. The temperature ranges from 16°C in winter months to 44°C in summer months.

Becuase of varied landscape the forest cover is found in different proportion in different areas. The plains associated with Subarnarekha basin here recorded considerable deforestation but the Dalma and Dhanjori highland areas are still clad with thick forest. High rains and summer temprature favour wild and luxuriant forest. The Sal trees are dominant in this area. Other trees are Gamhar, Mango, Jamun, Jack fruits, Kaaranj, Palas, etc. This area has gravel soil, eroded scarp soil in the higher areas and recent allunium in the plain areas.

This area provides dominance of culturable use of land in the plains and forest cover, waste land, mining, etc. in the hill areas. It may be noted that the plain areas of Dhalbhum is quite fertile. In this area farmers growAghani, Rabi and Bhadai crops. Among Aghani crops paddy is important. Farmers grow wheat (rabi crop) in plain areas. Oilseeds are also grown considerably in the area. The production of vegetables and seasonal fruits has remarkably increased due to urban demand. The area also has the facilities of irrigation by Chandil dam, Dih Barrage and Icha Dam.

This district has a good network of transport specially in the plains due to homogenous land and in the hilly areas due to presence of mines. The National Highway No. 6 passes through the South-eastern portion. The National High No.33 connecting Ranchi-Jamshedpur meets National Highway No. 6 at Baharagora. There are several state highways linking Adityapur-Kanda, Baharagora-Haldipokhar, Tatanagar-Bharbharia, etc. There is a good network of road inside Jamshedpur township which is about 240 km in length. Railways also connect almost all industrial centres. Jamshedpur is an important junction connecting Kolkata, Mumbai, Delhi, etc. Other rail routes include Tata- Gurimahisani-Badam Pahar, Kandra-Gamaharia hord, Barbil branch etc. Rivers are not navigable but during rainy season ferry service is available at some ghats. There is a good aerodrome at Jamshedpur. Tata Steel Authority has developed a separate air strip.

This district has several minerals found in Dharwarian system. Important minirals are iron ore, copper, manganese, uranium, etc. Copper is found at Chapari, Didheshwar and Dhobani mines. Kayanite mine is located at Lopso. Gold and silver are also found in Chapri and Sidheshwar alongwith copper. Mushbani is important for copper. Jadugora is noted for uranium. This district has a good number of industries. A few industries developed are of international and national repute. Important industries are TISCO, TELCO, TUBE Company, Tin Plate company, Hindustan Copper Ltd., Cable Company, Tata Pigment Ltd., Uranium Corporation of India, Glass factory at Kandra, etc.Virtually an industrial betl from Jamshedpur to Kandra has emerged.

This area is moderately and densely populated. Average density is 644 persons per sq.km. but in rural and urban areas density condiderably differ. Development blocks having very high density include Golmuri-Cum-Jugslai (1441) in which several townships are located. Development blocks having lower density include Dhalbhumgarh, Dumaria, Potaka where density is less than 350 persons per sq. km. The district has 1609 inhabited villages and 139 uninhabited village. There are 15 towns, out of which nine towns are located in only one development block of Golmuri-Cum-Jugsalai. Important towns are Jamshedpur (6,77,350), Jugsalai (49,660), Baghbera (78,356), Mango (22,380), Musabani (49,598), Ghatshila (40,624), etc. This district contains 4.86% scheduled caste and 28.51% scheduled tribe population. Important tribes are Bhumiy, Santhals, Kharias, Hos, etc. Important places are TELCO Ram Temple, Sakchi Budh Mandir, Kashidih Mandir, Kalibari Mandir, Golden Rankni at Potka, Jubli Park, Dharagiri Waterfall, 18 km from Ghatshila, Mata Shambhu Temple at Chitreshwar in Bahragora etc.

20. Ramgarh District:

Ramgarh district was carved on 12th September, 2007 from the Hazaribagh district. It lies at the centre of the Jharkhand state and known for mining, industrial and cultural hub. It is surrounded by Hazaribagh district in the north, Ranchi district in the south & east and Bokaro district in the west. It ranks 17th in terms of population (9,49,443 persons) and 23rd in terms of area (1341 sq. km.) as per the census of 2011 in the state of Jharkhand.

This district lies in the North Chotanagpur division and the history of this district is inseperable from the parent district of Hazaribagh. The Hazaribagh district came into existance in the year 1833 and later on Ramgarh became the centre of military cantonment. The name Ramgarh means "The Fort of God Ram". At present it has one sub-division named Ramgarh, 6 C.D. blocks, 257 Gram panchayats and 315 revenue villages. The 6 C.D. blocks are Patratu, Mandu, Ramgarh, Dulmi and Gola. There are altogether 23 towns out of which 01 is statutory town and the rest 22 towns are census towns.

This district is a part of Chotanagpur plateau which is characterised by uneven land surfaces. It has also steep scarps, residual hills and inter plateau valleys. The important physiographic regions of the district is Damodar trough which is bounded by Hazaribagh plateau in the north and Ranchi plateau in the south. The Barka Pahar (Marang Buru) is the highest peak (3442 feet) in the district situated along the Ramgarh - Ranchi border. Damodar is the main river which forms a major river basin comprising a number of tributaries. Important amongst them are Naikari, Bhera and Bokaro river. This district is endowed with a large and rich deposits of coal and coal bed methane. It also possesses various other minerals like limestone, fireclay etc. The coal formation lies mostly in the Damodar Valley which geologically belong to the Gondwana system.

The climate of this district is similar to that of Hazaribagh & Ranchi district. It receives about 1286 mm (average rainfall) of rainfall and about 85% falls during 4 months from June to September. The average temperature has been recorded at 25.2°C. The maximum & minimum temperature has been recorded at 44.2°C & 4.8°C respectively in the district.

The main occupation of the district is mainly agriculture. There are 3 main agricultural seasons in the district : Kharif, Rabi & Zaid. Rice, maize, ragi, fruits and vegetables are the main crops of the district. More than 44% of the total workers are engaged in the Primary sector. It is also one of the important industrial district of Jharkhand. Many industrial units have been established here like steel, sponge iron, cement, refractory & thermal power plants due to availability of coal, limestone, firclay etc.

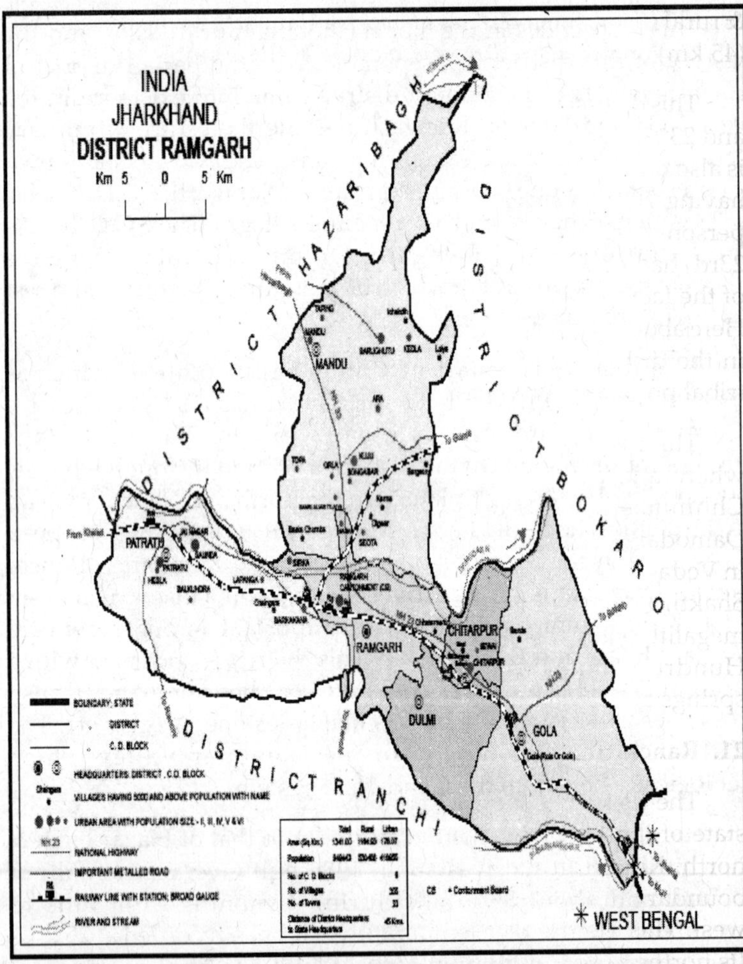

Source : Census of India 2011, Jharkhand, Series 21, Part XII 'B'.

Fig. 46

The district has a releatively a good network of railways and roads in comparison to other parts of the state. It is served by two railway zones : East Central Railway & South Eastern Railway. Barkakana is a railway sub-divisions under Dhanbad division of East Cental Railway zone. Presently, it is well connected to other parts of the state & country with railway. However, roads are the main means of transportation of passengers & goods. It is served by 02 National highways namely NH-33 and NH-23. Besides this SH-2 also passes through this district and it is supported by district & rural roads. Birsa Munda Airport at Ranchi is the nearest airport (45 km) and links to prominent cities of the country.

This district ranks 17th in terms of population (9,49,443 persons) and 23rd in terms of area (1341 sq. km) in the state of Jharkhand. It is also one of the densely populated district (ranks 3rd) in the state having 708 persons/sq. km. as against the state average of 414 persons/sq. km. However, the sex ratio is not favourable (ranks 23rd) having 921 females/1000 males in the state. This is because of the fact that a large section of population lives in urban areas. Here about 44.13% population lives in urban area as it has 23 towns in the district. Here majority of the population is non-tribal as tribal population accounts for only 21.19% of the district.

This district is also famous for natural, scenic, religios centres where a large no. of people visit evry year. The famous Maa Chhinmastika temple is situated at Rajrappa, where rivers Damodar & Bhairvi also forms a waterfall. This temple finds place in Vedas & Puranas and it has been recognised as strong source of Shakti. Besides, this district is also famous for china cemetry, megaliths sites. It is also dotted with many natural water falls like Hundru, Dhur-duria, Am-Jharia, Liril, Nimi, Dhara and hot springs at Gandhaunia.

21. Ranchi District :

The district of Ranchi has almost the central location in the state of Jharkhand. It is bounded by the districts of Ramgarh in north, Khunti in the south, Saraikela-Kharsawan and the state boundary in the east and Latehar, Lohardaga and Gumla in the west. This district also has to some extent physical demarcation. Its northern boundary is delimited by the presence of catchment area of the Damodar Basin and the higher plateau region lies in

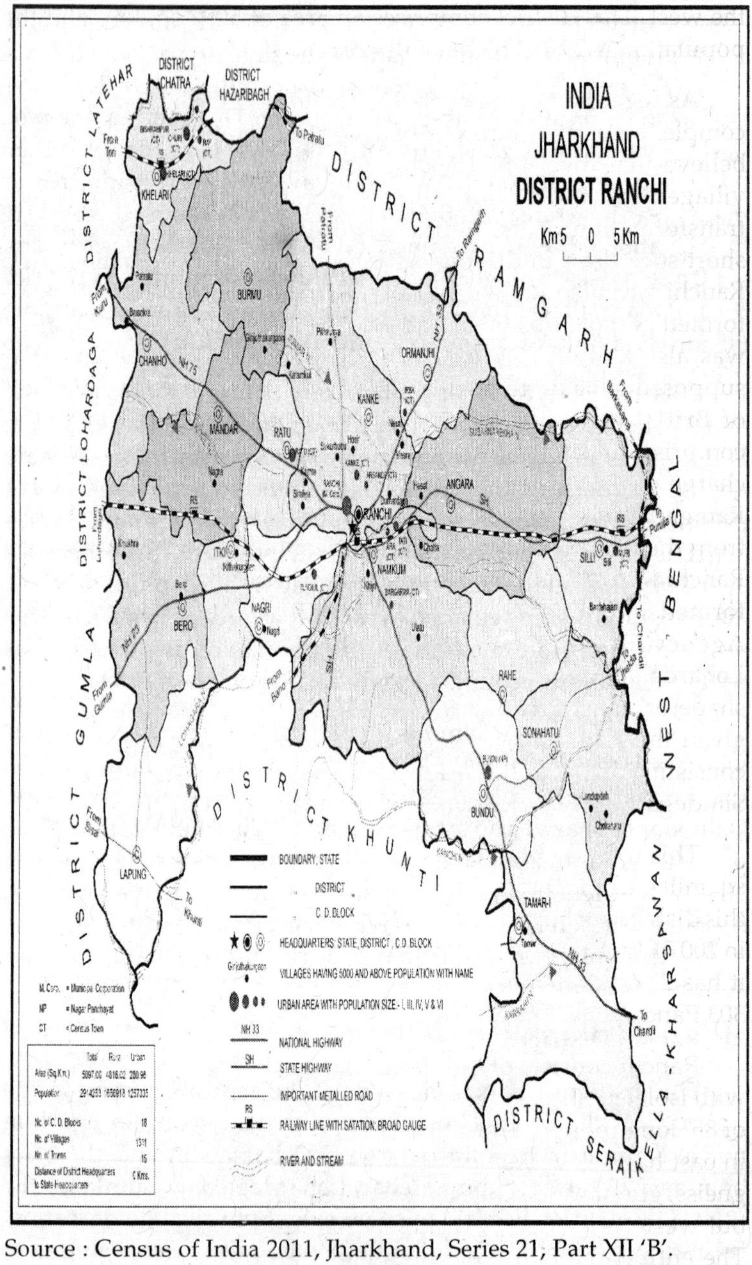

Source : Census of India 2011, Jharkhand, Series 21, Part XII 'B'.

Fig. 47

the west. This district comprises an area of 5097 sq. km. and the population is 29,14,253 persons (2011).

As regards the nomenclature and origin of the district it has complex accounts. The origin of the district is rather obscured some believe that the word Ranchi has been from Purani Ranchi, a village near Ranchi but others suggest that it is an anglicised transformation of the Mundari word 'Aranchi' which means a short stick used in driving cattle. Some people also believe that Ranchi was also known as Kokrah. In the Ain-i-Akbari Kokrah formed part of Suba Bihar and was the realm of Raja of Kokrah. It was also known to Mughals as Nappur or Coira Orissa. It is supposed to be original area of Nagbansi Raja. In the early phase of British rule this district with other adjoining territories comprised of a total area of 12004 sq. miles and was placed in the charge of the magistrate of Ramgarh in Hazaribag District but Ramgarh being outside the Ranchi Plateau it proved ineffective from the administrative point of view and in 1833 the district of Ranchi with the adjacent Parganas of Palamu and Dhalbhum was formed into a separate entity known as South-West Frontier Agency. This agency got abolished in 1854 giving rise to Lohardaga, Hazaribag and Manbhum districts. Finally it got the shape of Ranchi district in the year 1899 when Lohardaga was given the status of Subdivision and Ranchi became the district consisting of four subdivisions- Ranchi Sadar, Gumla, Khunti and Simdega.

This district remained intact till 1954 when an area of 118.2 sq. miles was transferred to Singhbhum Districts. Again in 1983 this district got bifurcated into Ranchi, Gumla and Lohardaga and in 2007 Khunti was made a district carving from Ranchi. Presently it has 2 sub-divisions namely Ranchi & Bundu, 22CD Blocks & 303 Panchayats.

Ranchi consists of tabular landmass. It has even flat surface with isolated hillocks known as Tongri. Certainly hills lying west of 85^0 longitude have elevation above 800 metres and those lying in east have elevation less than 75 metres. The land is formed of gneiss, granite and schists. Average elevation is above 650 metres but western portion is relatively higher than the eastern parts. The entire area is full of Tanrs and Dons on account of rolling topography. Tanrs are comparatively higher lands and Dons are the lower lands. Important rivers of the district are the Subarnarekha, South Koel and Sankh. Subarnarekha has important tributaries of like Kokro, Kanchi, Karkari, etc. The chief tributaries of South Koel are Karo, Bonai and Paras. The river Subarnarekha forms Hundru and Jonha waterfall and the river Kanchi forms Dasam waterfall. The entire district has red soil except a little in portion in the south-west where black soil is also found. This area has comparatively less undulating land surface. It has comparatively more human habitation coupled with more human interference on natural vegetation. This region has however, vegetal cover on 23% land. The forest is mainly confined in hilly areas. Important trees are Sal, Asan, Gamhar, Kendu, Semal, Mahua, Kathal, Jamun, etc.

The district is endowed with healthy climate. The tropic of capricorn passes over the district but the higher elevation reduces the temperature. Relative humidity also remains less so summer season is also not uncongenial. Certainly in the summer season, temperature rises upto 30.0^{0}C and in the winter season the same comes down to 17^{0}C. December is the coldest month with the mean daily maximum temperature of 22.9^{0}C and the mean daily minimum of 10.3^{0}C. The temperature in January and February is slightly higher than in December. After February temperature rises steadily and till May which is the hottest month. The mean daily maximum temperature in May is 37.2^{0}C. The highest maximum temperature recorded at Ranchi has been 43^{0}C and the lowest recorded has been 2.8^{0}C. Ranchi receives rainfall almost throughout the year, though the concentration is during the monsoon months from June to September. During the monsoon months the area receives about 82% of the annual rainfall. July and August are the rainiest months. Average annual rainfall of the district is 137.5 cm. In the months of July, August and September normal rainfall has been 35.5, 35.9 and 19.2 cm respectively. In the months of December and January the rainfall has been only 3.7 cm.

Relatively homogenous land surface has encouraged considerable use of culturable land. The lower areas provide suitable conditions for the production of paddy cultivation but the higher areas provide conditions for habitation, orchards and other uses and also for production of pulses, millets and vegetables.

Non agricultural uses of land are also important because of industrial and urban growth. There are important industries like H.E.C. at Ranchi, Cement Factory at Khelari, Aluminium factory at Muri, High Tension Insulator Factory at Namkum, Electrical Appliances at Tatisilbai, etc. There are lac factories scattered in the district. Other industries are Mahalaxmi Fibre Ltd., Nalanda Ceramic, Ranchi Distillery, Chotanagpur Engineering Works, Sri Ram Ball Bearing industries, etc. There are cottage industries like handloom. Saw mills are numerous in number based on local forest resource.

There is a good network of transport. Two national highways no. 23 & 33 pass through the district. A no. of state highways are also present in the district. Important National and State highways are Ranchi-Ramgarh-Hazaribag (91 km), Ranchi-Bundu-Tamar-Chandil (115 km.), Ranchi-Netarhat (154 km.), Ranchi-Chaibasa-Jamshedpur (202 km.), Ranchi-Daltonganj (163 km.), Ranchi-Gumla-Simdega, Rourkela (270 km). This district has also good connection with railways. There is a rail connection with almost important places like Kolkata, Delhi, Amritsar, Patna, Gorakhpur, Rourkela, etc. There is a direct rail route between Ranchi and Rourkela. There is a metre gauge between Ranchi and Lohardaga (67 km) which is mainly used for carrying goods especially bauxite. In this district rivers are not navigable because the land has got undulating slope. The aeroplane link stands between Ranchi and Kolkata, Patna and Delhi. Ranchi district is also well connected with other amenities of communications.

The spatial population in the district is uneven. It is larger in the eastern areas but the western areas contain smaller percentage of population. Ranchi district has the average density of 572 persons per sq. km. in 2011. This district has nine urban centres. Important urban-centres are Ranchi and Khelari urban agglomerations. Other towns are Itki, Bundu, etc. The area is inhabited by several aboriginals. Important tribals are Munda, Parhaiya, Kharia, Lohra, Mahli, Kharwar, Binjhia, Bedia, Asur, etc. Important languages and dialects spoken in this region are Hindi, Kuruk/Oraon, Mundari, Sadan, Panch Pargana, Nagpuria, Bengali, Kharia, etc. Urban and industrial centres have got considerable concentration of non-tribals who have come from neighbouring states and exhibit cross-culture. This district has

35.76% tribal and 5.25% scheduled caste population.

This district has some picturesque spots, cultural centres and historical sites where tourists visit. The waterfalls of Hundru, Jonha and Dasam are attractive tourist sites. Birsa Mrig Vihar has also become an important tourist site. There are also religious sites like Temple of Shiva and Durga at Mijothakuri (25 kms. north-west of Ranchi), Temple of Balram-Subhadra and Jagannath at Jagannathpur near Ranchi etc.

Important urban centres of the district are Ranchi, Kanke, Khelari, Bundu, Itki, etc.

22. Sahibganj District :

Sahibganj district is located in extreme north-east of the state as well as Santhal Pargana Commissionery. It is bounded by Katihar district in north, Dumka district in south, Malda district of West Bengal in east and Godda district in west. This district has also clear cut physical demarcation in north and east by the river Ganga. To the west boundary is formed by the Manjhure hills running from north to south. To the south boundary is formed by an irregular line of hills and highlands running from east to west which form the watershed between the Gumani and Bansloi river. This district extends from 24°43'N. Latitude to 25°18' N. Latitude and 87°27' E. Longitude to 87°57' E. Longitude and comprises an area of 2063 sq.km. It has a population of 11,50,567 (2011) persons. It consists of nine development blocks namely Sahibganj, Rajmahal, Talihari, Barharwa, Mandro, Udhwa, Pathna, Barhait and Borio. It measures about 60 km from north to south and 45 km from east to west. Its shape tapers in the north and broadens in the south.

The eastern portions of the district and a narrow strip along the Ganga to the north consist of alluvial and Diara land. The whole of western portion forming about three forth of the total area consist of hills, valleys and pleasant undulting land. To the north the hills extend in an unbroken line running parallel to the Ganga and leaving a narrow strip of level land between them and the river. To the east they run almost due north and south as far as Gumani river, except at one point where a wedge shaped block project to the east near Udhuwa Nala. All along their eastern face is a track of lowlying country streching down to the Ganga which is very liable to flood. The south-east of the district, however, on the borders of Pakur district, consist of undulting land with a gravelly soil.

There are two major ranges of hills, viz. the Rajmahal hills to the east and the Manjhwa hills to the west which run parallel to one-another from north to south at an average distance of about

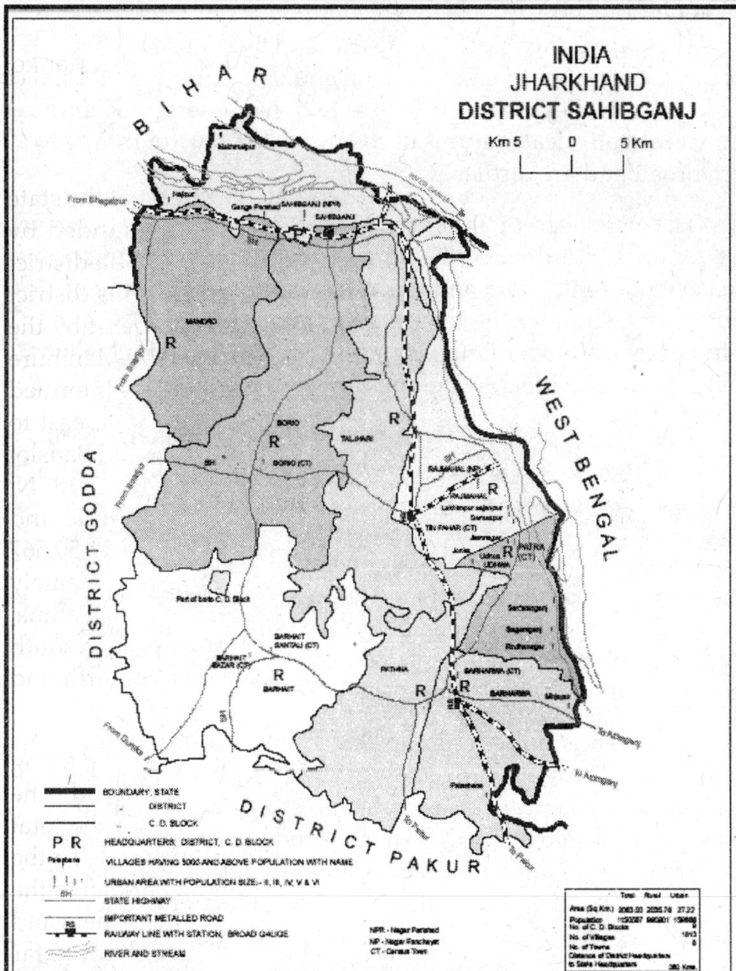

Source : Census of India 2011, Jharkhand, Series 21, Part XII 'B'.

Fig. 48

15 km. The Rajmahal hills extend as far as the point where the Gumani river debouches into the plains and the Manjhwe hills run south till they meet Gumani river near its source in the Godda district. These two ranges coalesce in the north and form a range running close to the river from east to west. South of the Gumani there is another irregular range of hills running more or less from east to west and forming the watershed between the Gumani and Bansloi rivers.

The river Ganga skirts the district and there are only two other of any importance like the Gumani and its tributary the Morel or Moran. The valleys formed by these two rivers are among the most noticeable physical features of the district. They are from 6 to 12 km broad and are fertile.

The nearness of this area to the humid Bengal, varied elevation etc. influence the climatic conditions. During winter season it becomes cold and records average temperature of 15°C but during summer the temperature ranges from 30°C to 40°C. The average annual rainfall is about 150 cm. The rainfall in winter season is negligible. Most of the rain occurs in rainy season.

The district presents two types of soils. The plains adjoining the hilly areas have clayey and sandy soils but the hilly areas have lateritic soil. At places gravelly soils are also found. Similar to soils, two clear cut regions emerge in terms of vegetative covers. The hilly and undulting areas have considerable vegetative cover but the plain areas have orchards and isolated trees. In the plain mango orchards are well developed but in the hilly areas forest has been depleted and bushes with isolated trees are found. Important trees are Sal, Mahua, Kathal, Jamun, etc.

This area is quite fertile in the plains and less fertile in the hilly areas. These two areas reveal distictive land use pattern. In the hilly areas there are areas under forest habitation of aboriginals, waste land caused by rocky exposure and soil erosion and patches of agricultural land in the valleys. On the other hand the plain areas are dominated by culturable land alongwith settlements, transport, orchards, etc. The plain areas are suitable for growing rice in Aghani season, maize and jute in Bhadai season, wheat and vegetables in the Rabi season. Hilly areas grow maize in Bhadai season and paddy in the Aghani season. During recent years people have started growing vegetables and seasonal fruits.

In terms of minerals and industries this district is underdeveloped. Important industry is Bihar Silicate Works. Small scale industries have developed. Some people have handloom industries. There are a few stone crushers. Industries related to making tin and packing materials have grown up at Rajmahal. China clay industry has also developed at Rajmahal.

This area is well linked with rail road and waterways. National highway No.34 passes by the eastern side of the district. The rail route connects Howrah and Bhagalpur. The headquarter of this district is also linked with a good metal road with Bhagalpur, Godda, Dumka, Pakur etc. There is a ferry service at Sakrigali Ghat across the river Ganga.

The population distribution in this district is not uniform. The average density is 588 (2011) persons per sq.km Development blocks having higher density include Rajmahal (1037), Barharwa (965) and the remaining development blocks have density less than the district average. This district has 1349 inhabited villages and 464 uninhabited villages. There are 8 towns - Sahibganj (88,214), Rajmahal (NP-22,514) and Boreo Bazar (CT-6964). This district has 6.29% scheduled caste and 26.80% scheduled tribe. Santhals and Pahari as are the main tribal population living in this district.

There is a Shiv Temple namely Ajgaibinath near Sahibganj. Rajmahal is the forgotten capital of West Bengal. It was once the capital of Viceroy's of Bengal. The remains of the old Mohamdan city are burried in ruins.

23. Seraikela-Kharsawan District:

The newly created district Seraikela-Kharsawan has been carved out from the parent district of Paschimi Singhbhum district on 1st April 2001. It is situated in the south-eastern part of the state. It extends between 22°29′26″N. Latitude to 23°09′34″N. Latitude and 85°30′14″E. Longitude to 86°15′24″E. Longitude. It is surrounded by Ranchi district & W. Bengal in the north, Purbi Singhbhum in the east, Odisha in the south & Paschimi Singhbhum & Khunti in the west. The total population of the district is 10,65,056 persons which is spread over an area of 2657 sq. km.

The history of Seraikela-Kharsawan is inseperable from the history of parent district of Paschimi Singhbhum and has registered many administrative changes. The parent district of Paschimi Singhbhum came into existance in 1990 after the seperation of old Singhbhum district, in which Seraikela & Kharsawan funtioned seperately as C.D. Blocks. Later on, Seraikela-Kharsawan were merged to form a seperate district

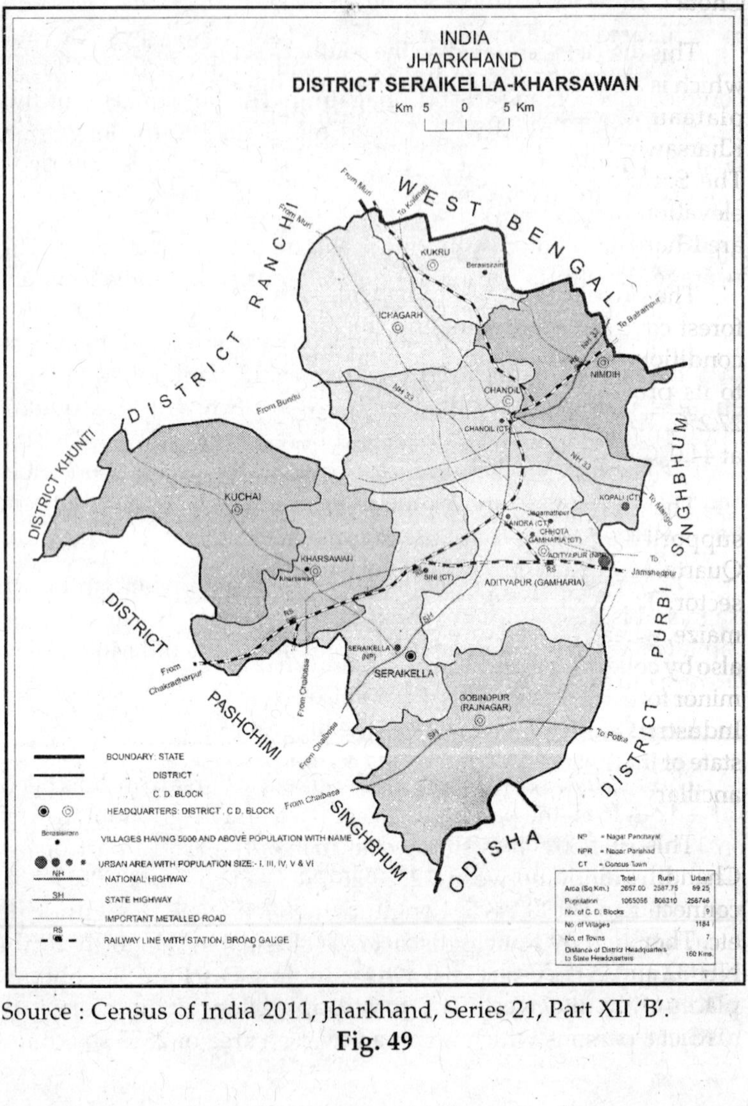

	Total	Rural	Urban
Area (Sq.Km.)	2657.00	2587.75	69.25
Population	1065056	806310	258746
No. of C. D. Blocks			9
No. of Villages			1184
No. of Towns			7
Distance of District Headquarters to State Headquarters			160 Kms.

Source : Census of India 2011, Jharkhand, Series 21, Part XII 'B'.

Fig. 49

entity on 01.04.2001. At present, Seraikela is the district headquarter which possesses 2 sub-divisions namely Seraikela & Chandil, 9 C.D. blocks, and 1184 villages. The 9 C.D. blocks Kuchai, Kharsawan, Adityapur, Seraikela & Gobindpur under Seraikela sub-division and Chandil, Ichagarh, Kukru & Nimdih under Chandil sub-division. Besides, there are 02 statutory towns namely Seraikela & Adityapur and 05 census towns namely Kandra, Sini, Chota Gamharia, Chandil & Kopali.

This district is situated on the southern scrap of Ranchi Plateua which is also known as Kolhan region. This area has hilly ranges, plateau & valleys which provide dense forest cover. The Kharsawan and Saranda hilly areas provide dissected topography. The Saranda hilly area consists of 700 hills and the general elevation ranges between 300 to 500 meter. The important rivers are Kharkai, Tikar, Shakha etc.

The proximity to Bay of Bengal, high relief and considerable forest cove have a dominating role in characterising the climatic conditions of this region. The total rainfall exceeds 1300 mm due to its proximity to Bay of Bengal. The average temperature is 27.2°C. The maximum & minimum temperature has been recorded at 44.0°C & 6.6°C respectively.

The economy of the region is mainly agrarian which is well supported by a large deposits of minerals like Kyanite, Asbestos, Quartz. More than 60% of the total workers are engaged in Primary sector. The Kharif & Rabi are the main season in which paddy, maize, wheat, oilseeds are grown. The tribals earn their livelihood also by collecting Kundu leaves, Mahua fruits & leaves, lac & other minor forest products. This district also boosts of having Adityapur Industrial Area which is one of the biggest industrial area of the state of Jharkhand. Here many big & minor industrial, engineering, ancillary units have been set up.

This district has comparatively a better railway network. Chandil, Adityapur and Sini are important railway stations and connect the district to Chakradharpur, Ranchi, Jamshedpur, Muri etc. There are 3 branches of railway network available here. The NH-32 and NH-33 passes through the district connecting distant places of the state & the country. It is well supported by SH and district roads.

It ranks 14th in terms of population as well as area in the state. The total population is 10,65,056 persons (2011 census) and the area is 2657 sq. km. So far as the population density is concerned, it is almost equal (401 persons/sq.km.) to that of the state average (414 persons/sq.km.). However, the sex-ratio of the district is comparatively better. These are 956 females/1000 males in the district against the state average of 949 females/1000 males. A significant number of people reside in urban areas here, due to the development of industrial centres & better communication links. About 24.29% (2011 census) population reside in urban areas. The tribal population constitute about 35.18% of the total population as per the 2011 census.

This district has several places of tourist interest. The district is well known for Seraikela Chhau, one of the three distinctive styles of Chhau dance. It is a unique amalgation of Oriya and tribal culture. Seraikela has become the 'Mecca' for the connoisseurs of music and dance. It is also famous for Chandil dam, Jayda temple, Maa Akarshini Devi Temple. Dalma Top is a beautiful place of natural attraction as it is situated at an altitude of 3000 feet from the sea level. This district also houses Dalma Wildlife Sanctuary, Chandil Dalma Elephant Reserve.

24. Simdega District:

The newly created Simdega district, earlier a sub-division of Gumla district was carved out on 30th April, 2001. It is istuated in the south western part of the state of Jharkhand and borders with Orissa and Chattisgarh state in the south & west respectively. It is surrounded by Gumla district in the north and west, Singhbhum in the east. It ranks 22nd in terms of population (5,99,578 persons) and 8th in terms of area (3774 sq. km.) in the state of Jharkhand.

The history of Simdega district is inseperable from the history of parent district of Gumla and has registered many administrative changes over a long priod of time. Gumla got a status of district on 18th May 1983 in which Simdega became a sub-division. Later on, it was carved out of Gumla district on 30.04.2001. Presently, it has one sub-division named Simdega, 10 C.D. blocks, 94 Gram Panchayats and 450 revenue Villages. The 10 C.D. blocks are Simdega, Kurdeg, Bolba, Thethaitangar, Kolebira, Bano, Jaldega, Pakartanr, Bansjore and Kersai. These is only one statutory town named Simdega.

This district consists of primarily Simdega Plateau which is more rugged and hilly with turbulent streams, steep slopes, high cliffs and narrow valleys. The major part of the landscape is formed of hills and undulating plateau. The major part of the geographical

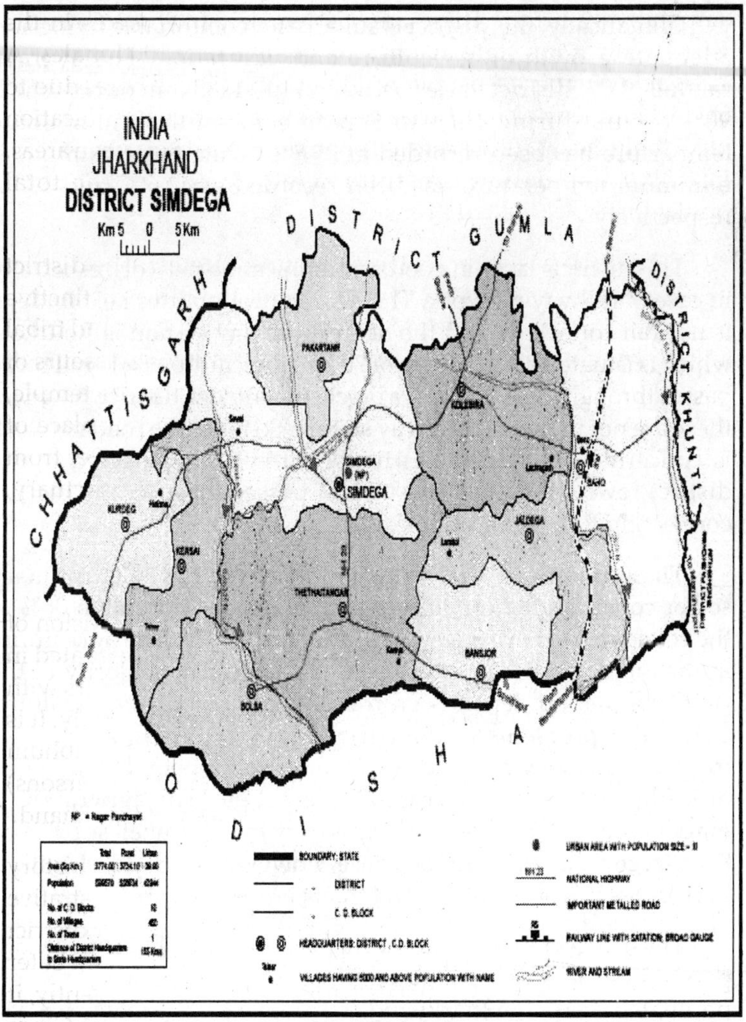

Source : Census of India 2011, Jharkhand, Series 21, Part XII 'B'.

Fig. 50

area of the district is covered with red laterite acidic soil. The upland is generally covered with morum and stone. The major rivers of the area are Sankh, Deo, Girwa and Palamara. It has about 32% forested land. The major trees are Sal, Bija, Gamhar, Jackfruit, Mango, Bamboo etc. and the important forest products are Tendu, leaves, coccon, sal seeds, lac, karanj etc.

Climatically, this district is suitable for comfortable living and adequate growth of agriculture and vegetation. The average rainfall of the district varies between 1100-1285 mm out of which 90% occurs during June to September month. The average temperture has been recorded at 25.8°C while the maximum & minimum temperature has been recorded at 43.6°C and 5.2°C respectively.

This districts lacks in good communication network especially in case of railway network. The district headquarter i.e, Simdega is not rail connected and the nearest railway station is Rourkela which is istuated at a distance of 70 km. Ranchi-Rourkela rail route passes through Bano & Jaldega blocks of Simdega district, however these are not important railway station. So far as the road network is concerned, all the blocks headquarter are connected to the district town Simdega. The NH-23 passes through the district connecting Ranchi and Rourkela.

The economy of the district is primarily based on primary sector consisting of agriculture and livestock. More than 86% of the total workers are engaged in Primary sector. The agriculture in the Simdega district is of primitive stage. The majority of the rural population depends on traditional method of agriculture. The major problems in the agricultural sector include lack of irrigation facilities, absence of scientific inputs, poor development of infrastructure, lack of capital, market etc. The agriculture is mainly rain fed and productivity is very low. It is well supported by a large number of milch cattle, however the yield of milk is very poor.

The district ranks 22nd in terms of population having 5,99,578 persons in the state of Jharkhand. It is one of the least densely populated district in Jharkhand state having 159 persons/sq. km as against the state average of 414 person/sq. km. However, the sex-ratio is very good and ranks 2nd in the state having 997 females/1000 males against the state average of 949 females/1000 males. This is primarily a rural district where only 7.16% population reside in urban area. The majority of the population is tribal (70.78%) as per the 2011 census. The major tribal groups in the district are Orans, Kharia and Mundas. A few families belonging to the primitive tribal group like Asur, Birhor etc. also reside in the district.

This district has tremendous potential for development of tourism and allied sector. Its natural scenic landscape with full of greenery coupled with pleasant climate may attract domestic as well as foreign tourists. Ram Rekhadam, Dangadi, Kelaghat, Bandurga, Birugarh, Rajadera, are the main places of interest which needs to be developed in terms of tourism infrastructure and service centres.

Bibliography

Ahmad, E. (1954) : Geographical Essays on India.

Ahmad, E. (1965) : Bihar : A Physical, Economic and Regional Geography, Ranchi University, Ranchi.

Basu, B.C. (1890) : Reports on the Agriculture of the District of Lohardaga, Calcutta.

Bikram, K. (1993) : Mal Pahariya Ranchi.

Bradley - Birt, F.B. (1903) : Chotanagpur, A little known province of the Empire, London.

Chaudhary, P.C. (1961) : Palamu : A District Gazetteer, Patna.

Chaudhary, P.C. (1964) : Dhanbad : Bihar District Gazetteer, Patna.

Chaudhary, P.C. 91965) : Santal Parganas, Bihar District Gazetteer, Patna.

Chakravarti, P.B. (1938) : Chotanagpur Raj : A brief sketch of its origin and status, Calcutta.

Datta, K.K. (1946) : Introduction to Bihar, Patna.

Dayal, P. (1953) : Bihar in Maps, Patna.

Diwakar, R.R. (1959) : Bihar Through Ages, Bombay.

Houlton, S.J. (1949) : Bihar : The Heart of India, Bombay.

Jha, J.C. (1964) : The Kol Insurection of Chotanagpur, Calcutta.

Jha, J.C. (1967) : The Bhumij Revolt, Calcutta.

Krishnan, M.S. (1955) : Geology of India and Burma, Madras.

Kumar, N. (1970) : Ranchi, Bihar District Gazetteer, Patna.

Lal Sota, C. and Oraon, P.C. (1993) : Adim Janjatia, Ranchi.

Mitra, P.K. (1956) : Mundari Folk Tales, Ranchi.

Mukharji, R.B.S.C. (1938) : Santal Parganas, Bihar District